D1162379

MAPPING THE NORTH AMERICAN PLAINS

MAPPING THE NORTH AMERICAN PLAINS

ESSAYS IN THE HISTORY OF CARTOGRAPHY

Edited by

Frederick C. Luebke

Frances W. Kaye

and Gary E. Moulton

PUBLISHED BY THE

UNIVERSITY OF OKLAHOMA PRESS, NORMAN AND LONDON

WITH THE

CENTER FOR GREAT PLAINS STUDIES, UNIVERSITY OF NEBRASKA–LINCOLN

BY FREDERICK C. LUEBKE

Immigrants and Politics: The Germans of Nebraska, 1880–1900 (Lincoln, Nebr., 1969)
Ethnic Voters and the Election of Lincoln (ed.) (Lincoln, Nebr., 1971)
Bonds of Loyalty: German Americans and World War I (De Kalb, Ill., 1974)
The Great Plains: Environment and Culture (coed.) (Lincoln, Nebr., 1979)
Ethnicity on the Great Plains (ed.) (Lincoln, Nebr., 1980)
Vision and Refuge: Essays on the Literature of the Great Plains (coed.) (Lincoln, Nebr., 1982)
Germans in Brazil: A Comparative History of Cultural Conflict During World War I (Baton Rouge, La., 1987)
Mapping the North American Plains: Essays in the History of Cartography (coed.) (Norman, Okla., 1987)

BY GARY E. MOULTON

John Ross, Cherokee Chief (Athens, Ga., 1978)
The Journals of the Lewis and Clark Expedition (ed.) (Lincoln, Nebr., 1983–)
The Papers of Chief John Ross (ed.) (Norman, Okla., 1985)
Mapping the North American Plains: Essays in the History of Cartography (coed.) (Norman, Okla., 1987)

Library of Congress Cataloging-in-Publication Data
Mapping the North American plains.

Chiefly papers presented at the symposium sponsored by the Center for Great Plains Studies of the University of Nebraska—Lincoln in April 1983.
Includes index.
1. Cartography—Great Plains—Congresses. 2. Great Plains—Maps—Congresses. 3. Great Plains—Discovery and exploration—Congresses. I. Luebke, Frederick C., 1927– . II. Kaye, Frances W. III. Moulton, Gary E. IV. University of Nebraska—Lincoln. Center for Great Plains Studies.
GA401.M37 1987 912'.78 86-25112
ISBN 0-8061-2044-4 (alk. paper)

The paper in this book meets the guidelines for permanence and durability of the Committee on Production Guidelines for Book Longevity of the Council on Library Resources, Inc.

All the chapters of *Mapping the North American Plains: Essays in the History of Cartography* save those by Ralph E. Ehrenberg and John B. Garver, Jr., are revisions of articles that appeared in several issues of *Great Plains Quarterly*, vols. 4 (1984) and 5 (1985).

CONTENTS

PREFACE

EXPLORATION, no matter how scientifically oriented or technologically involved, has been popularly viewed mostly as romantic adventure. From Renaissance mariners to contemporary astronauts, explorers have been remembered more for their experiences than for their accomplishments. In an effort to compensate for this imbalance, the Center for Great Plains Studies of the University of Nebraska—Lincoln sponsored the symposium *Mapping the North American Plains* in April 1983 to identify achievements in the cartography of the North American plains from earliest times to the present. Twelve scholars from the United States, Canada, and the United Kingdom presented papers on a variety of interrelated topics on Great Plains cartography.

The present volume consists of eight essays originally presented at the symposium that were first published in *Great Plains Quarterly*, plus two additional papers that have been written especially for this volume. They are supplemented by a catalog of the more than seventy historic maps that were assembled for the exhibition in connection with the symposium. Following a month-long showing in Lincoln, the exhibition traveled to Washington, D.C., where it was displayed in the Library of Congress in the autumn of 1983.

The symposium itself was the product of the imagination and energy of many persons, but especially of Brian W. Blouet, former director of the Center for Great Plains Studies, now head of the Department of Geography at Texas A&M University. The exhibition of maps was largely the work of Ralph E. Ehrenberg, assistant chief of the Geography and Map Division in the Library of Congress. The showing at the University of Nebraska was under the supervision of Jon Nelson, curator, Center for Great Plains Studies Collection of Western Art. The symposium was supported financially by the National Endowment for the Humanities, the InterNorth Foundation, and the University of Nebraska–Lincoln, especially its Research Council and Convocations Committee.

The great success of the exhibition was due essentially to Ehrenberg's expertise and energy. He served as its curator and subsequently prepared the catalog included in this book. His superb knowledge of American cartographic history and his acquaintance with map collections and their curators in the United States and Canada made it possible to assemble the outstanding materials that composed the final display.

I am pleased also to acknowledge the cooperation of the several libraries and archives that loaned rare and valuable maps for the exhibition; we are similarly indebted to their employees, who freely offered encouragement and valuable advice. Ralph Ehrenberg and the Library of Congress supplied the largest number of maps; other persons, and their institutions, who participated in preparing the exhibition are William H. Cunliffe, John Dwyer, and Chris Rudy-Smith, National Archives; Edward Dahl, Public Archives of Canada; Peter Bower, Provincial Archives of Manitoba; James Potter and Ann Reinert,

Nebraska State Historical Society; Carol Connor, Bennett Martin Library in Lincoln, Nebraska; Shirlee Smith, Hudson's Bay Company Archives of Winnipeg; Joseph Porter, Joslyn Art Museum in Omaha; Herman Viola of the Smithsonian Institution in Washington, D.C.; Barbara McCorkle and George Miles, Beinecke Library, Yale University; Robert Karrow, Newberry Library in Chicago; and Roman Drazniowsky, American Geographical Society Map Collection, University of Wisconsin–Milwaukee. The Nebraska showing also included loans from the private collection of Don L. Forke of Lincoln.

Members of the symposium committee were Brian W. Blouet, Stephen W. Cox, Frederick C. Luebke, Gary E. Moulton, Jon Nelson, and David Wishart, all at that time of the University of Nebraska–Lincoln. Special advice and participation came from Donald Jackson of Colorado Springs. Rosemary Bergstrom, formerly administrative assistant in the Center for Great Plains Studies, solved many problems and pursued scores of details with skill and fortitude. To all these people and institutions go our heartfelt thanks for their various contributions of money, loans of maps, wise counsel, and dedicated service.

The task of editing this book has been shared by three University of Nebraska faculty members who serve on the staff of the Center for Great Plains Studies. Gary E. Moulton, editor of the *Journals of the Lewis and Clark Expedition,* wrote the Introduction; Frances W. Kaye applied her considerable editorial skills to several papers before they appeared in *Great Plains Quarterly,* of which she is editor; and I provided general coordination for the project and guided it to publication by the University of Oklahoma Press. Each of us participated in the evaluation and editing processes.

FREDERICK C. LUEBKE

INTRODUCTION

Gary E. Moulton

CONCEPTIONS of unknown places are inevitably circumscribed by people's geographical experiences. Spanish explorers, French fur traders, and Anglo-American pioneers naturally perceived lands geographically similar to the ones from which they came when they penetrated regions beyond the mountains, forests, and deserts that separated them from the Great Plains of North America. Gradually, images established in the sixteenth and seventeenth centuries were reshaped by the experiences of the early probers into the region. New concepts founded in direct experience combined with ideas emerging from nascent geographical theory to influence the makers of the first maps of the plains.

The materials presented here illuminate the cartographic history of the Great Plains, one of the last great regions of the continent to be explored and mapped. Characterized by treeless grasslands traversed by shallow river valleys, the semiarid plains extend from Texas northward into the prairie provinces of Canada, and eastward from the Rocky Mountains to the humid, fertile lowlands of the Midwest.

The earliest maps produced by Europeans merely traced the borders of the Great Plains. On the east were the mighty Mississippi and Missouri rivers; above the mouth of the Platte River the latter was still a mystery and only a faint course on the maps of the eighteenth century. Beyond these rivers—the vast interior encompassing the plains, the Rocky Mountains, and the Great Basin—was terra incognita.

What little was known of this huge area rested on information gathered mostly from Indian sources. Accounts of Indian mapping are widespread in the records and literature of exploration and travel, and numerous examples of their maps in archives, museums, and map collections suggest that their influence was extensive and diverse. Many references to Indian geographic conceptions that reflected accurate knowledge of the interior lands were inaccurately translated to represent a compressed area with imaginary rivers and low-lying hills where mountain ranges should have been depicted.

By 1800 interest in the plains region had increased to the extent that exploration and mapping of the region began to push aside the earlier, fanciful notions of the land. The focus was the Missouri River. Spanish merchants using St. Louis as a base moved up the waterway in search of the lucrative Indian trade, while French traders operating under British companies expanded their markets with the Mandan and Hidatsa Indians who lived along the Missouri in present-day North Dakota. To establish a link between these markets had been a goal of the Spanish for some time, and in 1795 an expedition was sent to accomplish that purpose. Out of the exploration of James Mackay and John Evans in 1795–97 came a number of maps that gave the first accurate plotting of the Missouri River from St. Louis to the Mandan-Hidatsa villages and the first reliable description of the Great Plains.

With the purchase of Louisiana in 1803, the United States acquired lands of the Missouri Basin, including most of the Great Plains. The systematic mapping of this region commenced immediately. Under President Thomas Jefferson's guidance, the Lewis and Clark expedition set out to discover the dimensions of the newly acquired lands and explore the mysteries of the continental interior. During 1804, with Clark as the principal mapmaker, the explorers plotted the lower course of the Missouri River, set down its main geographic features, and depicted the hinterlands by interpreting the information of traders and Indians they met along the way. During their first winter spent at Fort Mandan, they queried more Indians and traders about the lands beyond that point and sent to Jefferson a grand map of their sightings and speculations. In 1805 and 1806 they crossed and recrossed the plains and mountains to the Pacific Ocean and continued their mapping project. After they returned to their homes, Clark rendered the results of their extensive surveys onto a great new map of the West. This was the beginning of a new generation of maps—maps that would accurately portray the American West because they were based on actual field sightings and acute topographic inference. The Missouri and its tributaries were set down with amazing skill, with only the upper reaches of the main stream's affluents charted conjecturally. The idea of a narrow, single ridge of mountains in the West was replaced by that of a complex cordillera with many ranges. One authority has called Clark's work "one of the most influential maps ever drawn."

In the first decades after Lewis and Clark, scientific and government-sponsored expeditions onto the Great Plains proliferated, and maps of many heretofore unexplored regions were generated. In 1805, Lt. Zebulon Montgomery Pike explored the Mississippi River to its supposed headwaters, and within the next two years he crossed the Great Plains of present-day Kansas and Colorado to reach the high peaks of the Colorado Rockies. Pike made numerous chartings of the central plains, plotted the course of the Arkansas River to its source in the Rockies, discovered the Royal Gorge and explored it, mapped the outstanding peaks of the region, and noted the sources of some great rivers of the West: the South Platte, the Colorado, the Rio Grande, and the Yellowstone. It had been Pike's intention, on his return trip, to pick up the headwaters of the Red River and follow its course back to the Mississippi. But he crossed into Spanish territory, where he was captured by Spanish soldiers. His expedition was thus cut short.

Following his return via Mexico, Pike gave the first negative reports about the plains, which he characterized as having "a barren soil, parched, and fried up eight months in the year." This conception of the plains was greatly strengthened by Maj. Stephen H. Long. In 1819–20 Long's party explored the Platte and South Platte rivers in Nebraska and Colorado and returned on a harrowing trip down the Canadian River in Oklahoma. Long's description of the country through which he passed is one of the most frequently quoted passages written by any explorer: "I do not hesitate in giving the opinion that it is almost wholly unfit for cultivation, and of course uninhabitable by a people depending upon agriculture for their subsistence." To these words Long added that incriminating epigram: "Great American Desert," written boldly across his map.

During the next several decades the Great Plains were mapped extensively. In 1842 John Charles Frémont surveyed the Oregon Trail to Walla Walla, and his maps and reports, published in 1845, were of great value to the pioneers and gold seekers who followed the western trail to California and Oregon. In the 1850s the desire to create a transcontinental railroad produced another burst of surveying and mapping. These Pacific railroad surveys, particularly the work of Isaac Stevens and G. K. Warren, added greatly to the geographic knowledge of the plains. After the Civil War four great western surveys were chartered by the federal government, and one of these, the Hayden surveys of 1867–78, covered large parts of the Great Plains. Hayden's reports and maps describe geology, topography, natural history, ethnology, land settlement, and mineral resources of the region. Through the remainder of the nineteenth and into the early twentieth century increasingly sophisticated maps of the plains were executed by government agencies: the Missouri River Commission in the 1890s and the Corps of Engineers are outstanding examples. Today, the U.S. Geological Survey keeps accurate topographic maps of the whole of the United States. Along with all the maps produced by the federal government, there were private and locally sponsored maps executed for railroad promotion, for local boosterism, for community plats, and for personal development.

Aspects of all these developments, from the early Spanish and French penetrations in the seventeenth and eighteenth centuries to the extensive governmental involvements of the nineteenth century, are examined here in detail. Each essay stands independently, but collectively the eleven articles and the catalog provide a comprehensive introduction to the cartographic history of the North American plains.

The first essay, "Exploratory Mapping of the Great Plains before 1800," by Ralph E. Ehrenberg, demonstrates that considerable cartographic work was going on in the pre–American period. The French were the first cartographers of the plains, and Ehrenberg discovers three periods of their activities: (1) from about 1670 to 1700, when the fringes of the plains were being depicted by Quebec and Paris cartographers; (2) from about 1700 to 1730, when French explorers ventured onto the southern plains; and (3) from 1730 to 1760, when the Vérendryes visited northeastern and central parts. After 1760, as France lost out in the New World, Frenchmen continued as explorers, but under Spanish and British sponsorship. British contributions to exploratory mapping concentrated on the northern plains and come largely from the two competing fur enterprises, the Hudson's Bay Company and the North West Company. Spain, motivated by the apparent encroachments of the British, encouraged expeditions and mapping on the upper Missouri River. Ehrenberg writes that the legacy of these seventeenth- and eighteenth-century efforts was that "their work provided a foundation for the more systematic mapping that was to follow" during the American period.

W. Raymond Wood's essay, "Mapping the Missouri River through the Great Plains, 1673–1895," enlarges and complements Ehrenberg's survey. His research reaches back to the 1673 maps of the French explorers Marquette and Jolliet, which show the Missouri as little more than a short, wavy line, and moves forward to the highly accurate maps done by the Missouri River Commission in the 1890s. During much of this period the Missouri River was the "Gateway to the West," and St. Louis was the base. The river cuts through the heart of the Great Plains and has influenced the whole area dramatically. Wood argues that despite its importance, the river has been neglected by scholars, who have focused on the overland trails instead. Wood divides his study into parts that relate to the successive control of the river by the competing powers of France, Spain, and the United States. For each period a seminal map or two emerged that inspired numerous copies and derivatives: for the French phase, Delisle's maps of 1703 and 1718; for the Spanish interim, Soulard's map of 1795 and the Mackay-Evans maps of 1796–97; and for the American era, the maps of Lewis and Clark.

John L. Allen's chapter, "Patterns of Promise: Mapping the Plains and Prairies, 1800–1860," is a review of the current state of scholarship on the American period. While covering familiar historic events of well-known expeditions, Allen also examines minor explorers, lesser known cartographers, scientists, and trained topographical engineers who set a pattern of exploration and mapping that gave promise of what was to be found on the great range of middle America. Through it all, Allen finds a thread of continuity: the cartography of the plains "was a mapping of the geography of hope and expectation rather than the geography of reality." His aim is to get at the thought behind the maps and discover the images they presented to those who studied them in the first half of the nineteenth century.

In "Indian Maps: Their Place in the History of Plains Cartography," G. Malcolm Lewis notes the paucity of scholarly study on maps of the North American plains drawn by Indians. He attributes this partly to the lack of materials for study and also to the scattered and hidden nature of the evidence. Nevertheless, these indigenous maps trace migration routes, reveal tribal history, exhibit mythical and religious elements, and display artistic characteristics, in addition to the usual geographic material. Lewis examines them in their historical and cartographic contexts and appeals for a greater appreciation of the influence they had on exploration and expansion. Knowledge of Indian maps will help researchers to understand those produced later by Euro-Americans—maps that drew on native intelligence and revealed perceptions of lands that predate European penetration of the plains.

The fifth essay is an excellent example of the scholarly research in Indian cartography that Lewis advocates. James P. Ronda, in "'A Chart in His Way': Indian Cartography and the Lewis and Clark Expedition," investigates the cultural interrelationships between natives of the plains and one group of explorers who needed indigenous maps. Ronda examines how Lewis and Clark approached the problem of mapping unseen lands by using Indian information. He shows how Lewis and Clark had to cope not only with language differences, but also with translating native concepts of land quality and configuration. As Ronda explains, the transfer of native knowledge to paper was a difficult process, not only because the Indians' "way" made use of different materials (animal hides or scratches in the dirt), but also because "Indian maps represented conceptions of distance, space, and time that were often fundamentally different from those commonly held by the bearded strangers." Indian maps were eagerly sought because they enabled the party to look ahead and to look beyond, that is, to alert them to the trail ahead and to reveal the nature of lands beyond their route. At several decisive points the captains used Indian testimony to guide them, and much of their mapping of areas that lay beyond the thin route of their traverse drew on information acquired from Indians.

Although the Lewis and Clark expedition was involved in a number of scientific investigations, mapping the new lands was a principal concern. For that purpose the men carried some of the best cartographic instruments of the day and used them extensively as they crossed the continent, but little has been written about this subject. The dearth of study of Lewis and Clark's astronomical observations is due partly to the difficulties of understanding the intricate procedures and antiquated instruments of the Corps of Discovery. Silvio A. Bedini brings the necessary expertise to the topic as he clarifies a highly technical subject in "The Scientific Instruments of the Lewis and Clark Expedition." By 1800 the science of astronomy had reached high levels of accuracy in determining latitude, but the calculation of a correct fix of longitude was yet to come, owing largely to the unreliability of the era's timepieces. Despite some training in "shooting the stars," Lewis may not have had sufficient skill in the task, and the difficult field conditions under which he worked exacerbated his problems. The captains were unable to overcome the obstacles, and their readings fell short of the accuracy for which President Jefferson had hoped.

John B. Garver, Jr., recognizes previously unheralded military men who mapped the American West but have not received the attention accorded the more illustrious explorers of the Corps of Topographical Engineers. In his essay, "Practical Military Geographers and Mappers of the Trans-Missouri west, 1820–1860," Garver considers the work of these "practical" surveyors and geographers, including Isaac McCoy, Enoch Steen, Nathan Boone, Philip St. George Cooke, Clifton Wharton, and Seth Eastman. From 1828 to 1832 McCoy made more than a half-dozen excursions into Kansas and Oklahoma. Garver finds McCoy's maps a marked improvement over Long's 1823 map, especially since he showed the courses of the Arkansas, Kansas, and Platte rivers more accurately. Moreover, McCoy extended the line of arable land about one hundred miles west of Long's delineation. Under orders to consider the topography of western lands more carefully, dragoon expeditions led by Henry Dodge in 1835, Nathan Boone in 1843, and Clifton Wharton in 1844 produced maps as well as reports. The reconnaissance maps were helpful in showing general locations of Indian tribes, previously uncharted lands, and possible routes for future travel. Garver emphasizes that these are just a few of the many military officers who were involved in mapping the trans-Missouri West and who expanded contemporary knowledge of the Great Plains.

The rectangular survey system with its base lines, principal meridians, townships, ranges, and sections, is one of the most striking characteristics of the settlement pattern of the Great Plains. Ronald E. Grim's essay, "Mapping Kansas and Nebraska: The Role of the General Land Office," examines the work of this federal bureaucracy. Grim is principally interested in the field work and reports on the personnel and their tenure, tools, tasks, and results. The surveying began shortly after the establishment of the territories in 1854 and progressed rapidly until the Civil War, during which time progress was greatly retarded. After the war, surveying increased dramatically, reaching a peak in Kansas in 1871 with 7.1 million acres plotted. Grim concludes that the General Land Office's cartographic activities represented a new phase in the mapping of the Great Plains. It was a change from earlier portrayals, which were limited to the basic elements of the landscape, to comprehensive, large-scale topographic mapping of the region. Land Office maps thus document a transitional phase in the cartographic history of the plains.

The Canadian plains are the subject of the two final essays. Richard I. Ruggles's chapter, "Mapping the Interior Plains of Rupert's Land by the Hudson's Bay Company to 1870," is concerned with cartographic activities of British fur trading factories along the Hudson Bay shoreline onto the plains of Manitoba and Saskatchewan. Having made a thorough study of the impressive collection of maps in the Hudson's Bay Company Archives in Winnipeg, Ruggles uses these primary sources extensively for his essay. He counts about ninety maps of the plains made between 1755 and 1870, or about 11 percent of the manuscript total. Ruggles identifies two major periods of mapping: an exploratory period from 1755 to 1815 with explorers such as Peter Fidler; and a diversified period from 1815 to 1870, which produced many cadastral maps showing details related to settlement. The Hudson's Bay Company used the maps to identify locations for developing trading strategy, transport routing, and a general understanding of the company's territory. They were a means to discover more efficient routes to trading customers in the interior. The company ended its cartographic efforts in 1870 as it surrendered its rights to Rupert's Land to the British government.

The final essay in this volume, "Mapping the Quality of Land for Agriculture in Western Canada," by James M. Richtik, brings the cartographic history of the plains into the twentieth century. Richtik finds that the search for fertile soil in the Canadian plains predates agricultural settlement. Early mapmakers of the nineteenth century, such as John Palliser and Henry Youle Hind, allowed American perceptions of the plains as the "Great American Desert" to influence their work, as they tended to describe the Canadian prairies in negative terms. Yet both Palliser and Hind described the land as having a greater potential for agriculture than the "true desert country" south of the border, while others found much more to praise. Although nineteenth-century cartographers and surveyors offered varied opinions on the soil quality of the plains, their assessments were often speculative and based on unsystematic classifications. Twentieth-century assessments anticipated extensive settlement by World War I veterans. Many of the maps of this period were overly optimistic; not surprisingly, much of the land was later abandoned. Not until the 1950s were systematic and accurate soil classifications made. Richtik found a low correlation between these maps and earlier ones, and the same must be said for the most recent classifications. Nevertheless, a certain consistency emerges. Since value and the perceptions of quality are tied to the expected use of land, Richtik concludes that we might expect classifications to continue to evolve as scientific knowledge increases and perceptions of land value change.

These essays offer much more than factual information about the West. Each points to new areas for research or to the reevaluation of existing ideas. Ehrenberg and Wood reveal the wealth of resources available for studying the French and Spanish period of Great Plains mapping. Allen provides a thesis to grapple with, as well as a review of well-known explorations. Lewis and Ronda not only suggest the possibilities for fruitful study of indigenous maps, but also provide important findings from their own investigations of such material. Ronda and Bedini demonstrate that we can return to such familiar figures as Lewis and Clark and still find fresh ideas. In addition to such prominent explorers of the early nineteenth century, less well known military figures are shown by Garver to have done significant cartographic work on the frontier. Similarly, Grim outlines the role of surveyors employed by the General Land Office in laying the lines for future frontier development. The Canadian plains also offer vast resources for further study, as is demonstrated in the essays by Ruggles and Richtik. Taken together, these contributions reveal the extent and quality of the mapping that was done, especially in the exploratory and early settlement periods, and effectively demonstrate the importance of cartography for the history of the plains region.

MAPPING THE NORTH AMERICAN PLAINS

1. EXPLORATORY MAPPING OF THE GREAT PLAINS BEFORE 1800

Ralph E. Ehrenberg

THE GREAT PLAINS as a geographic entity did not form a part of the European image of North America before the middle of the seventeenth century. Despite journeys to the fringe of the vast interior plains of the North American continent by Hernando de Soto, who reached eastern Oklahoma, and Francisco Vásquez de Coronado, who discovered the high plains of Kansas in the early 1540s, the region remained unknown to European mapmakers. As a consequence, from the time of the first publication of the earliest known printed map of North America by Giovanni Matteo Contarini and Francesco Rosselli of Florence in 1506 to major French explorations of the continental interior, the leading mapmakers of Europe left the interior of their maps on North America blank; designated the area terra incognita; filled it with fanciful mountains, rivers, and forests; or decorated it with ornate cartouches (see Chapter 11, Map I.1).

The only features relating directly to the North American plains that appeared on printed maps during this period were a few place names ascribed to secondary sources rather than maps made from actual observations. The first European explorers to reach the Great Plains were Spaniards who began moving northeastward from Mexico soon after the Aztecs were subdued in 1521. In search of *Quivira,* a mythical kingdom of gold, Coronado crossed the Great Bend of the Arkansas River and reached as far north as McPherson County, Kansas, then inhabited by Wichita Indians. In 1542 his expedition returned to Mexico City by way of an ancient Indian trade route, which later became famous as the Santa Fe Trail. Although Coronado's original maps have not been found, European cartographers adopted the place name *Quivira* for this region as early as 1556 and placed it within a geographic reference based on imagination and fancy current to that day (Chapter 11, Maps I.2 to I.4).

Following Coronado's expedition, Spanish interest in the plains subsided until the 1580s, when Don Juan de Oñate colonized New Mexico and subsequently led an exploratory party from Santa Fe northeastward as far as central Kansas. The earliest extant map of a portion of the Great Plains based on actual observation dates from this expedition. Compiled by Enrico Martínez in 1602 for use by the viceroy in Mexico City, it had no impact on European cartography, since it was not made available to other mapmakers.[1] Now preserved in the Archives of the Indies in Seville, Spain, it was based on information supplied by Juan Rodríguez, a soldier who had accompanied Oñate. With surprising accuracy the map depicts the Rio Grande flowing from Taos to the Gulf of Mexico. Martinez located the headwaters of the *Rio de la Madalena* (Canadian River) and the *Rio del Robredal* (Arkansas River) immediately to the east of Taos. A dashed line shows the route of Oñate's expedition from Mexico City to the Rio Grande and then onto the southern plains. It continues northward to his destination, a large Indian village on the Walnut River, a branch of the Arkansas, approximately twenty-five miles southeast of the present Wichita, Kansas. This unique map further notes that the region is flat, that it contains many animals called the "cattle of Ciuola [buffalo]," and that the Indians reported that gold could be found along one of the streams entering the Arkansas.

At least four additional expeditions were made to the southern plains from New Mexico during the remainder of the seventeenth century, but little of this information reached the cartographers of northern Europe, who continued to distort the geography of the region. Their most obvious error was to show the *Rio del Norte* (Rio Grande) flowing southwesterly into the Gulf of California rather than southeasterly into the Gulf of Mexico (Chapter 11, Map II.1). Not until the 1680s did Spanish knowledge of the Rio Grande reach commercial mapmakers of northern Europe. The source for this new information was Diego de Peñalosa, a former governor of New Mexico, who went to Paris in 1672 after being condemned by the Inquisition and banished from New Spain.[2] Peñalosa's map is still in the naval library in Paris, where it was consulted by French and Italian cartographers (Chapter 11, Map II.3). It depicts the *Rio del Norte* flowing in a southeasterly direction from Taos. Along the eastern bank of the river a chain of mountains is depicted, and further to the east the names *Quivira* and *Apache* are shown.

Despite increased Spanish exploration of the western border of the southern and central plains from 1696 to 1727, which reached as far as the North Platte, it was left to French explorers and mapmakers to reveal to the world the first geographical images of the North American plains. Spanish knowledge of the region was restricted for administrative and military purposes.

French Exploratory Mapping

The concept of the North American plains as a distinct geographical entity first took form during French exploration of New France. Three periods of French exploratory mapping of the continental interior can be discerned.[3] During the initial period (ca. 1670–1700), missionaries, Canadian *voyageurs,* and explorers traversed the central basin of the continent from Hudson Bay to the Gulf of Mexico, gathering geographic information for cartographers in Quebec and Paris. Because of their lack of direct, first-hand geographical information and proper surveying equipment, they reported grossly generalized and greatly distorted features. During the second period (1700–30), French adventurers and merchants from Louisiana first ventured onto the southern plains in search of trade and treasure. They sent reports and sketch maps back to France, where they were carefully analyzed and summarized by the famous family of French cartographers, Claude and Guillaume Delisle, whose printed maps of the region influenced later mapmakers for the remainder of the century. The third period is associated with the expeditions of Pierre Gaultier de Varennes, Sieur de la Vérendrye, and his sons, who ranged over large areas of the northeastern and central portions of the North American plains between 1731 and 1749 in search of a waterway to the Western Ocean. For the first time, a major section of the plains was mapped from direct observation. With the fall of New France in 1760, official French exploratory mapping of the northern plains ceased, but French traders employed by Spanish authorities continued to make important contributions to the mapping of the central plains during the 1790s, particularly in the trans-Missouri region.

1. Early French Period, 1670–1700

On the eve of French penetration of the continental interior, European cartographers had a general image of the inland periphery of North America but still no direct knowledge of the North American plains. This image was summarized by Nicolas Sanson on his map *Ameriqve Septentrionale,* which was published in Paris in 1650 (Chapter 11, Map II.1). Copies of Sanson's 1650 map and his revisions of 1656 and 1669 were carried into the interior by later French missionaries and explorers as they ventured westward beyond Lake Superior.

French exploration and mapping of the New World during the second half of the seventeenth century coincided with the shift of the center of mapmaking from Holland to France. While the Dutch establishments of Blaeu, Hondius, and Jansson were famous for their decorative wall maps and atlases during the first half of the seventeenth century, the work of their successors, the French mapmakers, became known for their scientific precision. Under the patronage of Louis XIV and the direction of his minister of finance, Jean-Baptiste Colbert, the Royal Academy of Sciences, established in 1666, made notable advances in surveying, geodesy, and cartography. The most significant French developments were the accurate measurement of an arc of meridian, leading to a more accurate determination of the dimensions and figure of the earth, and the perfection of a method for determining longitude. These two advances placed mapmaking on a firmer scientific basis.

In an effort to extend French authority over North America, Jean-Baptiste Talon, the intendant of New France, initiated France's exploring and mapping program of the continental interior in the latter half of the seventeenth century. In 1673 Talon sent Louis Jolliet, a French Canadian with some experience as a surveyor and cartographer, and Jacques Marquette, a French Jesuit missionary, down the Mississippi River to determine whether it flowed into the Pacific Ocean or the Gulf of Mexico. Although they journeyed only as far as the mouth of the Arkansas River, the two explorers deduced that the Mississippi River continued in the same southerly direction to the Gulf of Mexico. In addition to making this momentous discovery, confirmed a decade later by René-Robert Cavelier, Sieur de la Salle, Jolliet and Marquette first described the Missouri River, which they called by its Indian name, *Pekitanoui* (Muddy River); associated the Missouri River with the interior plains; and suggested the Missouri River as a water route to the Pacific Ocean.

The expedition of Jolliet and Marquette produced two maps: Marquette's surviving sketch map of Lake Superior, Lake Michigan, and a portion of the Mississippi,[4] and a map drawn from memory by Jolliet in 1674. The latter became the basis of Thévenot's map of the Mississippi valley in 1681, the first printed map showing a portion of the Great Plains (Chapter 11, Map II.2).

Jolliet's original map also served as a prototype for several other maps that depicted parts of the Great Plains. The earliest and most complete copy is known as the Buade map in reference to the designation of the Mississippi River as the River Buade, the family name of Frontenac, the governor of New France. It is titled *Nouvelle Decouverte de Plusieurs Nations dans la Nouvelle France En L'annee 1673 et 1674.*[5] The draftsman is unknown. The Buade map shows a general outline of North America bisected by the Mississippi River, which flows from three imaginary lakes west of Lake Superior to the Gulf of Mexico. Along the Missouri River the cartographer included names of several Indian tribes later identified with the Great Plains (fig. 1.1).

From 1678 to 1700 the cartography of New France relating to the Great Plains was dominated by the work of Jean-Baptiste Louis Franquelin. A trader and teacher of navigation, Franquelin served in Quebec as the official court cartographer for the governor-general.[6] In this position he had direct access to the sketches, information, and field notes prepared by western missionaries and traders. Franquelin's maps were never published and hence are not well known.

Franquelin's map of 1678 was the first attempt to show the Dakota or Sioux country west of Lake Superior. Entitled *Carte Gnlle de la France Septen-Trionalle,* it was based on Jolliet's map of 1674 and Sanson's map of 1650; however, the Mississippi River is projected much farther northward into Canadian territory some 250 miles north of Winnipeg, and a lake located to the northwest of Lake Superior is identified as *Lac des Assinibouels* (Assiniboine). Along the east side of the course of the Mississippi River north of Lake Superior, Franquelin lists eight Indian tribes associated with the Great Plains. These include the *Ihanct8a* (Yankton, a major division of the Western Dakota or Sioux); *Pintoua* (Tintonha, Dakota word for "land without timber," that is, prairie); *8apik8ti* (Wahpeton, a Dakota tribe); and *Chaiena* (Cheyenne, a tribe allied with the Dakota who at that time lived in the vicinity of the Minnesota River and Red River of the North).[7]

From 1684 to 1687, Franquelin compiled several more manuscript maps of the newly discovered western region of New France, incorporating material from La Salle's explorations to the mouth of the Mississippi River in 1682, Chevalier de Troyes's expedition to Hudson Bay in 1686, and Pierre-Charles le Sueur's discoveries in Minnesota.[8]

On his *Carte de la Louisiane* (1684) and *Carte de l'Amérique Septentrionalle* (1686 and 1688) Franquelin introduced two major geographical misconceptions derived from La Salle's map which were to

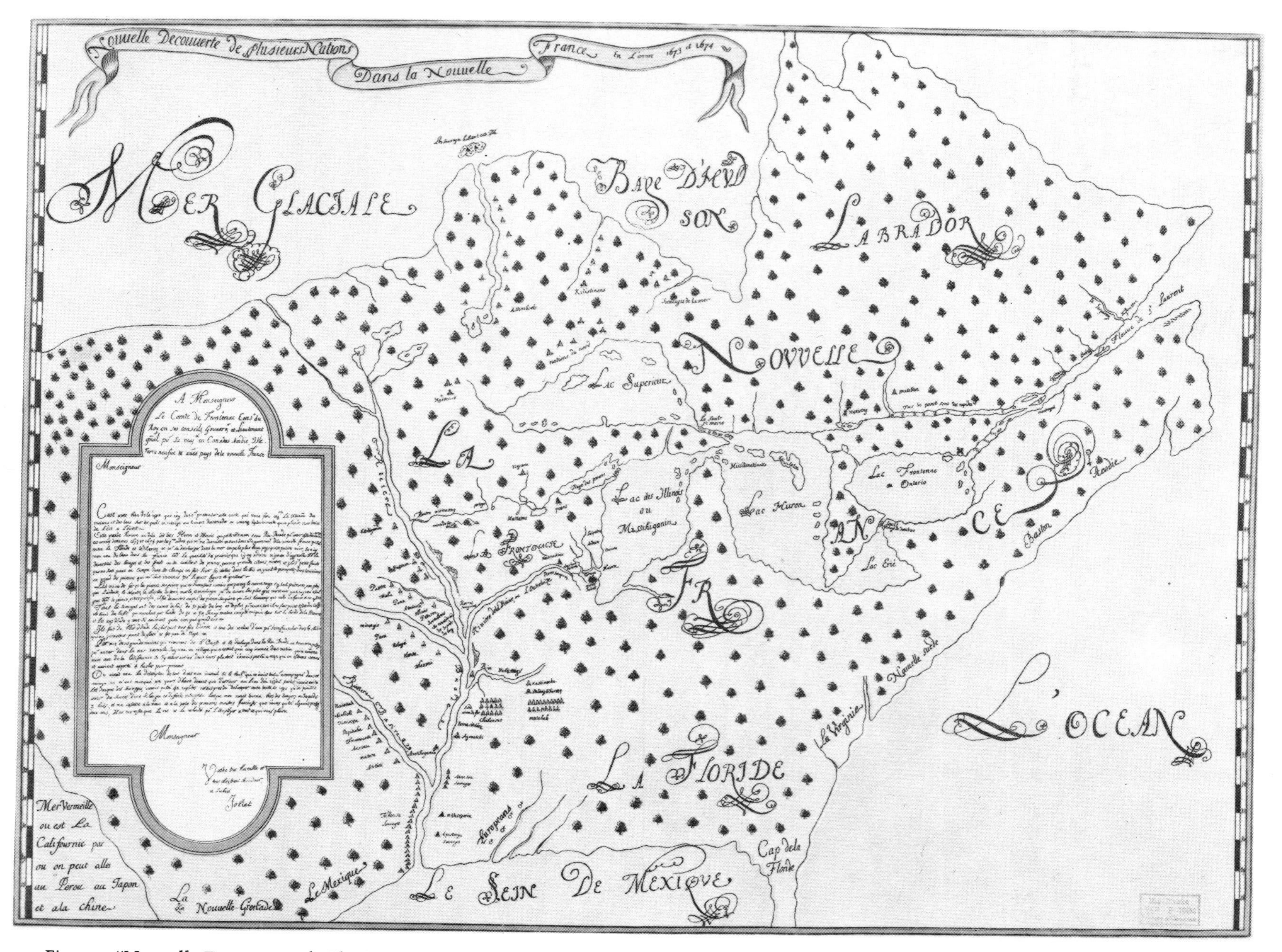

Fig. 1.1. "Nouvelle Decouverte de Plusiers Nations Dans la Nouvelle France En L'annee 1673 et 1674." The "Buade Map" based on Louis Jolliet. (Library of Congress)

have a profound impact on European cartography for the next twenty years (fig. 1.2). Franquelin's map of 1684 is a copy of La Salle's draft, which was prepared for the Marquis de Seignelay, Colbert's son and successor.[9]

In the first instance, La Salle had placed the mouth of the Mississippi River on the far western coast of the Gulf of Mexico some six hundred miles west of its true location, entering the Gulf from a westerly direction rather than from the north, thereby launching, in the words of Louis DeVorsey, "the greatest geographical hoax in the history of North American explorations."[10] This concept was not corrected until 1703, when the true course of the lower Mississippi River began to appear on European maps. As the first explorer to travel the full length of the Mississippi, La Salle was misled by contemporary maps that failed to depict a delta along the Gulf Coast. He was also misguided by Indians, who reported that all the creeks and rivers of the region ran eastward. Finally, his own desire to lead a colonizing expedition to the Mississippi by sea blinded his reasoning. In an effort to achieve financial backing for his return to the Gulf Coast, La Salle sought the support of Seignelay, who was more interested in obtaining the treasures of New Spain than in colonizing. Therefore, La Salle apparently distorted his map to show that the mouth of the Mississippi was closer to the fabled gold and silver mines of New Spain than originally believed. La Salle "found a great welcome with M. de Seignelay," the Spanish ambassador to England wrote in 1687, "to whom he presented a map marked with latitudes and longitudes that made his proposals plausible. On his word they believed in him and in his map. He made them believe what he wanted them to believe."[11]

The second major geographical misconception was La Salle's description of the Missouri River as having its source near Santa Fe, leading later mapmakers and explorers to conjecture that the mines of New Spain could be reached directly by the Missouri River. Plains Indians encountered by La Salle and his party provided the explorers with their information of the region west of the Mississippi. He compiled a list of some of the interior tribes, placing the Pawnee about five hundred miles from the Mississippi and estimating the navigable course of the Missouri at more than one thousand miles. "The Indians assured us that this river is formed by many others and that they ascend it for ten or twelve days to a mountain where they have their source," noted La Salle's chaplain, Father Zenobius Membré, "and that beyond this mountain is the sea where great ships are seen, that [the Missouri] is peopled by a great number of large villages, of several different nations; [and] that there are lands and prairies and great cattle and beaver hunting."[12]

Franquelin's delineation of the Missouri River, depicted as a complex network of streams, was derived from a young Pawnee captured by La Salle, who mistook the upper Platte, his home, as the main stream of the Missouri.[13] This misconception led La Salle and Franquelin to show the Missouri River as flowing almost directly eastward from the mountains just east of Santa Fe.

Below the Missouri, La Salle and Franquelin delineated the courses of the two major river highways as flowing eastward through the southern plains. For the first time the Arkansas and Seignelay (Red) are shown as extensive rivers with their headwaters near the Rio Grande. The Arkansas is depicted in a gradual west-northwest orientation, while the Red curves more sharply to the north. The mouths of both rivers, however, are located too far westward because of La Salle's gross distortion of the lower course of the Mississippi.

The next contribution, and the first map to depict the upper Mississippi River region based on actual observations, was compiled and drafted in 1682 by Abbé Claude Bernou and M. Peronel. It was based on information obtained by the French soldier and fur trader Daniel Greysolon Duluth, who entered the Sioux country west of Lake Superior in late 1678 and by the following summer had reached "the great village of the Nadouecioux called Izatys" on the shore of Lake Mille Lacs.[14] Although this map was never published, Duluth's geographical model of the upper Mississippi was incorporated in maps published separately a year later by Roussel and Vincenzo Coronelli and was still used as late as 1727, when it appeared on Henry Popple's map of North America. Bernou, compiler of La Salle's official report of his discovery of the Mississippi, was well versed in the geography of western exploration; nothing is known of Peronel, the draftsman.

Fig. 1.2. "Carte de l'Amérique Septentrionale" (1688) by Jean-Baptiste Louis Franquelin. (Library of Congress; manuscript copy of the original map in the Archive du Dépôt des Cartes de la Marine, Paris)

Roussel's *Carte de la Nouvelle France et de la Louisiane* was drawn to accompany Louis Hennepin's *Description de la Louisiane,* which was published in Paris on 5 January 1683 and was widely circulated throughout Europe.[15] Roussel followed Bernou closely with respect to the upper Mississippi, but while Bernou did not show the lower course of the Mississippi River, Roussel depicted it with a dotted line in its approximate location. The region west of the Mississippi River is left blank.

Coronelli's delineation of North America, on the other hand, shows the full length of the Mississippi River.[16] In 1680, Cardinal César d'Estrées, impressed by a pair of globes he saw in Parma, Italy, persuaded the author, Coronelli, to go to Paris to construct a pair of giant terrestrial and celestial globes for Louis XIV. Prepared during his stay in Paris from 1681 to 1683, Coronelli's globes, each measuring fifteen feet in diameter, were compiled from the most up-to-date material. Bernou, who was also sponsored by Cardinal d'Estrées, provided Coronelli with La Salle's official report, which Bernou had compiled sometime in the spring of 1683. In addition to the globes, Coronelli issued a map of North America in two sections under the title *America Settentrionale* for inclusion in his great atlas, *Atlante Veneto,* published in four volumes in Venice in 1690–96 (Chapter 11, Map II.3).

On both Coronelli's terrestrial globe and map the upper course of the Mississippi is derived from the Duluth-Bernou model, while the lower course is based on the La Salle–Franquelin model. Despite access to La Salle's report, Coronelli neglected to depict the explorer's geographic concept of the Missouri River system, relying instead upon Sanson's earlier imaginary rendition of the region.

2. Middle French Period, 1700–1730

European knowledge of the Great Plains was expanded by the work of Claude Delisle and his son, Guillaume, which represents the second period of French mapping of this region. From 1700 to 1718, the Delisles compiled a series of distinct maps of New France and Louisiana which greatly influenced later cartographers. Claude Delisle was the leading geographer of his time and is credited with reforming cartography. He was aided by his son, whose name appears as the author of the printed maps. As royal geographers and members of the French Royal Academy of Sciences, the Delisles had a unique opportunity to examine and copy the maps and reports sent to Paris from the Mississippi valley. Their voluminous papers, maps, and extracts were collected by the Archives du Service Hydrographique and now reside in the Archives Nationales.

Three Delisle maps were printed. The first, in 1700, entitled *L'Amerique Septentrionale,* closely follows Franquelin's map of 1699 with respect to the headwaters of the Mississippi and the Arkansas and Red rivers in the south. The Missouri River, on the other hand, is depicted as a single stream, with its course almost due east. The lower course is once again given the name *Pekitanoni* (a variant spelling), and it is connected by a dotted line to the upper course, whose source is in a mountain region just north of Taos on the Rio Grande. Two states of the 1700 map exist in printed form. In the first state, the lower course of the Mississippi follows the La Salle–Franquelin model; in the second state, the lower course is positioned some eight degrees to the east.[17] The latter innovation was derived from the maps and memoirs of Pierre le Moyne, Sieur d'Iberville, governor of Louisiana, who had explored the lower Mississippi in 1699 and 1700. This repositioning of the lower course contributed to the accurate placement of the mouths of the two major water routes leading to the Great Plains from the southeast—the Arkansas and Red rivers. "All charts [of this region] hitherto made," noted Iberville, in reference to La Salle's earlier work, "have been drawn by people who do not know the degrees of latitude and longitude nor how far places are from one another and who do not count the turns and twists of the way."[18]

In 1703 the Delisles published their famous two-sheet map of North America culminating seven years of intense research (Chapter 11, Map II.5). The northern half, entitled *Carte du Canada ou de la Nouvelle France et des Decouvertes qui y ont été faites,* displayed New France north of the confluence of the Missouri and Mississippi rivers; the southern section, entitled *Carte du Mexique et de la Florida,* showed the region south of the upper Mississippi and was the most accurate map then published with regard to the Mississippi River and the Gulf Coast. For this work, the Delisles relied heavily upon the maps and reports of Franquelin; Jean le Sueur; La Salle's Italian lieutenant, Henry de Tonti; and Iberville.[19] An English version of the two maps was issued by John Senex in 1710.

The rather scanty knowledge of the Missouri River valley and the southern plains as portrayed by the Delisles was enlarged during the next fifteen years as French traders, soldiers, and churchmen began to push up the Red, Arkansas, and Missouri rivers. Attracted by the rich markets of Santa Fe, alluring stories of gold in northern Mexico, Indian trade, and the hope of finding a route to the Western Sea (and thus the route to Japan, China, and India), Canadian *voyageurs* began exploring the Red River about 1700. By 1714 they had reached Spanish colonies on the Rio Grande.

Farther north, the first organized expedition ascended the Missouri River in 1702, reaching a point somewhere along the present Iowa-Nebraska state line. Between 1706 and 1708, several more expeditions brought back sketchy accounts of the broad river valley. The first systematic exploration of the Missouri that has survived was that of Étienne Veniard, Sieur de Bourgmont, who navigated the river as far as the Platte in 1714 and may have journeyed up the Platte through Nebraska and into eastern Wyoming. Additional information was gathered by missionaries and soldiers, notably Father Le Maire and Sieur Vermale, a French cavalry officer, whose information and maps found their way to the Delisles. Le Maire, in particular, drew influential maps of Lower Louisiana that contained important geographical information on the Red and Arkansas rivers.[20]

The Delisles incorporated much of this new information in a map they published in 1718, entitled *Carte de la Louisiane et du Cours du Mississipi*, which included a more accurate rendition of the lower Missouri and the Osage rivers, the introduction of the Kansas and Platte rivers, and the extension of an elongated Red River, protruding in a northwesterly direction deep into the southern prairies. In addition, the French belief that one could reach New Spain by traveling up the Missouri, first reflected on the maps of Franquelin, was supported by the Delisles, who depicted the headwaters of the Missouri in close proximity to those of the Rio Grande and a chain of mountains paralleling the Rio Grande. Although the Delisles had abandoned this notion on their 1703 map of Mexico and Florida, they reintroduced it in 1718, probably on the basis of information obtained from Le Maire. Because of their unequaled reputation within the mapping community, the Delisles' conception of the central and southern portion of the North American plains influenced later cartographers even when more recent information was available. Prominent mapmakers who relied upon the 1703 and 1718 models for the central and southern plains included Henry Popple (1727), John Mitchell (1755), Jacques Nicolas Bellin (1755), Philippe Buache (1763), William Faden (1777), Thomas Kitchin (1787), and Aaron Arrowsmith (1797) (Chapter 11, Map III.7).

Spanish expeditions as far north as the Platte River spurred further French activity. In 1718–19, Jean-Baptiste le Moyne, Sieur de Bienville, commandant-general of French Louisiana, sent two expeditions westward in an effort to contact the Padoucas and clear the way for trade with the Spanish on the Rio Grande. These journeys were led by employees of the French Company of the Indies. Jean-Baptiste Bénard, Sieur de la Harpe, ascended the Red River, crossed overland to the junction of the Canadian and Arkansas rivers, and explored the region to the northwest in what is now Oklahoma.[21] He was accompanied by Sieur de Rivage, "the famous surveyor whom I brought with me from France," who meticulously recorded compass directions and corrections for river windings.[22] The second expedition, led by Claude-Charles du Tisné, crossed overland from Kaskaskia near the junction of the Missouri and Mississippi to the headwaters of the Osage River and then southwest to the Wichita villages in southeastern Kansas situated on the Verdigris River, a tributary of the Arkansas.[23]

These two expeditions resulted in a more accurate image of the region south of the Missouri, particularly the relationship of the Osage, Arkansas, and Red rivers, which were displayed on several manuscript maps of the western province of Louisiana between 1720 and 1725. All have a similar title, *Carte Nouvelle de La Partie de L'Ouest de La Louisianne, Faite sur les Obervations & decouvertes de M. Bénard de la Harpe, l'un des Commandans audit Pays*, but there is some difference in content. These maps were based on La Harpe's journal as well as his field sketches and maps, which were drawn by him or under his direction.[24]

The La Harpe maps can be divided into two different types. The first is preserved in at least three variants in the naval library at Paris.[25] It was compiled in 1720 by Beauvilliers, an engineer to the king and member of the Royal Academy of Science. While it covers essentially the same geographical area as the Delisles' 1718 map west of the Mississippi, the Arkansas and Red River systems are shown more accurately.

The second variety is attributed to Jean de Beaurain.[26] The original is found in a one-volume manuscript copy of the memoirs and journals of La Harpe, Du Tisné, Iberville, and Bienville compiled by Beaurain about 1723–25 for presentation to the king. Entitled *Journal historique Concernant l'Establissement des Français à la Louisianne,* it is now housed in the Library of Congress. Beaurain, geographer to Louis XV, prepared many maps and atlases but is best known for his invention of a perpetual almanac for civil and ecclesiastical needs that was published in 1724. This map covers a larger area than the first type, extending from the Mississippi valley west to the Pacific Coast and from the Gulf Coast north to the Minnesota River (fig. 1.3). It reflects La Harpe's geographical concepts of 1723–25, which incorporated the results of his later Arkansas River expedition in 1722 and discussions with French Canadians in 1722–23.[27]

Although these maps were never published, La Harpe's geographical concepts of the Red, Arkansas, and Osage river systems eventually appeared in printed form when Jean-Baptiste Bourguignon d'Anville issued his noted map of North America in 1746. D'Anville, a scholar who collated written texts and contemporary maps, continued the French tradition of cartographic excellence initiated by the Delisles. La Harpe's concepts of the region were also later used by Dr. John Sibley, who explored the Red River in 1803 for the U.S. War Department when he discovered a bound folio of the *Journal historique* in Natchitoches.[28]

Although French fur traders and trappers continued to traverse the southern plains—Bourgmont, for example, crossed the Kansas plains in 1722–23, and the Mallet brothers made the first overland trek from the Missouri River to Santa Fe in 1739—no new geographical information of this region reached the European map publishing community until the nineteenth century.[29]

3. Late French Period, 1730–1760

The third major period of French exploratory mapping of the North American interior plains occurred in the 1730s. After beginning in the south and moving progressively northward as the western tributaries of the Mississippi were explored in preliminary fashion, the French westward movement shifted to the Canadian prairies, bypassing the upper Mississippi and Missouri rivers. The latter remained unexplored and therefore unmapped from 1700 to 1763, since the Fox Indians west of Lake Michigan and the Sioux of the upper Mississippi discouraged French *voyageurs* and explorers from entering that region. Instead, Frenchmen pushed around this Indian barrier, following the boundary lakes and streams to the northwest. French exploration of the northwest was also stimulated by their wish to check the flow of western furs to the English, who had gained sovereignty over Hudson Bay in 1713, and by their continuing search for a water route to the Pacific Ocean.

The search for an overland water route to the Pacific Ocean was associated with two related geographical concepts that had their origin early in the western movement: the Sea of the West and the River of the West. Seventeenth- and eighteenth-century European geographers theorized that the Pacific Ocean could be reached by two waterways crossing the western half of the North American continent, separated either by an elevated area of land or a great lake. As explorers and traders moved farther westward, the Sea of the West, in the words of Jean Delanglez, who traced the early history of this concept, receded beyond the horizon like a mirage, and the river leading to it increased in length.[30] Father Claude Dablon, the noted seventeenth-century missionary-geographer of New France, believed that this sea took the form of a large inlet, "a sort of North American mediterranean, with its outlet on the Pacific."[31] In response to these concepts, maps of North America portrayed large interior lakes or inlets well into the eighteenth century.

The search for the illusory River of the West was a major factor underlying the pioneer exploration of

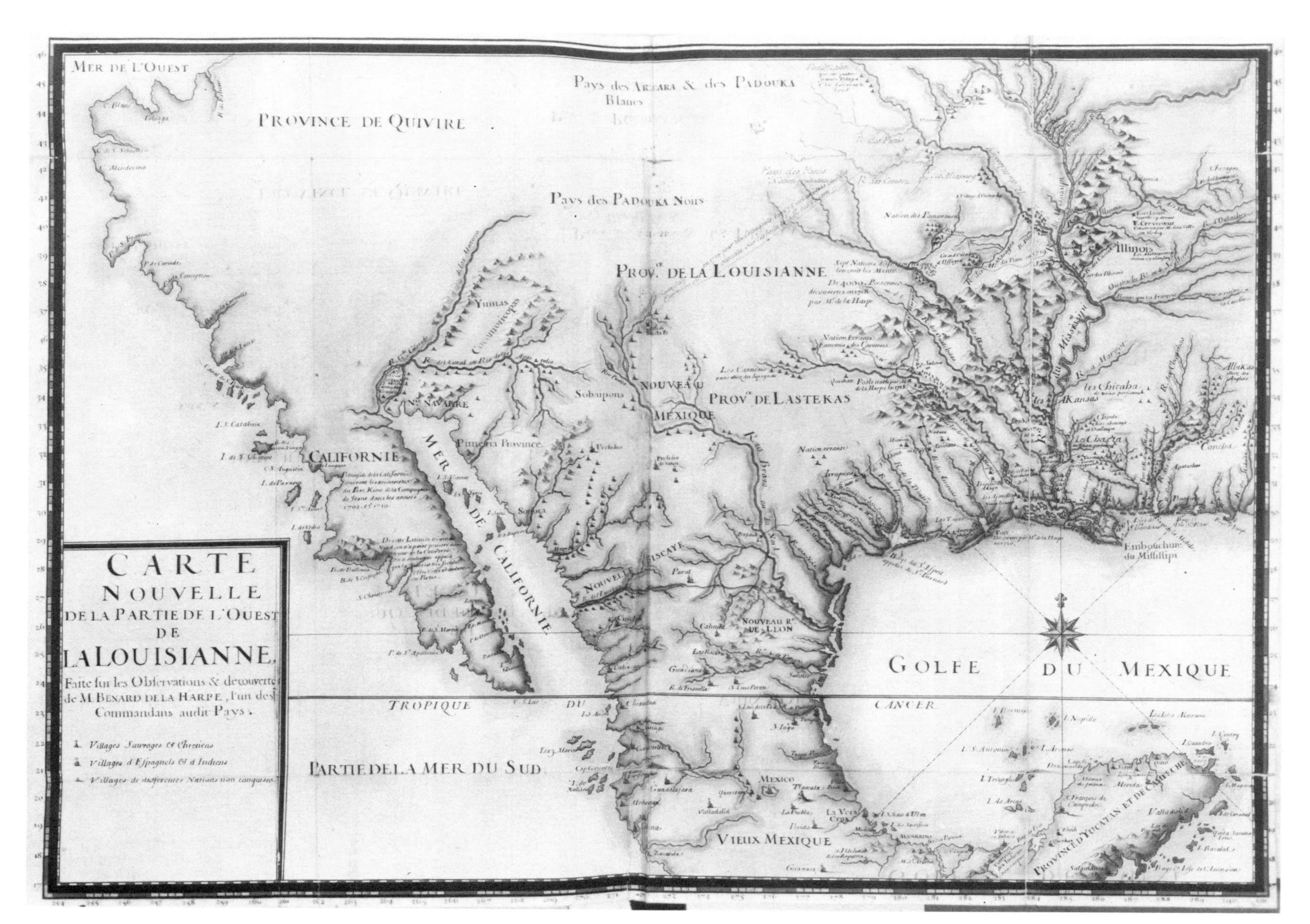

Fig. 1.3. "Carte Nouvelle de la Partie de l'Ouest de la Louisianne" (1723–25), based on Bénard de la Harpe, attributed to Jean de Beaurain. (Library of Congress)

the North American plains. It induced French explorers to push onto the great prairies of the West. The first reference to the River of the West is found in a detailed account of North America in the *Jesuit Relation* of 1659–60.[32] The concept was translated to cartographic form as early as 1684, when it appeared on Franquelin's map entitled *Carte de la Louisiane ou Des Voyages du Sr. De La Salle & des pays qu'il a découverts depuis la Nouvelle France jusqu'au Golfe Mexique, les annees 1679.80.81 & 82*. Franquelin displayed a river flowing southwesterly to the Gulf of California from a lake located near the headwaters of the Rio Grande and the Platte (labeled the Missouri). In 1697, Louis de la Porte de Louvigny, a French officer serving at Mackinac, summarized French geographical lore of the West in his manuscript map of the Mississippi valley entitled *Carte du Fleuue Missisipi*.[33] Louvigny depicted the headwaters of the same river with a note indicating that this river flowed to the Sea of the West. The notion of a River of the West was further reinforced by the Delisles' map of 1703, which showed not one but two rivers whose courses led westward—Lahontan's Long River and Le Sueur's "grande Rivière nommee Meschasipi" (Chapter 11, Maps II.4 and II.5).

In 1720 the French government sent the Jesuit priest Pierre-François-Xavier de Charlevoix on a fact-finding mission to the western frontier to inquire about the Western Sea. As a result of countless interviews with frontier military commanders, missionaries, Indians, and *voyageurs*, Charlevoix concluded that the Western Sea could be reached either by ascending the Missouri River (which he favored) or by establishing an outpost among the Assiniboines north and west of Lake Superior from which future expeditions could be sent westward. The latter course was chosen by Jean-Frédéric Phélippeaux, Comte de Maurepas, the French minister of marine and colonies under Louis XV. Pierre Gaultier de Varennes, the Sieur de la Vérendrye, was selected for this assignment.

From 1731 to 1743, La Vérendrye and his sons launched major exploratory expeditions to the Canadian prairies and the upper Missouri River valley extending as far west as the Black Hills. Most significantly, the Vérendryes recorded their explorations on several important maps that were incorporated in printed maps by the leading French and English mapmakers of the day.

A native-born Canadian, La Vérendrye was directed by the governor general of New France to establish a chain of trading posts northwest of Lake Superior along the border lakes region and to use the profits gained from trading to cover the cost of his search for the Western Sea. By 1733, his eldest son had reached the Red River and established a post on Lake Winnipeg. Three years later, La Vérendrye built Fort Rouge at the forks of the Red and Assiniboine rivers and Fort La Reine further up the Assiniboine, near the present city of Portage la Prairie. From the latter point, La Vérendrye explored the Great Plains in two different directions. He crossed the plains in 1738–39 in a southwesterly direction which led to the Mandan villages on the Great Bend of the Missouri River, thus becoming the first French explorer to visit that region. In 1742–43, two of his sons journeyed westward beyond that point to northeastern Wyoming and the Black Hills in a further attempt to locate the long-sought Western Sea. Unsuccessful in his southwestern expeditions, La Vérendrye turned northwestward in 1741 to explore Lake Winnipegosis as far north as the lower Saskatchewan. Although he failed to locate the River of the West, La Vérendrye explored and mapped the first major segment of the northern plains.

The earliest recorded contemporary map of a portion of the Great Plains associated with La Vérendrye was a composite of Cree Indian maps prepared at his request in 1728–29. Several versions of this map survive, showing the river and lake network west of Lake Superior. In 1830, La Vérendrye noted that the first version of his map was drawn by "Auchagah, a savage of my post, greatly attached to the French nation."[34] The map portrays, in an elongated configuration, the rivers, lakes, and portages of the border lakes region that connect Lake Superior with Lake Winnipeg. A mythical River of the West is depicted as flowing directly westward from Lake Winnipeg to the *Montagne de Pierre brilliantes* (Rocky Mountains).

At least three similar Indian birch bark maps were sent to Quebec, where Gaspard-Joseph Chaussegros de Léry, a prominent military engineer in New France, made a composite copy for transmission to the minister of marine and colonies in Paris along with La Vérendrye's report of 1730. Chaussegros de Léry also reduced and superimposed Auchagah's "River of the West" on a printed copy of Guillaume Delisle's 1730 map of North America, leading La Vérendrye to conclude that the River of the West flowed to an inlet on the California coast supposedly discovered in 1603 by the Spaniard Martin Aguilar, a member of the Vizcaino expedition.[35] The composite map, which greatly compressed the region west of Lake Winnipeg, was published separately as an inset in 1754 by Philippe Buache on his *Carte Physique des Terreins les plus élevés de la Partie Occidentale du Canada* (fig. 1.4). Buache was the son-in-law of Guillaume Delisle and succeeded him as the royal geographer. Like his predecessor, Buache was attached to the Ministry of Marine and Colonies, which provided him direct access to all of the reports and maps sent from New France. The Indian composite image of the border lakes and the River of the West also found expression on Jacques-Nicolas Bellin's map of 1743, entitled *Carte de l'Amerique Septentrionale,* which was published in 1744 in Charlevoix's popular *Histoire et description générale de la Nouvelle France,* and d'Anville's 1746 map, *Amerique Septentrionale.*

The second major map produced by the Vérendrye group was probably drawn by Christophe Dufrost de la Jemeraye, La Vérendrye's nephew and second-in-command, following his expedition to Lake Winnipeg and the lower course of the Assiniboine River in 1733.[36] Like the former map, it was redrafted by Chaussegros de Léry in 1734 and sent to Paris. Entitled *Carte d'une Partie du Lac Superieur avec le Découverte de la Riviere depuis le Grande Portage A jusqu'au Lac Ouinipigon, ou on a construit le fort Maurepas–le fort,* it is the first map of the northern plains based on actual observation to show the lower courses of the Assiniboine River and the Red River of the North. The Red River of the North, which forms the eastern edge of the northern plains, is identified as the Maurepas River, in honor of the French minister of marine and colonies, but the map also includes a note that reads "the Savages call it Miscouesipi which signifies the Red River." West of the Red River of the North, the geography of the river systems is noted as having been "traced according to the report of the Savages." The *Riviere St. Pierre* (Upper Assiniboine) and *Riviére St. Charles* (Souris-Assiniboine) are shown as forming one stream which enters Lake Winnipeg from the southwest. A section of the Great Bend of the Missouri, shown in the southeastern corner of the map, is characterized as the *Riviere de l'Ouest,* thereby confirming that La Vérendrye had revised his notion that a river flowed directly westward from Lake Winnipeg to the Western Sea. Further evidence that this river is the Missouri is the associated phrase *Village des Ouachipouenne* (village of "the Sioux who go underground"), the Monsonis Indian word for the Mandans. A trail identified as the "Route of the Warriors" links Lake Winnipeg to the Missouri River, and ultimately to the waterway to the Pacific Ocean.

A third map by La Vérendrye's group, compiled by Cree and Assiniboine Indians to aid in the search for the western sea, was sent to Quebec in 1737. The map is entitled *Carte contenant Les Nouvelles découvertes de l'ouest En Canada, Mers, Riviéres, Lacs, et nations qui y habittent en L'année 1737.*[37] This map may have been redrawn by La Vérendrye's youngest son, Louis-Joseph, who had accompanied him on the 1735–36 expedition. Louis-Joseph was a trained draftsman who had been sent to Quebec by his father during the winter of 1734–35 to learn "mathematics and drawing so that he may be able to make a correct map of the countries we shall pass through."[38] Although the geographical features are grossly generalized, it is a significant document, for it places for the first time on a single map the four major river systems of the central and northern plains: the Missouri River (now identified as the "Grand River of the Nation of the Couhatehalle," the Cree Indian name for the Mandans), which is shown for the first time as flowing to the south; the Red River of the North; the Assiniboine; and, for the first time, the Saskatchewan (identified as the *R[iviére] Blanche,* or White River). The source of the Saskatchewan is depicted as a lake situated in a range of western highlands, from which a second river flows westward to an unknown sea.

The most influential map produced by La Vérendrye's expedition is titled *Carte des nouvelles decouverts dans l'Ouest du Canada et des nations, qui y habitent* (Chapter 11, Map II.6). It was

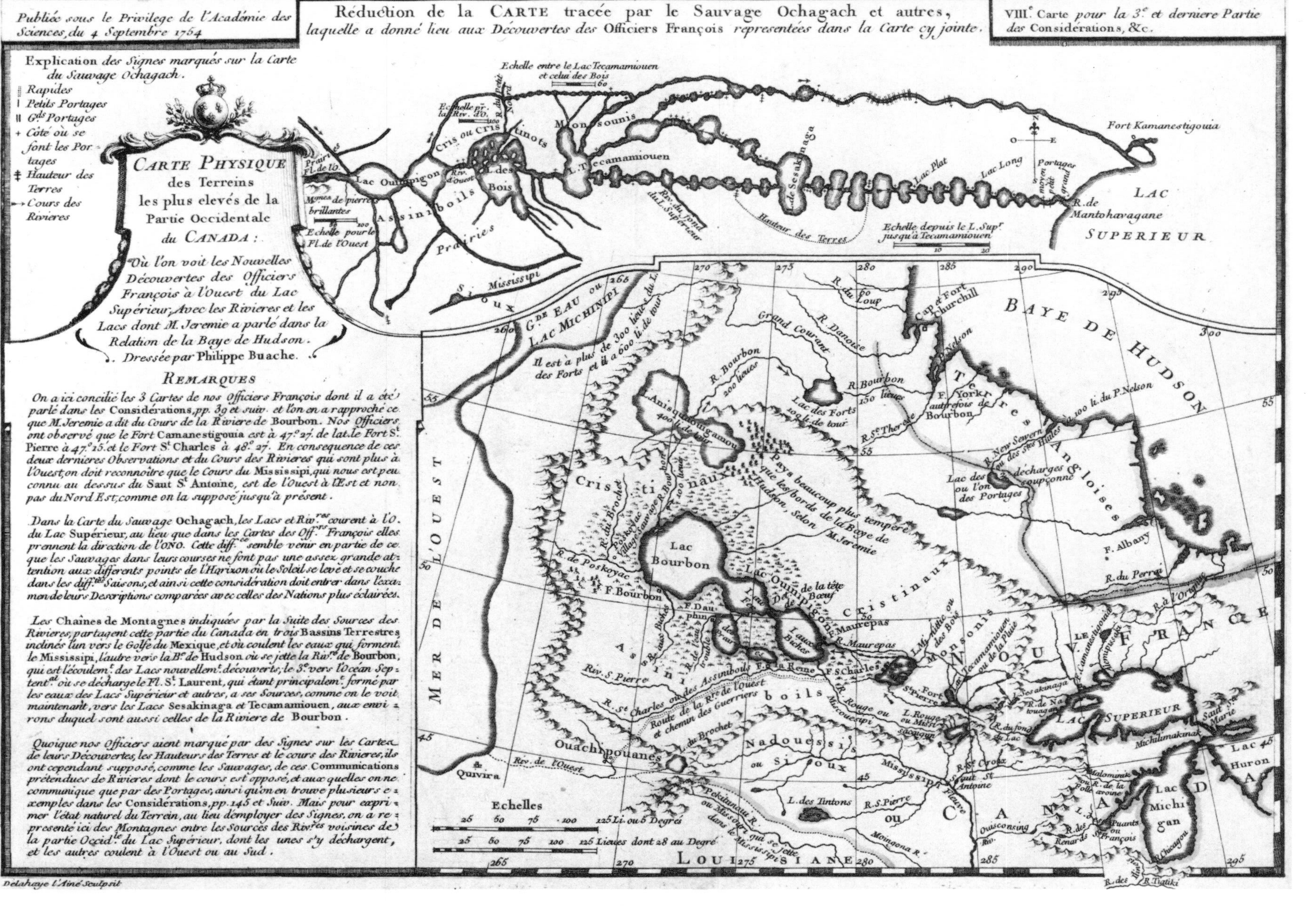

Fig. 1.4. "Carte Physique des Terreins les plus elevés de la Partie Occidentale du Canada" (1754) by Philippe Buache. (Library of Congress)

based on the memoirs of La Vérendrye and first prepared about 1740, with some information added in 1750, for Roland-Michel Barrin de la Galissoniere, commandant-general of New France. Although La Vérendrye believed that the map was "very imperfect," it nonetheless summarized the explorer's geographical concept of the northern and central plains. This map incorporated information from his journey southwestward across the prairies to the Mandan villages on the Great Bend of the Missouri River, which he reached 3 December 1738,and that of his third son, François, who reached the Saskatchewan in 1739. The western rivers that are portrayed on the 1737 map are now omitted, but the relative relationship of the Souris-Assiniboine and Missouri rivers is depicted more realistically, and the lower Saskatchewan is placed in its proper location. *Lac des Prairies* is portrayed as curving around the southwestern shoreline of Lake Winnipeg. Although somewhat exaggerated, this crescent-shaped Lake Manitoba is displayed more accurately than in earlier renderings.

La Vérendrye's concept of the northern and central plains as depicted on this map appeared in a number of printed French and English maps beginning with Jacques-Nicolas Bellin's *Carte de L'Amerique Septentrionale,* published in 1755. An engineer by training, Bellin served as senior hydrographer to the king of France for fifty-one years beginning in 1721. He became well known for his hydrographic atlases. In this position, Bellin also was in charge of the Dépôt des Cartes et Plans de la Marine, that great archive of Franch discovery and exploration that served as the final repository for La Vérendrye's maps.

Bellin's map of 1755 provides the most comprehensive European image of the plains before the fall of New France in 1760. For his rendering of the Mississippi and the lower Missouri rivers, Bellin followed the earlier works of the Delisles and d'Anville. For the north and central plains, Bellin copied La Vérendrye's configuration. On the line representing the Assiniboine River, however, Bellin added that it was believed that this river extended to the Western Sea, thereby perpetuating the notion that the Pacific Ocean and the continental interior were connected by a direct water route. Near the headwaters on the Assiniboine River, he included the phrase "Mountains of Shining Rocks according to reports of Indians," first recorded on La Vérendrye's Indian map of 1728–30. A segment of the middle course of the Missouri River, which was reached by La Vérendrye in 1738, is depicted, but not joined to the lower channel of the river; it is simply identified as the Mandan River. The only apparent connection that the cartographer made between the Mandan and Missouri rivers is a note along the Mandan River that indicates that "one does not know if this is the source of the Missouri." Bellin's map served as the model for subsequent British maps of North America, including those of Thomas Jefferys, the most productive English cartographer of the eighteenth century (1762); Jonathan Carver (1778); and Thomas Kitchin and John Harrison (1787) (Chapter 11, Map III.1).

English Exploratory Mapping

British contributions to the exploratory mapping of the North American plains focused primarily on the north and can be divided into two parts: (1) discovery maps prepared by surveyors and mappers associated with the Hudson's Bay Company, which eventually found expression in Aaron Arrowsmith's published maps of North America; and (2) manuscript maps prepared by American traders associated with the North West Company, a rival fur-trading group.

Although the Hudson's Bay Company was the first of the two great eighteenth-century North American fur-trading companies to encourage surveying and mapping of the plains, the earliest important maps of large segments of the region were prepared by American traders working for the North West Company. Following the collapse of French power in Canada in 1760, independent "pedlars" from Montreal began to build trading posts on rivers and lakes leading to the northern prairies. Following the earlier French route from Lake Superior up the border lakes and rivers to Lake Winnipeg, the first of these traders, James Finley, reached the Saskatchewan River no later than 1767. By 1784 these free traders had formed the North West Company. Two of these early Montreal-based fur traders, Alexander Henry and Peter Pond, prepared maps of the region.

Henry compiled the first map to define the natural limits of a portion of the northern plains. A native of New Jersey, Henry traded on the Saskatchewan River during the winter of 1775–76, when he accompanied a party of Assiniboines to their winter camp. Upon his retirement in 1796, Henry provided a first-hand account of the western fur trade in his *Travels and Adventures in Canada and the Indian Territories between the Years 1760 and 1776*, which was published in New York in 1809. This work provides one of the earliest descriptions in English of the area. "The Plains, or, as the French denominate them, the Prairies, or Meadows," Henry observed, "compose an extensive tract of country, which is watered by the Elk, or Athabasca, the Sascatchiwaine, the Red River and others, and runs southward to the Gulf of Mexico."[39]

His map, entitled *A Map of the North West Parts of America . . .*, was completed just before his departure for England and France in 1776 and presented to Lord Dorchester, then governor of Canada (Chapter 11, Map III.2). In addition to delineating the eastern boundary of the Great Plains, Henry's treatment of the configuration and placement of Lake Winnipeg, Cedar Lake, and the Saskatchewan River represents a major improvement over earlier works.

The most important contribution to the cartography of the western interior since La Vérendrye's maps of the 1740s, however, was the work of Peter Pond, an early associate of Henry. Born in Milford, Connecticut, Pond explored and traded on the St. Peters River (Minnesota River), Lake Dauphin, and the North Saskatchewan; pioneered the trade to the Lake Athabasca region northwest of the Saskatchewan; and aided in the development of the North West Company. Despite his considerable accomplishments, Pond remained a controversial figure in the fur trade and was accused of murdering two rival fur traders and killing another in a duel.[40]

Pond compiled three basic continental maps.[41] The first was presented to Congress in March 1785, the second was given to Lord Hamilton in April 1785, and the third was prepared for presentation to the empress of Russia in 1787 (Chapter 11, Map III.4). These maps survive in various copies and apparently one printed version. The last was published in *The Gentleman's Magazine* of March 1790. Pond's maps, encompassing all of the interior from Hudson Bay and Lake Superior west to the Pacific Ocean and from the Arctic Circle south to the Missouri River, provided the most comprehensive treatment of the general hydrographic system of the region until 1797. His delineation of the Manitoba lakes and the Red River of the North represent major improvements over earlier maps. His map was based on charting the canoe routes. For direction he used a compass, and for distance he "adopted those of the Canadian canoe men in leagues . . . sketching off the lake shores the best he could." The result was a fairly accurate rendition of the geographical features, David Thompson noted some years later, "but by taking the league of the canoe men for three geographical miles" (I found they averaged only two miles) he increased his longitude so much as to place the Athabasca Lake at its west end near the Pacific Ocean.[42]

A more systematic mapping program was undertaken by the Hudson's Bay Company. Chartered by the English crown in 1670, the company was established on the southwest shore of Hudson Bay to open the fur trade with western Indians and to discover the Northwest Passage. Following the French defeat, the Hudson's Bay Company moved aggressively to counter the St. Louis–based French-Spanish traders pushing up the Des Moines and Missouri rivers and the Montreal-based Canadians driving westward from Lake Superior and the border lakes. Company traders were sent southwestward down either the Nelson or Hayes rivers from their base at York on the Hudson Bay to the upper Saskatchewan and its tributaries. Although an employee of the Hudson's Bay Company was the first English trader to travel up the Saskatchewan River in 1766, independent Montreal traders such as Henry and Pond had cut substantially into their trade by the 1700s.[43] By following Indians to their hunting camps rather than waiting for them to come to fixed trading posts on the Saskatchewan, the Montreal Canadians had a decided advantage. To meet this challenge, the Hudson's Bay Company directed that surveys and maps be made of the unknown region west and southwest of Hudson Bay. Several important maps relating to the northern plains were then prepared for study by company officials, but they were not made public until near the end of the century.

Among the earliest of such maps was Andrew Graham's *A Plan of Part of Hudson's Bay, & Rivers, Communicating with York Fort and Severn.* Compiled about 1774, it is the first map to show the North and South Forks of the Saskatchewan River based on information obtained from a Hudson's Bay Company trader, Matthew Cocking, who had visited the Forks of the Saskatchewan in 1772–73.[44]

More scientific maps of the region were required, however, and in 1778 the secretary of the Hudson's Bay Company noted that the company intended to send to Hudson Bay "three or more Persons well skilled in the Mathematics and in making Astronomical Observations." These persons would carry the title of "Inland Surveyors."[45] The first person chosen for this position was Philip Turnor, who, in the words of Thompson, "was well versed in mathematics, was one of the compilers of the nautical Almanacs and a practical astronomer."[46]

From 1778 to 1779, Turnor surveyed the route from Hudson Bay to Hudson House, a company trading post on the North Saskatchewan River beyond present Prince Albert. He spent the next seven years surveying and trading along the southern coast of Hudson Bay. In 1787 Turnor returned to London, where he compiled a number of maps based on his surveys, including two relating to the Saskatchewan trip. After Turnor's contract as inland surveyor was renewed, his next assignment was to find the exact location of Lake Athabasca. It seems likely that Turnor and company officials consulted Peter Pond's 1785 map for information on that region.[47] Alexander Dalrymple, the noted nautical surveyor, had been asked by agents of the company to compare a copy of Pond's map with the charts of Captain James Cook and concluded that Lake Athabasca was situated only one hundred miles from the Pacific Ocean and that this lake would provide a short route to Asia.[48]

Turnor's second stay in North America was brief, but he succeeded in his mission and during this period trained two apprentices, Peter Fidler and Thompson, to carry on his work. Back in London by 1792, Turnor began work on his large general *Map of Hudson's Bay and the Rivers and Lakes Between the Atlantick and Pacifick Oceans.* This manuscript map, which measures 5 feet 10 inches by 8 feet 4 inches, incorporated all of his surveys as well as information obtained from other traders and Indians.

Fidler and Thompson added substantially to the map of the northwestern interior begun by Turnor. Altogether, they surveyed and mapped some 20,300 miles of waterways, 16,000 miles of which had never before been surveyed.[49] Fidler, who was to succeed Turnor as inland surveyor in 1792, joined the Hudson's Bay Company as a laborer in 1788. He accompanied Turnor as his assistant on the expedition to find and map a shorter route to Lake Athabasca from 1790 to 1792. In addition to important work in northern Manitoba, Saskatchewan, and Alberta, Fidler was the first to map the precise courses of the headwaters of the North and South Saskatchewan (1792–93) and the headwaters of the Assiniboine (1795). He also compiled a composite Indian map prepared by the Blackfoot Indian Chief Ac ko mok ki (The Feathers) in 1801.[50] The latter provided for the first time a fairly accurate representation of the major hydrographic features of the headwaters of the Missouri River and adjacent plains (Chapter 11, Map III.6).

Thompson, the third member of the Hudson's Bay Company triad of eighteenth-century surveyors, was sent to Hudson Bay at the age of fourteen in 1784 and was taught by Turnor in 1789–90. While Thompson is best known for the map of his 1811 expedition down the Columbia River to the Pacific Ocean and his 1813 master map of the western interior, he also surveyed the upper course of the North Saskatchewan as early as 1793–94 and a portion of the upper Missouri River in 1798.

In 1797 Thompson left the Hudson's Bay Company for the North West Company because the former failed to employ fully his skill as an explorer and cartographer. His first assignment for the North West Company led, in the words of Victor G. Hopwood, to "one of the world's great feats of surveying."[51] From 1797 to 1798, Thompson surveyed the traditional trade route from Lake Superior to Lake Winnipeg. Then, turning southwestward, he followed the Swan and Assiniboine rivers to the Souris, where, like La Vérendrye before him, he crossed the plains to the Mandan-Hidatsa villages on the Great Bend of the Missouri. Returning to the Assiniboine in 1798, he journeyed up the Red River of the North, crossed over to the headwaters of the Mississippi, and then went down the St. Louis River to Lake Superior. Outfitted with "a sextant of ten inches radius, with quick silver and parallel glasses, an excellent achromatic telescope, a lesser for common use, drawing instruments, and two thermometers," Thompson determined for the first time the latitude and longitude of the Great Bend of the Missouri (Chapter 11, Map III.5).[52]

The new geographical information of the northern and central plains acquired by Turnor, Fidler, and Thompson (before 1797) was sent to the London Committee of the Hudson's Bay Company, where it was soon incorporated into maps of North America published by Aaron Arrowsmith. Like the earlier commercial publishing houses of Delisle and Bellin in Paris, which were given access to the invaluable sketch maps of French explorers in the Dépôt des Cartes et Plans de la Marine, Arrowsmith was the first commercial English cartographer to enjoy this privilege with the Hudson's Bay Company. Arrowsmith acquired his knowledge of the map publishing business from John Cary, with whom it appears he apprenticed. His high standards for cartographic workmanship came from Alexander Dalrymple, hydrographer to the Admiralty, with whom he served as an assistant in 1795.[53] It was, perhaps, through Dalrymple that Arrowsmith received his entrée to the Hudson's Bay Company archives.

Beginning with his 1795 map *The New Discoveries in the Interior Parts of North America*, Arrowsmith established a new measure of excellence for published maps of the American West. He based his work only on original surveys or information obtained from direct observations; sources of the surveys or information were noted on the map; areas for which no information was available were left blank. Constantly revised until after 1850 as new information was obtained, this map went through at least nineteen editions or alterations (Chapter 11, Map III.7).[54] Thus, Arrowsmith's 1795 map with additions to 1796 embodied Turnor's survey of the lower Saskatchewan (1778–79) and his master map of the region north of the Saskatchewan (1794); Thompson's survey of the North Saskatchewan (1794–95); and Fidler's surveys of the headwaters of the South Saskatchewan (1792–93) and the upper Assiniboine and Swan rivers (ca. 1795). The lower Assiniboine (attributed to a Mr. Grant, 1784–91) and a segment of the Missouri River (derived from Bellin?) are misplaced and inaccurate, but they still demonstrate the relationship between these two major rivers.[55] A dotted line connecting the Little Missouri River (Souris) with the Missouri River segment contains the following description: "12 Hours Journey on Horse Back." Arrowsmith's next map, issued in 1802, corrected these errors by incorporating Thompson's surveys of 1797–98. The 1801 Ac ko mok ki model of the headwaters of the Missouri and South Saskatchewan (shown in part by dotted lines to symbolize the lack of adequate surveys) is also embodied in the 1802 map.

Spanish-Sponsored Exploratory Mapping

During the last decade of the eighteenth century, as English surveyors were establishing a cartographic framework for the northern plains, Spanish-sponsored expeditions based on direct observation simultaneously contributed to the geographic knowledge of the central plains. Motivated primarily by the fear of English encroachments along the upper course of the Missouri River, Spanish authorities in New Orleans and St. Louis encouraged and aided three major expeditions up the Missouri. In an effort to establish trade with the Plains Indians, they sought to expand their knowledge of Spanish Louisiana and to discover a route to the Pacific Ocean.

Spanish knowledge of the Missouri River basin was limited. Although the Delisles' *Carte de la Louisiane* (1718) provided a fairly accurate image of the lower Missouri to the mouth of the Platte, little was known of the region north of that point. But other maps of French or English derivation were unavailable to Spanish officials. Nevertheless, when the governor was asked for a map of his province in 1785, he provided an account of river geography and Indian tribal locations that advanced beyond the French maps of the 1750s. He also was able to define the limits of the southern plains:

> All this vast country beginning with the Arcanzas River to the shore-that-sings [Great Falls of the Missouri] is an immense meadow cut up and watered by the above mentioned rivers. This meadow land is bounded on the west by the mountains of New Mexico, on the east by the shore of the Mississippi; to the south the river San Francisco de Arcanzas; to the north

its boundaries are unknown because the Indians who have given this information do not go beyond the shore-that-sings."[56]

The first Spanish subject to venture beyond the Platte was Jacques d'Eglise, who reached the Mandan-Hidatsa villages on the Great Bend of the Missouri in 1792. However, no effort to map this region was made until two years later, when a group of St. Louis merchants formed the *La Compagnie de commerce pour la Decouverte des Nations du haut du Missouri,* commonly known as the "Missouri Company."

In preparation for the first expedition, its leader, Jean-Baptiste Truteau, was furnished on June 30, 1794, with a "plan . . . in order that he may know his whereabouts."[57] This "plan" appears to be the first new map of the central plains since Delisle's map of 1718. It was compiled by Antoine Soulard, the first surveyor-general of Upper Louisiana, at the request of Baron de Carondelet, Estevan Miró's successor as governor-general of Louisiana.

Although the original map has apparently disappeared, copies in Spanish, French, and English survive.[58] A copy of the English version was used by Lewis and Clark (Chapter 11, Map II.8). The map encompasses the Great Plains from the Saskatchewan River south to the headwaters of the Arkansas and Rio Grande. The delineation of the rivers of the northern part was derived from James Mackay, a Montreal-based fur trader who claimed to have visited the Mandans in 1787 and wintered on the Saskatchewan in 1786 and 1787, and who apparently was associated with Robert Grant on the Assiniboine. Mackay's portrayals of the Saskatchewan and Assiniboine are similar to Pond's maps, but his configuration of the Manitoba lakes and the addition of the Souris River represent an advance in the treatment of this region. With respect to the central plains, the Missouri-Platte watershed is portrayed in a new manner, with many secondary streams shown for the first time. The Platte River is depicted more realistically than earlier maps, with two upper forks almost touching the Rocky Mountains. The Grand Detour (a huge meander in the Missouri above the mouth of the White River, portrayed for the first time) and the Great Bend of the Upper Missouri are merged, resulting in an exaggerated loop that gives the upper course of the Missouri a unique alignment.

Soulard's model of the central plains, easily recognizable by his distinctive treatment of the upper Missouri, was incorporated in printed maps for some twenty years. His map was redrawn by Samuel Lewis, engraved by Benjamin Tanner, and published in 1804 by J. Conrad of Philadelphia under the title *Louisiana.* Another engraving was made by Francis Shallus and published in 1805 by Mathew Carey for his *American Atlas.* Hebron's engraved *Map of North America* inaccurately credited the Soulard delineation of the Missouri River to Meriwether Lewis. The latter appeared in *Pinkerton's Modern Atlas,* published in London in 1812.

The second important contribution to the map of the Great Plains during the 1790s was prepared by or for Gen. Georges Henri Victor Collot, a skilled engineer who reconnoitered the Ohio and Mississippi rivers at the request of the French minister to the United States. During a brief stay in St. Louis in 1796, he apparently ascended the Missouri as far as the Osage River. More importantly, it appears that Truteau prepared an account of his ascent of the Missouri River for Collot, who later incorporated its information into his book *Voyage en Amérique Septentrionale* and on his map of the region.[59] Subsequently arrested in New Orleans for conspiracy, Collot turned up in 1797 in Philadelphia, where he offered the Spanish minister copies of all the reports and plans that he made on his journey to the West, including "the plans of all the American forts on those frontiers, state of the country and other notices which will be profitable to Spain, either keeping Louisiana or in case of making an exchange with France."[60]

Collot's final map in its English version is titled *Map of the Missouri; . . . and of the elevated Plain . . .* (fig. 1.5). Apparently prepared in Paris, it summarizes on one sheet the most recent geographical information of the western interior then available. The delineation of the northern plains was taken directly from Arrowsmith's map of 1795 and that of the Platte and Kansas watersheds from Soulard's map of the same year. Soulard's upper course of the Missouri, however, was realigned to fit with Arrowsmith's segment of the Missouri. Several new tributaries of the Missouri, notably the Cheyenne and White rivers, were derived from Truteau's description. A large northeast-flowing tributary depicted west of the Mandan villages is suggestive of

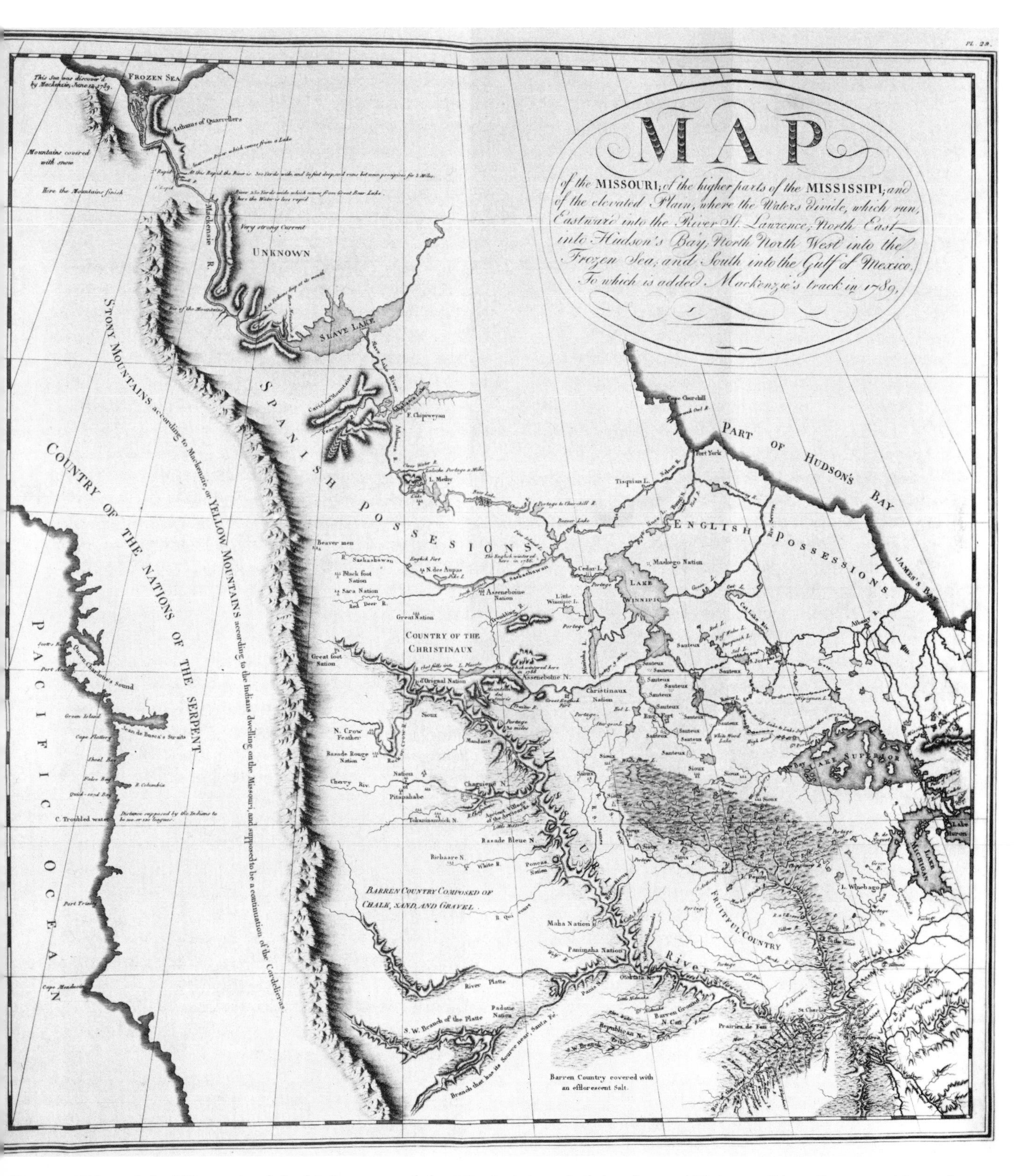

Fig. 1.5. "Map of the Missouri; of the higher parts of the Mississipi; and of the elevated Plain, . . ." (1796) prepared by or for Georges Collot. (Library of Congress)

the Yellowstone. Its identification was probably derived from Truteau, who had interviewed Cheyenne Indians, and a Frenchman named Menard, who had resided among the Mandans for some fourteen years.[61] Although Collot's map and book were compiled in 1796 or shortly thereafter, they were not published until 1804 and not released until 1826.

Following the unsuccessful efforts of Truteau and his successor Lecuyer to reach the upper course of the Missouri, the Missouri Company employed James Mackay as "principal explorer and director of the Company's affairs in the Indian Country."[62] Assisted by John Evans, a Welshman who came to America in 1792 in search of a tribe of Welsh Indians believed to be situated on the Missouri, the expedition left St. Louis in August, 1795. After wintering at Fort Charles, near present Sioux City, the exploring party split up, with Mackay reconnoitering the interior of Nebraska and Evans continuing upriver with orders "to discover a passage from the sources of the Missouri to the Pacific Ocean."[63] Evans was further instructed to "take care to mark down your route and distance each day, whether by land or water; in case you will be short of ink, use the powder, and for want of powder, in the summer you will surely find some fruit whose juice can replace both." Mackay and Evans returned separately to St. Louis in 1797, the latter failing to push beyond the Mandan-Hidatsa villages.

Although Mackay and Evans did not reach the Pacific, they provided the first accurate renderings of the upper course of the Missouri based on astronomical sightings and compass readings, and by carrying their survey up to the Mandan-Hidatsa villages, they finally brought into contact the two exploratory mapping frontiers of the interior plains. For the first time it was possible to connect on a map the major river systems of the central and northern plains with reasonable accuracy. While nothing is known about the early training of the two explorers, both Mackay and Evans demonstrated some knowledge of surveying. Zenon Trudeau, lieutenant-governor for Upper Louisiana, noted that Mackay "was instructed in the matter" and, following his return to St. Louis, Evans surveyed and mapped land grants in the Cape Girardeau area under the direction of Soulard.[64]

Two different maps of the region resulted from this expedition. Evans compiled a detailed six-sheet traverse chart of the Missouri River from Fort Charles to the Great Bend of the Missouri plus a single sheet showing a rudimentary sketch of the river from the Mandan-Hidatsa villages to the Rocky Mountains (Chapter 11, Map II.9). The latter also contains an early rendition of the Yellowstone River, which appears to have been derived from an Indian map acquired by d'Eglise, who passed it on to Mackay and Evans.[65]

The second map depicts the Missouri and its tributaries from the Mississippi to the Yellowstone River and was apparently prepared by Mackay. Work began on this map before Evans returned from the upper course of the Missouri, for on 26 May 1797 it was mentioned in a letter to Carondelet: "*Monsieur* Mackay made a very good map of the Missouri from its mouth as far as the *rivière Blanche* which includes about three hundred and fifty leaues. You will receive this map as soon as it is *distinct.*"[66] It may be this version that later served as the model for François Perrin de Lac's *Carte du Missouri Levée ou Rectifiée dans toute son Etendue* of 1802, which was published in his *Voyage dans les deux Louisianes* (Paris, 1805).[67] In addition to showing the Missouri only as far north as the White River, Perrin de Lac's map depicts Mackay's route of exploration in some detail in Nebraska and contains copious descriptive notes concerning that region. Particularly revealing is a note near the headwaters of the Loup and Platte rivers pertaining to a "Great desert of drifting sand without trees, soil, rocks, water, or animals of any kind, excepting some little varicolored turtles, of which there are vast numbers." This early reference to the Sand Hill region of Nebraska foreshadowed the later notion of the Great American Desert, which would come to play such an important role in geographical lore of the western interior. When Mackay's map was finally forwarded to Carondelet's successor in New Orleans, Governor-General Manuel Gayoso de Lemos, on 16 January 1798, it was described as depicting the Missouri "as far as the Mandan nations."[68] Thus, it appears that following the return of Evans, Mackay expanded his map, incorporating Evans's observations and chart.

Although the original maps have not been found, copies of both the Mackay map and the Evans chart were acquired in 1803 by William Henry Harrison (then governor of Indiana Territory and later ninth president of the United States), who forwarded them to either Thomas Jefferson or William Clark for planning and mapping the first leg of the Lewis and Clark expedition.[69] The Mackay-Evans model of the Missouri was also incorporated into a manuscript map of the North America interior by an unknown cartographer in about 1797 (Chapter 11, Map II.10). In addition, this map is noteworthy for its treatment of the upper course of the Missouri, which is depicted with its source in New Mexico in conformance with the Spanish concept, embodied in Carondelet's report of 1794, that the Mississippi and Missouri rivers encircled "the kingdom of Mexico . . . from the gulf almost to the South Sea."[70]

Whereas the map of the northern plains and, to some extent, the central plains began to take form during the late eighteenth century under the impulse of commercial and government interests in London, Montreal, and St. Louis, the image of the southern plains remained sketchy and obscure. Spanish authorities in New Orleans, Santa Fe, and St. Louis continued to base their images of this region on conjecture and vague Indian accounts rather than direct observation. The one exception was the work of Pedro Vial, a naturalized Frenchman who had lived for fourteen years among the Indians on the upper Red River. Commissioned by Domingo Cabello, the governor of Texas, Vial traversed the rolling plains of western Oklahoma and the Texas Panhandle from 1789 to 1793, establishing three new routes that would connect Santa Fe southeastward with San Antonio, eastward with Natchitoches on the Red River, and northeastward with St. Louis. The last was destined to become the Santa Fe Trail, one of the great caravan routes of the Great Plains. In the process, Vial compiled at least three maps, two of which have survived.

Following his first journey from San Antonio to Santa Fe in 1786 and 1787, which took him overland to the Red River and then west across the uncharted Llano Estacado (Staked Plains) of the Texas Panhandle, Vial enclosed a map with his report of 5 July 1787, which has disappeared. Shortly thereafter, however, he prepared a second map for the governor which portrayed the entire trans-Mississippi West from the upper course of the Missouri River to the Gulf of Mexico. This map is titled *Mapa et tierra qe. yos. pedro Vial taingo tranzitau en St. Tafee este dia 18 de octubre de Lann, 1787.*[71] Although this map is highly generalized, it nevertheless is a remarkable cartographic document. In the north, the Missouri River is delineated from St. Louis to the Rocky Mountains. While the major western tributaries of the Missouri such as the Platte and Kansas are not developed, the *mandanes* (Mandan) and *Riqura* (Arikara) villages are shown some five years before they were visited and described by the first Spanish-sponsored expeditions from St. Louis. This suggests, as Loomis and Nasatir have noted, that Vial may have reached the upper Missouri by way of Canada before going on to Spanish Territory.[72] The southern portion of the map somewhat resembles La Harpe's map of 1720, but the upper course of the Red River and its northern tributary, the Washita, are better placed (although their headwaters are positioned about five degrees too far to the north). Two types of Indian settlements are distinguished: encampments of nomadic Comanches, located on the upper courses of the Red River, are identified by small pictorial clusters of teepees, while more permanent villages of agricultural groups (*"pueblo de los yndios"*) associated with the Missouri and Red rivers are portrayed by a small hemisphere with a circle to symbolize an earthen or straw lodge.

The third map associated with Vial was compiled in Natchitoches in late 1788 following his arrival from Santa Fe. The expedition, designed as part of a program to open a route from Louisiana to California and Sonora, produced a detailed route map and travel diary, the latter prepared by Francisco Xavier Fragoso. The map shows the route crossing the Staked Plains, which Fragoso described as "so extensive that one sees only sky and plain," down the Palo Duro Canyon and the Prairie Dog Town Fork of the Red River (named *Rio Blanco* on the map) and the Red River proper to the village of the Taovayas or Jumanes (Spanish form of Shuman, a Pueblo Indian group) near present Ringgold, Texas, and then continuing overland to Natchitoches. Rivers and Indian settlements along the trail are identified. The Cross Timbers, a noted landscape feature that marks the eastern edge of the southern plains, is shown crossing the route just north of the Trinity

River. Fragoso characterized this band of stunted timber, extending from the Brazos River in Texas north as far as the Arkansas River, as "a very beautiful forest of oak which they say is more than 200 leagues long and only 3 wide."[73]

Legacies

This survey of the mapping of the Great Plains during the seventeenth and eighteenth centuries suggests that the contributions of these early mapmakers were not insignificant. Although their cartographic legacy was primarily in the form of a composite map of the hydrographic features, their work provided a foundation for the more systematic mapping that was to follow in the wake of Jefferson's initiatives in the southern and central plains and English efforts north of the Canadian border.

A second legacy was the use of native maps and geographical descriptions, a source that continued to be exploited by later American and English explorers. From Jolliet and Marquette's first depiction of the lower Missouri to Arrowsmith's portrayal of the upper Saskatchewan and Missouri, Indian maps played a vital role in the early mapping of the Great Plains. Related to the use of Indian maps was the decision by mapmakers to retain original native place names for important hydrographic and natural features.

Another contribution of these pre-1800 mapmakers was the process of transmitting, revising, preserving, and dispersing the newly acquired geographical information and maps through official or commercial archives. The Dépôt des Cartes et Plans de la Marine in Paris and the archives of the Hudson's Bay Company in London provided models for later United States and Canadian mapping agencies that served a similar function of making manuscript field notes and maps available to commercial publishers. The latter, in turn, carried the image of the Great Plains to a wider public, providing them with their first visual impressions of this vast region. Finally, in the last decade of the eighteenth century, efforts to train selected fur traders and explorers such as Fidler, Thompson, Evans, and Mackay in surveying and mapping presaged the training of Lewis and Clark and later explorers.

Notes

1. Martinez's map is reproduced in Carl I. Wheat, *Mapping the Transmississippi West, 1540–1861*, vol. 1, *The Spanish Entrada to the Louisiana Purchase, 1540–1804* (San Francisco: Institute of Historical Cartography, 1957), opposite p. 29. See also George P. Hammond and Agapito Rey, *The Rediscovery of New Mexico, 1580–1594* (Albuquerque: University of New Mexico Press, 1966), p. 63.

2. Peñalosa's map is reproduced in Wheat, *Mapping*, 1: 44.

3. The best summary of the French mapping of North America is Conrad E. Heidenreich and Edward H. Dahl, "The French Mapping of North America in the Seventeenth Century," *Map Collector* 13 (December 1980): 2–11, and "The French Mapping of North America, 1700–1760," *Map Collector* 19 (June 1982): 2–7.

4. Reproduced in Sara Jones Tucker, *Atlas: Indian Villages of the Illinois Country*, Illinois State Museum Scientific Papers, *vol. 2, no. 1* (Springfield: State of Illinois, 1942), pl. 4.

5. Reproduced in Emerson D. Fite and Archibald Freeman, *A Book of Old Maps Delineating American History from the Earliest Days Down to the Close of the Revolutionary War* (Cambridge: Harvard University Press, 1926; reprint, Dover Publications, 1969), p. 160.

6. Jean Delanglez, "Franquelin, Mapmaker," *Mid-America* 25 (January 1943): 29–53.

7. Jean Delanglez, *Hennepin's Description of Louisiana* (Chicago: Institute of Jesuit History, 1941), pp. 134–49; John R. Swanton, *The Indian Tribes of North America*, Smithsonian Institution Bureau of American Ethnology, Bulletin 145 (Washington, D.C.: GPO, 1952), p. 279. Franquelin's 1678 map is reproduced in John Warkentin and Richard Ruggles, *Manitoba Historical Atlas* (Winnipeg: Historical and Scientific Society of Manitoba, 1970), p. 39.

8. Delanglez, "Franquelin," pp. 54–74.

9. La Salle's draft has not been found, but Jean Delanglez believes that "there was no essential difference" between La Salle's map, Franquelin's map of 1684, and a similar map completed by Minet, the engineer who accompanied La Salle, in 1685. A copy of the draft was sent to Rome by Seignelay. See Jean Delanglez, "The Cartography of the Mississippi, part 2, La Salle and the Mississippi," *Mid-America* 31 (January 1949): 48–49. A section of Franquelin's map of 1686 is reproduced in Warkentin and Ruggles, *Manitoba Historical Atlas*, p. 47; Franquelin's 1688 map and Minet's 1685 map are found in Tucker, *Indian Villages*, pls. 7 and 11a.

10. Louis DeVorsey, Jr., "The Impact of the La Salle Expedition of 1682 on European Cartography," in *La Salle and His Legacy: Frenchmen and Indians in the Lower Mississippi Valley*, ed. Patricia K. Galloway (Jackson: University Press of Mississippi, 1982), p. 70.

11. Quoted in Delanglez, "Cartography of the Mississippi," part 2, p. 39.

12. Quoted in Gilbert J. Garraghan, "The Emergence of the Missouri Valley into History," *Illinois Catholic Historical Review* 9 (April 1927): 310.

13. Raphael N. Hamilton, "The Early Cartography of the Missouri Valley," *American Historical Review* 39 (1933–34): 651.

14. Reproduced in Tucker, *Indian Villages*, pl. 8; quotation is from John B. Brebner, *The Explorers of North America, 1492–1806* (New York: Macmillan, 1933), p. 351.

15. Reproduced in Fite and Freeman, *Old Maps*, p. 172.

16. Jean Delanglez, "The Cartography of the Mississippi: The Maps of Coronelli," *Mid-America* 30 (October 1948): 257–84.

17. Reproduced and described in Seymour I. Schwartz and Henry Taliaferro, "A Newly Discovered First State of a Foundation Map, *L'Amerique Septentrionale*," *Map Collector* 26 (March 1984): 2–6.

18. Philip Lee Phillips, ed., *The Lowery Collection. A Descriptive List of Maps of the Spanish Possessions Within the Present Limits of the United States, 1502–1820, by Woodbury Lowery* (Washington, D.C.: GPO, 1912), p. 221.

19. Jean Delanglez, "The Sources of the Delisle Map of America 1703," *Mid-America* 25 (1943): 275–98.

20. Jean Delanglez, "Documents—M. Le Maire on Louisiana," *Mid-America* 19 (April 1937): 124–43.

21. Mildred Mott Wedel: "J.-B. Bénard, Sieur de la Harpe: Visitor to the Wichitas in 1719," *Great Plains Journal* 10 (Spring 1971): 37–69.

22. Quoted in Brebner, *Explorers*, p. 337.

23. Mildred Mott Wedel, "Claude-Charles Dutisné: A Review of His 1719 Journeys," *Great Plains Journal* 12 (1972): 5–25, 147–73.

24. Mildred Mott Wedel, "The Bénard de la Harpe Historiography on French Colonial Louisiana," *Louisiana Studies* 13 (1947): 49.

25. Wheat, *Mapping*, 1: 68–69. A copy is reproduced on pl. 101, opposite p. 70.

26. Wedel, "De la Harpe Historiography," pp. 48–49. A reduced reproduction is found in W. P. Cumming, S. E. Hiller, D. B. Quinn, and G. Willems, *The Exploration of North America, 1630–1776* (New York: G. P. Putnam's Sons, 1974), p. 170.

27. Wedel, "De la Harpe Historiography," p. 65.

28. Ibid., pp. 37–38.

29. Henri Folmer, "The Mallet Expedition of 1739 Through Nebraska, Kansas and Colorado to Santa Fe," *Colorado Magazine*, September 1939, pp. 163–73; Henri Folmer, "Étienne Veniard de Bourgmond in the Missouri Country," *Missouri Historical Review* 36 (April 1942): 279–98.

30. Jean Delanglez, "A Mirage: The Sea of the West, Part I," *Revue d'Histoire de l'Amérique Française* 1 (December 1947): 381.

31. Ibid.

32. Ibid., p. 357.

33. Tucker, *Indian Villages*, p. 6; reproduced on pl. 14.

34. Lawrence J. Burpee, ed., *Journals and Letters of Pierre Gaultier de Varennes de La Vérendrye and His Sons*, Champlain Society Publication XVI (Toronto: The Champlain Society, 1927; reprint, New York: Greenwood Press, 1968), 52–53. Two versions of the Cree maps are reproduced in Warkentin and Ruggles, *Manitoba Historical Atlas*, p. 73.

35. Burpee, *Journals*, p. 64.

36. Warkentin and Ruggles, *Manitoba Historical Atlas*, p. 63, map reproduced on p. 77; and Martin Kavanagh, *La Vérendrye: His Life and Times* (Brandon, Manitoba: Martin Kavanagh, 1967), pp. 109–10, translated copy of map, chart 17.

37. A reduced copy of this map with English translation is found in Kavanagh, *La Vérendrye*, chart 18.

38. Burpee, *Journals*, p. 194.

39. Alexander Henry, *Travels & Adventures In Canada and the Indian Territories Between the Years 1760 and 1776*, ed. James Bain (Boston: Little, Brown, 1901), p. 267.

40. Harold A. Innis, *Peter Pond: Fur Trader and Adventurer* (Toronto: Irwin & Gordon, 1930), p. 142.

41. Henry R. Wagner, *Peter Pond: Fur Trader & Explorer* (New Haven: Yale University Library, 1955), pp. 26–32. Facsimiles of all three maps are included as separate sheets.

42. David Thompson, *Travels in Western North America, 1784–1812*, ed. Victor G. Hopwood (Toronto: Macmillan of Canada, 1971), p. 151.

43. J. B. Tyrrell, *Journals of Samuel Hearne and Philip Turnor* (Toronto: The Champlain Society, 1934), pp. 5, 68.

44. Warkentin and Ruggles, *Manitoba Historical Atlas*, p. 94. A reduction of this map is on p. 95.

45. Tyrrell, *Journals*, p. 61.

46. Thompson, *Travels*, p. 104.

47. Tyrrell, *Journals*, pp. 70–87.

48. Thompson, *Travels*, p. 151.

49. James G. MacGregor, *Peter Fidler: Canada's Forgotten Surveyor, 1769–1822* (Toronto: McClelland and Steward, 1966), p. xvii. One of Fidler's maps is reproduced in Warkentin and Ruggles, *Manitoba Historical Atlas*, p. 103.

50. D. W. Moodie and Barry Kaye, "The Ac Ko Mok Ki Map," *The Beaver: Magazine of the North*, Spring 1977, pp. 4–15.

51. Thompson, *Travels*, p. 147.

52. Ibid., p. 153. For a good treatment of Thompson's surveying instruments, see D. Smith, "David Thompson's Surveying Instruments and Methods in the Northwest, 1790–1812," *Cartographica* 18, no. 4 (1981): 1–17.

53. Coolie Verner, "The Arrowsmith Firm and the Cartography of Canada," *Canadian Cartographer* 8 (June 1971): 1; G. S. Ritchie, *The Admiralty Chart: British Naval Hydrography in the Nineteenth Century* (New York: American Elsevier, 1967), p. 19; Herbert George Fordham, *John Cary: Engraver, Map, Chart and Print-Seller and Globe-Maker, 1754 to 1835* (Cambridge University Press, 1925; reprint, Wm. Dawsons & Sons, 1976), 18.

54. Verner, "Arrowsmith," p. 5.

55. The "Mr. Grant" cited by Arrowsmith was probably Robert Grant, a North West Company trader who built a post on the Qu'Appelle about 1787 and was later in charge of the company's Red River Department. Grant returned to

Scotland in 1794. See Gordon Charles Davidson, *The North West Company* (Berkeley: University of California Press, 1918), p. 46.

56. Ibid., pp. 526, 531. Miro describes the location of the "shore-that-sings" as being a large cascade on the upper Missouri River.

57. "Clamorgan's Instructions to Truteau, St. Louis, June 30, 1794," *Before Lewis and Clark: Documents Illustrating the History of the Missouri, 1784–1804,* ed. Abraham P. Nasatir (St. Louis: Historical Documents Foundation, 1952), 1: 245.

58. The discussion of Soulard's map is based on Aubrey Diller, "Maps of the Missouri River Before Lewis and Clark," in *Studies and Essays in the History of Science and Learning Offered in Homage to George Sarton on the Occasion of his Sixtieth Birthday, 31 August 1944,* ed. M. F. Ashley Montagu (New York: Henry Schulman, 1946), pp. 507–508; Aubrey Diller, "A New Map of the Missouri River Drawn in 1795," *Imago Mundi* 12 (1955): 175–80; and Wheat, *Mapping,* 1: 157–59.

59. Annie Heloise Abel, ed., *Tabeau's Narrative of Loisel's Expedition to the Upper Missouri* (Norman: University of Oklahoma Press, 1939), p. 15, n. 30.

60. Noel M. Loomis and Abraham P. Nasatir, *Pedro Vial and the Roads to Santa Fe* (Norman: University of Oklahoma Press, 1967), p. 133.

61. Nasatir, *Before Lewis and Clark,* 1: 91.

62. A. P. Nasatir, "John Evans, Explorer and Surveyor," *Missouri Historical Review* 25 (April 1931): 225.

63. Ibid., p. 441.

64. Nasatir, *Before Lewis and Clark* 2: 545; Nasatir, "John Evans," pp. 591–93.

65. Nasatir, *Before Lewis and Clark* 2: 415. An excellent description of Evans's maps and redrawings of each sheet are found in W. Raymond Wood, "The John Evans 1796–97 Map of the Missouri River," *Great Plains Quarterly* 1 (Winter 1981): 39–53.

66. Nasatir, *Before Lewis and Clark* 2: 520.

67. For the connection between Mackay's map and Perrin du Lac's map, see F. J. Teggart, "Notes Supplementary to any Edition of Lewis and Clark," *Annual Report of the American Historical Association for the Year 1908* (Washington, D.C.: GPO, 1909), 1: 188–89. Perrin du Lac's map is reproduced in Wheat, *Mapping* 1: opposite p. 159.

68. Nasatir, *Before Lewis and Clark* 2: 545.

69. For a discussion of the complex provenance of these maps and related correspondence, see Donald Jackson, ed., *Letters of the Lewis and Clark Expedition with Related Documents, 1783–1854* (Urbana: University of Illinois Press, 1962), pp. 135–36, 140, 155, 163; and John Logan Allen, *Passage Through the Garden: Lewis and Clark and the Image of the American Northwest* (Urbana: University of Illinois Press, 1975), p. 141, n. 50. Both maps are reproduced in Gary E. Moulton, ed., *Atlas of the Lewis and Clark Expedition* (Lincoln and London: University of Nebraska Press, 1983), pls. 5, 7–12.

70. Nasatir, *Before Lewis and Clark* 1: 254.

71. This map is reproduced in Wheat, *Mapping,* 1: opposite p. 126, and in Carlos E. Castaneda, "Communications between Santa Fe and San Antonio in the Eighteenth Century," *Texas Geographic Journal* 5 (1942), opposite p. 160. A general description is found in Loomis and Nasatir, *Pedro Vial,* pp. 384–85.

72. Loomis and Nasatir, *Pedro Vial,* p. 387, n. 39a.

73. The map is reproduced in Wheat, *Mapping,* 1: opposite p. 127; the travel diary and itinerary are found in Loomis and Nasatir, *Pedro Vial,* pp. 327–48.

2. MAPPING THE MISSOURI RIVER THROUGH THE GREAT PLAINS, 1673–1895

W. Raymond Wood

For decades, "the Way West" referred not to any kind of overland trail but to the channel of the Missouri River. St. Louis became famous as the gateway to the West because it was the port of entry to the vast western domains drained in part by this mighty stream. Considering the extensive scholarship devoted to such land routes as the Oregon, Santa Fe, and Overland trails, it is curious that the equally important role of the Missouri River as an artery of exploration has been neglected. Only three works have made any real attempt to offer such a history, two of them popular.[1] The third, by Abraham Nasatir, is a short but heavily documented history of the river from its discovery in 1673 until 1805, when the course of the stream was finally explored in its entirety by Lewis and Clark.[2] Even so, the emphasis in Nasatir's study is on the two decades spanning the years 1785 to 1804.

An article by Raphael Hamilton is the only general study that describes and illustrates the history of the mapping of the Missouri River, but a host of published papers significantly augment Hamilton's work.[3] Our purpose here is to draw these scattered sources together in a brief narrative for the period from 1673 to 1895. By the latter date the entire course of the river was known and accurately mapped in detail.[4]

From first to last the mapping of the river was inspired principally by commercial interests. For the first century and a half of the Missouri's modern history, Indian trade, especially for furs, dominated the reasons for mapmaking. Maps made during the next seventy-five years, on the other hand, were stimulated in large part by the needs of those using the steamboat to trade with and settle the West. The latter period ended about 1902 with the dissolution of the Missouri River Commission, a federal unit charged with improving the navigational capabilities of the river.[5]

This study of the mapping of the Missouri River requires that we consider the entire reach of the stream from the time of its discovery by European explorers. The first crude maps of the Missouri, as well as most later general maps, depicted Native American tribal locations and other details on the Great Plains proper. These details extended well to the north and west of the mouth of the Kansas River, which, in some cultural schemes, marks the approximate eastern boundary of the Great Plains on the lower Missouri River. We are not concerned here with the precision of such details as Native American tribal locations, but rather with the developing exactness of the representation of the river itself through nine generations, or general stages, of Missouri River mapping. Each generation depicted a basic design of the river's configuration, and each ended as new data permitted significant refinement of that particular conformation.

The French Period: 1673–1770

The first-generation maps are those of Marquette and Jolliet. Although there are hints that the Spanish in the Southwest had learned from Indians of the existence of the Missouri River as early as 1541, Europeans did not actually lay eyes upon the stream until more than a century later.[6] In late June 1673, Father Jacques Marquette and Louis Jolliet and their party passed the mouth of the Missouri on their way down the Mississippi River. Marquette, a Jesuit, and Jolliet, a frontiersman, provided a graphic description of the mouth of the Missouri River as they passed it—not surprisingly, since it would have been discharging its spring floodwaters into the Mississippi at the time. They named the Missouri the *Pikistanouï*, a name that survived in various spellings (and, no doubt, pronunciations) for several decades.[7]

The cartographic documentation of the Missouri by this expedition included only the position of the mouth of the river. Unfortunately, the originals of Jolliet's map were lost at Lachine Rapids, a few miles from Montreal, on his return home in 1674. He produced a copy of the map from memory, however, and his superiors sent it on to France. It too has been lost, although it was copied by several European cartographers before it disappeared. Father Jean Delanglez has reconstructed a prototype of the lost map (called the Jolliet "X" map) using five such copies.[8] On these maps, as well as on the autograph map Marquette produced in 1673–74, the Missouri is shown simply as a short stub of a stream entering the Mississippi from the northwest.[9] The relationships between the various Marquette and Jolliet maps and their derivatives (as reconstructed by Father Delanglez) illustrate the kind of "genealogy" that must be prepared as the basis for a critical interpretation of these early maps.

For decades, Marquette and Jolliet's speculation that the Missouri River would provide a route to New Mexico fueled French interest in that stream as a means of reaching Mexico and its silver. As Bernard DeVoto said, "This idea was to confute government, diplomacy, and military strategy till the Great Valley became American, and to confuse geographical thinking till Lewis and Clark got home."[10]

The stublike depiction of the Missouri River persisted on copies of the lost Jolliet "X" map and on derivatives of the Marquette map produced by the French mapmakers Franquelin, Randin, and Bernou as late as the 1680s.[11] On a few maps of the period (such as the Coronelli 1688 map), the Missouri bears the name *Riv. des Ozages*, after one of the principal tributaries of the lower Missouri River and the important Indian tribe living along it in what is now western Missouri.[12] A few maps made as late as 1700 continued to show no real improvement over the Marquette and Jolliet sketches, in spite of the passage of time and the increasing number of French explorers, traders, and priests living near the mouth of the Missouri, some of whom penetrated a short distance up the river.

Less than a decade after Marquette and Jolliet's passage, René-Robert Cavelier, Sieur de la Salle, also explored the Mississippi River, this time between the mouth of the Illinois River and the Gulf of Mexico. Assisted by his lieutenant, Henry de Tonti, La Salle and his party arrived at the mouth of the Missouri on 14 February 1682 as they moved downstream. When the expedition approached the mouth of the Mississippi, La Salle took possession of the basin of the Mississippi River in the name of Louis XIV, naming the country *Louisiane* in his honor.[13]

After his return, La Salle produced a sketch of his impressions of the country of New France, apparently based in part on information obtained from a boy who is believed to have been a Wichita Indian slave.[14] This information was probably augmented by data from other French explorers and priests who were familiar with the area of his map. Frenchmen were certainly in the Missouri valley by this time, for according to Pierre Margry, two French *coureurs de bois* were captured by the Missouri Indians and taken to their village in 1680 or 1681, about a year before La Salle's visit to the mouth of the Missouri River.[15]

La Salle's map (now lost) was passed along to Jean-Baptiste Louis Franquelin in Paris when Franquelin served La Salle as a draughtsman in 1684 (fig. 2.1).[16] This map provided the second generation of Missouri River charts and continued to be produced by Franquelin for many years. Although they were never published, Franquelin's maps became widely known, for by 1686 he had become the royal hydrographer in Canada. The maps based on La Salle's data are readily identifiable by a distinctive and bizarre rendering of the lower reaches of the Missouri River, best described as "braided," with three immense "islands" depicted between what appear to

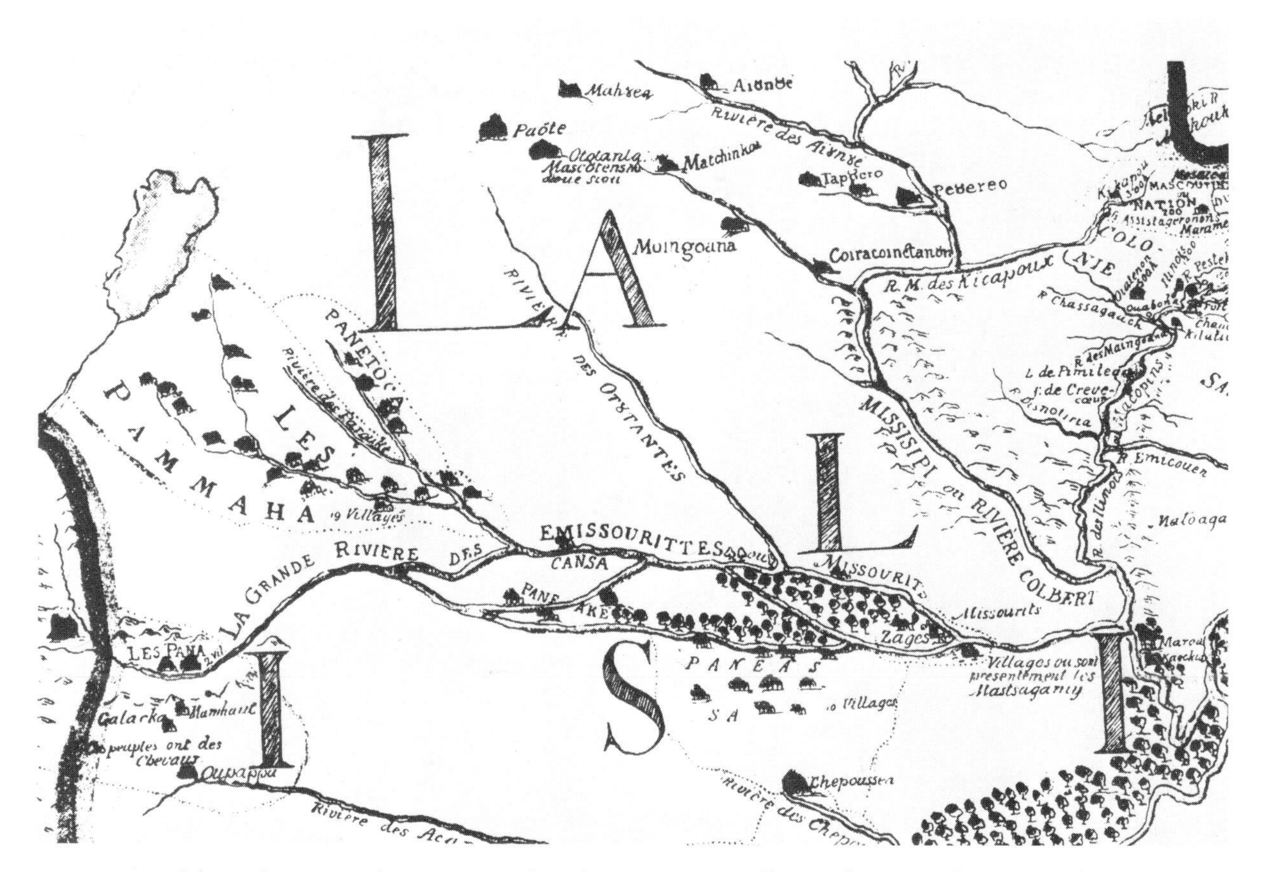

Fig. 2.1. Detail from the Franquelin 1684 map, based in part on La Salle's explorations. (From Temple, Supplement to *Atlas: Indian Villages of the Illinois Country*)

be the lands of the Missouri and the Kansa Indians. The Kansas River was apparently mistaken for the Missouri proper, for the Pawnees (*Panimaha*) are shown on one of three northwesterly affluents of what is called, on the 1684 map and some other versions, *La Grande Riviere des Emissourittes.* The rider . . . *ou des Ozages* appears on the 1699 version.[17]

North of and parallel to the *Grand Riviere,* or Kansas River, is a stream that John Champe identified as the Platte River, with affluents to the north that he believed to represent the Loup forks, since these streams bear village symbols for the Pawnee. French information of the time obviously did not extend much farther upriver than the Platte, because the river divides north of that stream (as identified by Champe): the west fork may represent the Niobrara, and the right fork, the Missouri.[18]

Other contemporary maps duplicate Franquelin's distinct configuration for the lower Missouri River. One of them is a chart by Minet, a French engineer who accompanied the La Salle expedition to the vicinity of the mouth of the Mississippi in 1685. Minet obviously used a La Salle map for the Mississippi and Missouri rivers.[19]

The next, and third, stage in representing the Missouri River is the 1703 Delisle map of North America, which shows the area of concern to us more realistically than any preceding attempt.[20] The map is a landmark in the history of mapping the river if for no other reason than that the sources Delisle used for the map are listed in a document in the Archives Service Hydrographiques in Paris; in most cases there is no record of the actual sources used to produce a given map of this period. The map in question was actually made by Claude Delisle, and not by his son, Guillaume, as the map cartouche claims.[21] Sources for this map include La Salle's map of the Mississippi River as copied by Franquelin in 1684, and Louis de la Porte de Louvigny's 1697 map of the Mississippi.[22]

The 1703 Delisle map is based in large part on the explorations of Pierre-Charles le Sueur. In 1702, under Le Sueur's guidance, Delisle prepared a set of five maps entitled *Carte de la Riviere de Mississippi Sur le memoires de Mr. le Sueur.*[23] These five sheets were basic to the development of the Delisle 1703 maps. The data came from Le Sueur's 1700 ascent of the Mississippi River. He traveled up the river from its mouth as far as the Minnesota River, thence to found a trading post, Fort l'Huillier, on the Blue Earth River. Le Sueur reached the mouth of the Missouri on 13 July 1700; the source for his data on the Missouri is uncertain but was probably derived from local traders or priests living near its mouth. Father Gabriel Marest, a French Jesuit, had settled at the mouth of the Des Pères River (now within the city limits of St. Louis), accompanied by a band of Kaskaskia Indians, in about 1700, and he may have been there at the time of Le Sueur's passage. There may also have been traders at the Tamaroa Indian village across the Mississippi River from Marest's settlement near present-day Cahokia, Illinois.[24] Since Le Sueur reached the Tamaroa village in June and did not pass the Missouri River mouth for another two weeks, there was ample time for him to gather data on the area.

On Delisle's 1703 map the Missouri River is shown as flowing almost directly southeast, in a nearly straight line from a point of origin near the Omaha Indians (*les Maha*), curving to the east only at the mouth of the Osage River (fig. 2.2). The 1702 prototype carried no detail beyond the course of the Missouri River itself, save for the *R. des Ozages* and, near its upper reaches on another sheet, two streams on which the Omaha and Iowa Indians lived.[25] Curiously enough, the general configuration of the river is reasonably accurate for its course as far north as the Great Bend near the Mandan villages, but on Delisle's map this part of the river is compressed by about half, so that it is shown as beginning at a point west and south of the headwaters of the Des Moines River.

The fourth-generation maps are those of Delisle and Mitchell. The finest map of the lower Missouri River to be produced prior to the late 1700s was a product of the exploration of the Missouri by Etienne Vèniard de Bourgmont in 1714. The map illustrates his explorations of the Missouri from its mouth to the Platte River. This manuscript map, although drawn by the famous French mapmaker Guillaume Delisle, did not have the impact one might expect, even on later maps by Delisle himself.[26] In any event, the chart is the earliest map resulting from the observations of a traveler on the Missouri, but it was made almost a half century after the river's discovery by European explorers (fig. 2.3).

The most important and influential map of the French period was the 1718 *Carte de la Louisiane* by Guillaume Delisle. The draft version, dated May 1718, was not changed for the Missouri and its tributaries when it was published the following month.[27] Bourgmont's 1714 data permitted Delisle to delimit the lower reaches of the Missouri River more accurately, and in fact, the course of the stream is well represented as far upriver as the present Nebraska–South Dakota boundary, although the Kansas and Platte rivers are badly distorted (fig. 2.4).

Delisle's map was plagiarized and reproduced essentially in its original form in several languages from the time it was issued until the 1790s.[28] Though John Mitchell's map of 1755 introduced numerous refinements and an improved configuration for the Missouri River below the Kansas River, Mitchell's famous chart illustrates little that is new for the Missouri valley.[29] A great number of maps followed Mitchell's map from the date of its publication until the end of the century, sometimes vying in popularity with variants of the Delisle map of 1718.

The Spanish Period: 1770–1804

France ceded Louisiana to Spain in the secret Treaty of Fontainebleau in November 1762, although residents of Louisiana did not learn of the transaction until late in 1764. The first Spanish officials did not arrive in lower Louisiana for another two years, and it was 1767 before the Spanish actually made their presence felt in upper Louisiana, with the construction of Fort Don Carlos near St. Louis. French con-

Fig. 2.2. Detail from the Delisle 1703 map, based in part on Le Sueur's explorations. (From Tucker, *Indian Villages of the Illinois Country*)

trol of upper Louisiana was not surrendered to Spain formally until 1770.[30]

St. Louis had been founded by the French in mid-1764, a century after the discovery of the mouth of the Missouri. The first settlement on the Missouri River itself, St. Charles, was made in 1769, and the small town of La Charette was founded almost thirty years later, in 1797. Although it was only a few miles from St. Charles, La Charette remained the settlement that was farthest upstream until the time of Lewis and Clark.

Informal trade with tribes on the Missouri continued, although the Spanish made efforts to limit trade to those who were licensed for the purpose. In 1794, the Baron de Carondelet, Louisiana's governor-general, and Jacques Clamorgan oversaw the founding of the Company of Discoverers and Explorers of the Missouri (better known as the Missouri Company). This company of local merchants was intent on exploiting the fur resources of the upper Missouri River.Clamorgan, the director of the company, planned a series of forts on the river and hoped eventually to extend the chain west to the Pacific Ocean. The first exploration of the river by the new company took place in the fall of 1794, when Jean Baptiste Truteau ascended the Missouri as far as present-day central South Dakota.[31]

No maps of the region he was to explore, however, were available to him. As late as 1785, Esteban Miró, the Spanish governor-general of Louisiana in New Orleans, had no general maps of Spanish Louisiana.[32] Nine years later, in November 1794, it

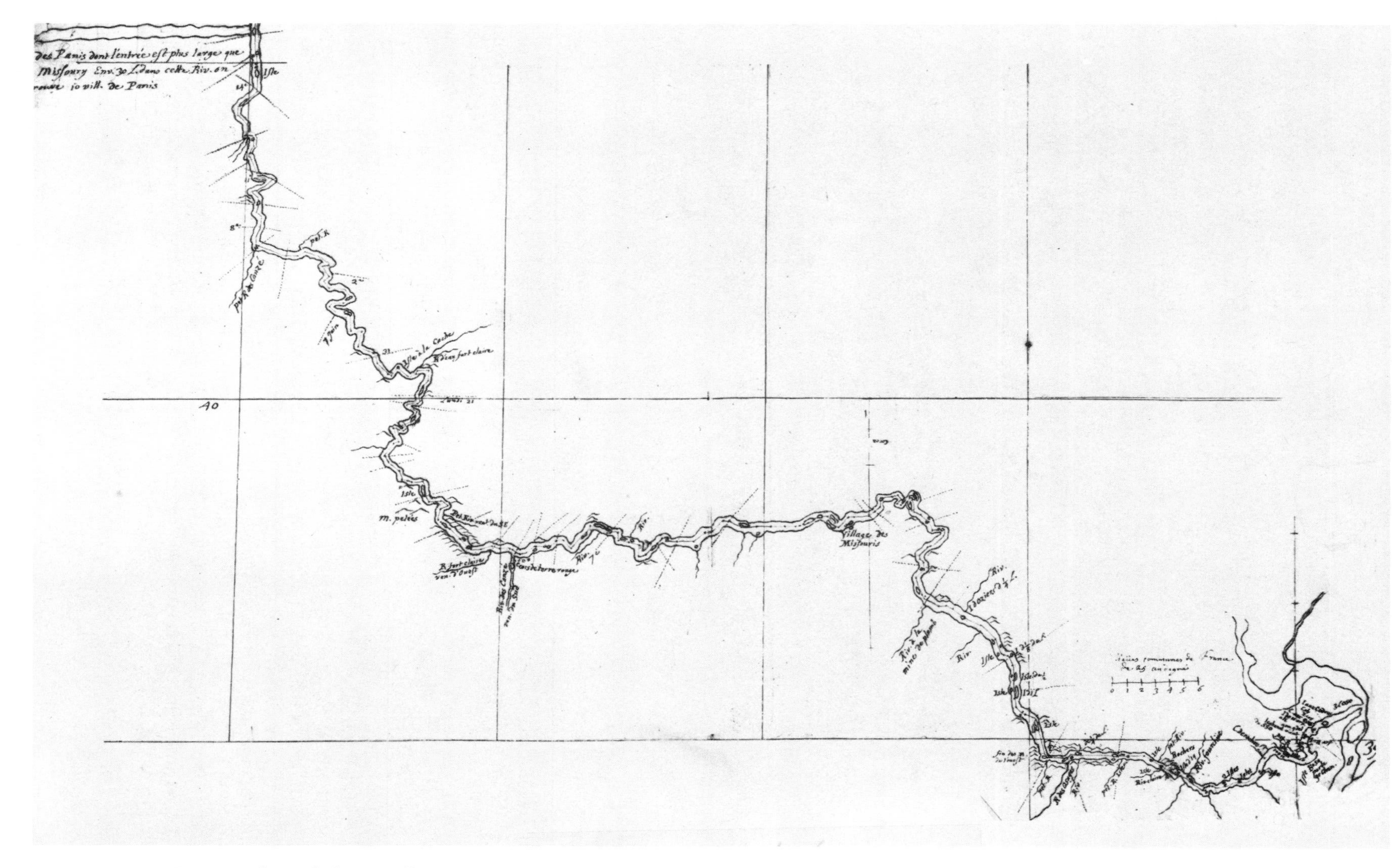

Fig. 2.3. The Delisle map illustrating Bourgmont's explorations in 1714. (Service Historique de la Marine, Chateau Vincennes)

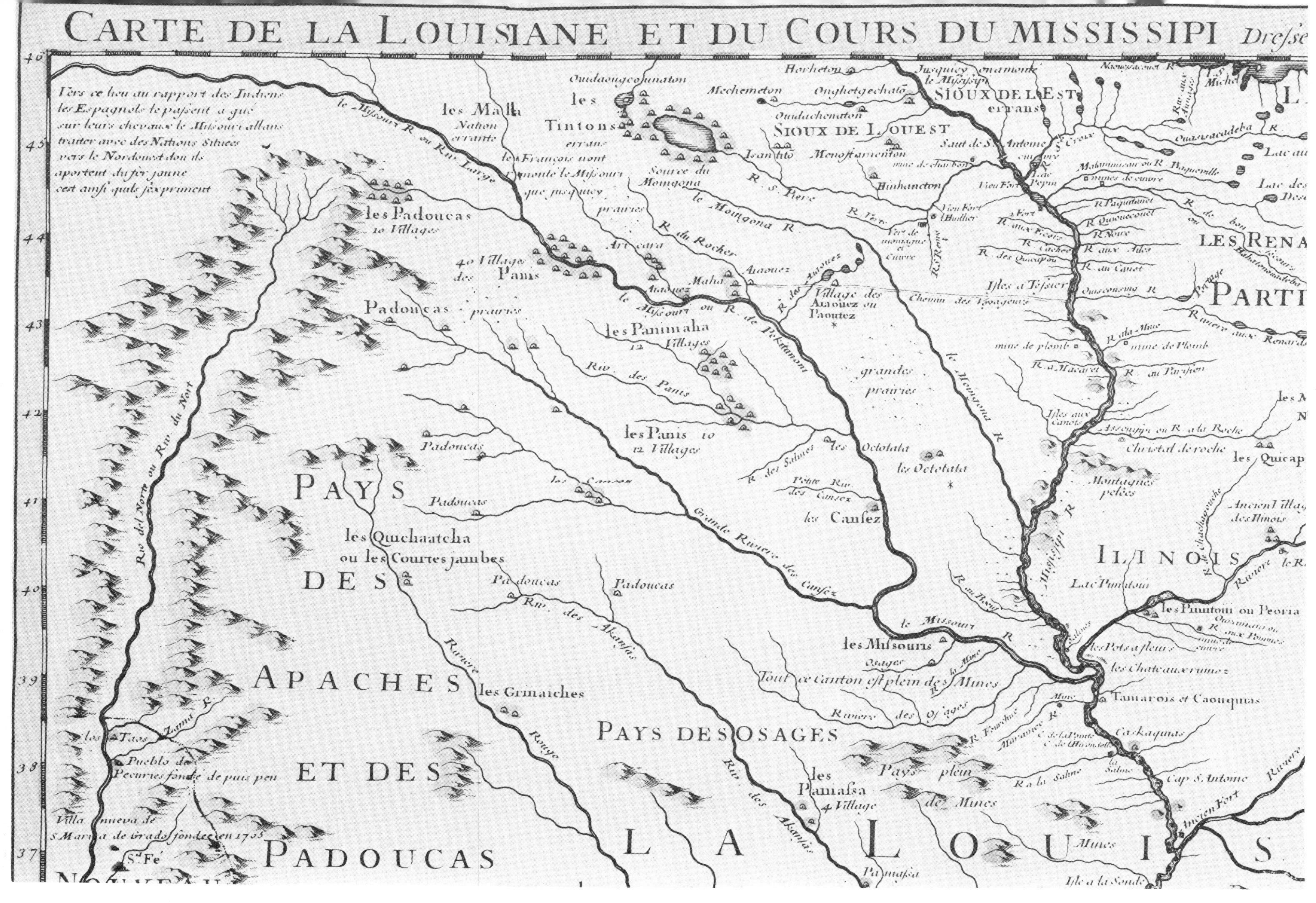

Fig. 2.4. Detail from the engraved Delisle 1718 map. (From Paullin's *Atlas of the Historical Cartography of the United States*)

was his successor, Governor-General Carondelet, who found it necessary to order "a wholly new map prepared for the information of Jean Baptiste Truteau's expedition up the Missouri River." This chart, drawn by Antoine Soulard, provided the fifth generation of Missouri River maps. Although the original map has been lost, no fewer than three copies of it are still extant.[33] Aubrey Diller has observed that the map was "virtually the first original and independent map of the river since Delisle's famous" map of 1718.[34] The chart owes a great deal to Canadian traders, one of whom appears to have been James Mackay, for the characteristics shown on the upper reaches of the Missouri.[35]

The Soulard map is distinguished by its characterization of the "Grand Detour" of the Missouri River in present central South Dakota as an immense U-shaped bend many times its actual size, below which the river is charted relatively precisely. Above the bend, however, the river is almost wholly speculative. Maps based on the Soulard chart carry the distinctively exaggerated Grand Detour and are easily recognized. The Samuel Lewis map of 1804 and, in turn, its derivatives (such as the maps used to illustrate Patrick Gass's account of the Lewis and Clark expedition) are thereby identifiable as such.[36]

Truteau's travels, which may have carried him to modern central South Dakota above the Grand Detour, resulted in no maps of his own. Data from his expedition were, however, incorporated into the narratives and maps of others. The account and maps of General Victor Collot dating to 1796, for example, owe a great deal to Truteau.[37]

Except for the Soulard map, the Spanish produced no significant charts of the Missouri River during the time they ruled upper Louisiana—that is, until 1797. Furthermore, the Soulard map was inaccurate, even for the lower Missouri River, which by now was well traveled. The sixth-generation maps were based on the explorations of James Mackay and John T. Evans. In 1795, Mackay and Evans were sent up the Missouri by the Missouri Company to help open the area to Spanish traders. A map usually referred to as the "Indian Office map," but almost unanimously credited to James Mackay, was produced in St. Louis in 1797 following the return of the Mackay-Evans expedition.[38]

The Indian Office map is a very exact rendering of the river from St. Charles, Missouri, to the Mandan villages. The person responsible for the lower part of the map is unknown, but it is certain that a map produced by John Evans in 1796–97 of the river from the mouth of the Big Sioux River to the Mandan villages was the basis for the upper part of the chart. It is in fact possible that Evans supplied the sketches for the entire map.[39] In any event, the Indian Office map and the map produced by John Evans were used in the construction of no less than ten secondary maps in French, Spanish, and English.[40]

One such map was produced in Paris in 1802 by a French merchant of New Orleans, James Pitot, on the basis of sketches he received from Barthélemy Lafon. Several variants of this map, originally described by Carl I. Wheat simply as "The Mississippi, 1802," have been published.[41] The Missouri River as far upstream as the Mandan villages is well represented. A Soulard-type Grand Detour was retained, but it was pushed upriver into terra incognita, so that the Mandan and the nearby Arikara villages are shown on its eastern or downstream margin (fig. 2.5). This general area marks the locale where the Missouri River makes the great turn to the south that later led to its designation as the "Great Bend," a term sometimes confused with "Grand Detour."

Louisiana was secretly transferred from Spain to France in 1800, but before any effective French control could even be contemplated, Napoleon resold the territory to the United States in May 1803. It was not until the following March, in 1804, that the Spanish formally relinquished control of upper Louisiana to the French, and the next day the United States took possession of the region.[42]

The American Period: 1804–1895

The seventh generation of maps are those produced by Lewis and Clark. Our knowledge of the maps generated by the two captains during their 1804–1806 expedition has been significantly updated by several recent studies.[43] Before the expedition left Camp Dubois near St. Louis, the captains had accumulated the latest and most reliable charts of the Missouri River then extant anywhere, including copies of those by Soulard, Evans, and Mackay. During the expedition itself, Clark (occasionally assisted by Lewis) produced maps of their entire route

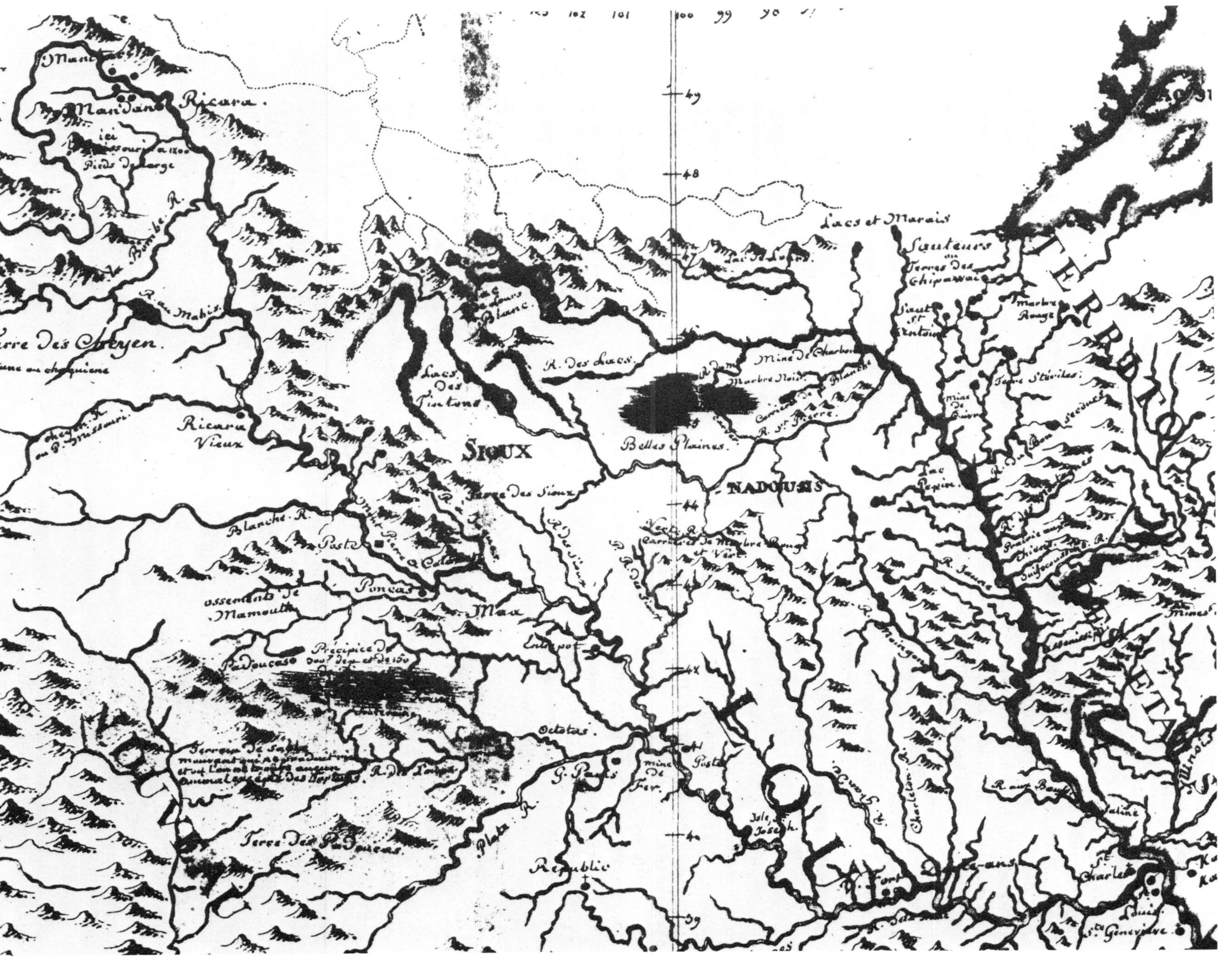

Fig. 2.5. Detail from the Pitot/Lafon 1802 map. (Service Historique de la Marine, Vincennes)

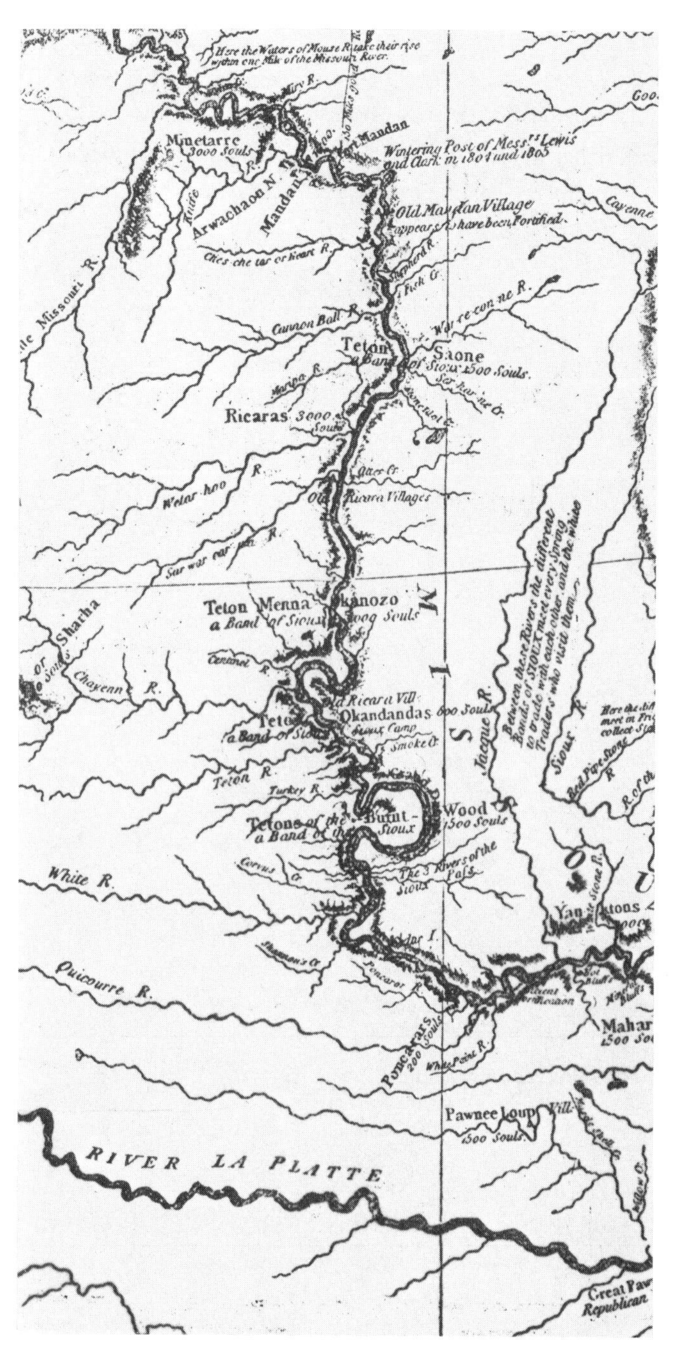

Fig. 2.6. Detail from the Samuel Lewis engraving of William Clark's map of 1810. (From Biddle, *History of the Expedition*, 1814)

across the continent. There was no public release of these maps until 1814, with the publication of Nicholas Biddle's *History of the Expedition* (fig. 2.6).[44] The general map in that volume, engraved by Samuel Lewis, was prepared from an 1810 chart by William Clark that synopsized the expedition's many detailed route maps.

No less than four members of the Lewis and Clark entourage kept journals that have survived. Only one of them (Gass's) was illustrated by a map, and it was no more than a version of the pre–expeditionary Samuel Lewis map of 1804—a map showing no improvements for the upper Missouri over its predecessor from the Spanish era, the 1795 Soulard map. One original map was prepared with the assistance of Robert Frazer, a member of the expedition, for a book that was never published. This curious map leans heavily on the 1796–97 Evans map, apparently having been drawn by a French cartographer in St. Louis in 1807.[45]

The Clark map engraved by Samuel Lewis remained the standard chart of the Missouri River for more than forty years after its publication by Biddle. The advent of steamboat navigation on the river, however, brought on an avalanche of new and improved maps.

A craft called the *Yellow Stone* was the first steamboat to ascend the Missouri River as far as the Yellowstone River, completing the round trip in 1832. Not long after this memorable voyage, charts of the river by military passengers began to appear as steamers made more and more frequent trips upstream. These charts provide the eighth generation of Missouri River maps.

One of the first of this group was a map prepared by Joseph N. Nicollet and his assistants for the Corps of Topographical Engineers. It was made in 1839 during an ascent of the river on the steamboat *Antelope* from the Gasconade River, in present-day Missouri, to Fort Pierre, in central South Dakota. Nicollet's detailed sectional map includes geological data, the tortuous path of the steamboat up the river channel, Indian camps along the river, and nightly stopping places along the route.[46]

One of Nicollet's fellow passengers for part of this journey was a Jesuit, Father Jean-Pierre De Smet. This priest, later to become famous on the western frontier, boarded the boat at a point near modern

Council Bluffs, Iowa, for an excursion of about 360 miles upstream. His destination was a Dakota Indian camp at the mouth of the Vermillion River, near present-day Elk Point, South Dakota. Father De Smet left a detailed map of this thirteen-day expedition. The precision and detail of this chart, far more elaborate than any of his other manuscript maps, is easily explained: it is a close copy, although not a tracing, of part of Nicollet's map of this part of the Missouri.[47] De Smet's map, however, is more useful historically than Nicollet's charts in that he added data to his version that were ignored by Nicollet—the locations, for example, of many Indian camps and villages of the period near present-day Omaha, Nebraska.

In the next fifteen to twenty years, military cartographers produced several general maps that brought the mapping of the river to a higher standard of precision. Isaac I. Stevens's three-part map appeared in 1855: sheet 2 shows the Missouri River from Fort Pierre to its headwaters.[48] In 1853, Lt. A. J. Donelson, a member of Stevens's party then exploring a railroad route from St. Paul to the Pacific Ocean, surveyed the Missouri River from its mouth to a point just west of Fort Union (near the modern boundary between North Dakota and Montana), where he met with Stevens's main party. Unfortunately, Donelson's maps were "mostly lost afterwards on the Isthmus of Panama, and [Stevens's] map was made from incomplete notes."[49] Stevens's map west of Fort Union, however, was more accurate; with its more precise rendering of the Missouri's route through what is today Montana, it represents an important contribution to the mapping of the river.

In 1855 and 1856 Lt. Gouverneur K. Warren mapped parts of the Missouri River on charts that were basic to the production of his general map of 1857, which showed the Missouri from its mouth to its source. The latter map was based on all previous surveys and explorations known to Warren, and was acclaimed by Wheat as superior to all of its predecessors in detail.[50] Warren's 1856 manuscript map of the Missouri River from the Big Nemaha to the Big Muddy River, sixty miles west of Fort Union, was the first significant eyewitness map of that portion of the river to be produced since Clark made his own charts, since most of Donelson's charts made three years earlier had been lost. Made to the same scale as a fifteen-minute United States Geological Survey quadrangle, Warren's manuscript maps show many details of historic significance, such as those in the vicinity of Fort Clark, North Dakota (fig. 2.7). Many of these details were deleted from his general map of 1857.[51]

The surveys of Capt. W. F. Raynolds and Lt. H. E. Maynadier of the topographical engineers in 1859 and 1860 charted western South Dakota and southern Montana. Their "Map of the Yellowstone and Missouri Rivers," published in 1860, set a new standard for the upper Missouri River.[52] The stage was now set for the next step in mapping; modern cartographic standards were about to arrive.

The ninth-generation maps were engraved sectional charts. In the 1860s, sectional maps were made for the upper Missouri River, illustrating the channel and valley of the river in great detail and designed expressly for the use of steamboat pilots. One of the first such charts, made in 1867 by Maj. C. W. Howell, was a manuscript map in seventeen sheets showing the Missouri between the Platte River and Fort Benton, Montana. His sketches, "made from the Pilot House of a steamboat, while in motion," were collated and arranged on the basis of Raynolds's 1860 map.[53]

By 1890 the Missouri River Commission had completed the task of secondary triangulation for the Missouri River. Modern mapping had arrived, and the river was now mapped with precision from its mouth to Three Forks. Two important sets of maps date from this period: the Missouri River Survey maps, dated 1892, and the Missouri River Commission maps, published in 1892–95 in eighty-four individual sheets. Engineer O. B. Wheeler described some of the conditions the surveyors faced in completing the secondary triangulation near Fort Benton, Montana, in 1889:

> The season was unusually dry, and the river never known so low. The smoke from the mountain forest and prairie fires was very dense, for the Indians were burning the prairies that the buffalo bones could the more easily be secured for market at the railroad stations.[54]

With the publication of these beautifully and delicately engraved charts, the Missouri River was at last revealed in detail, including channel depths,

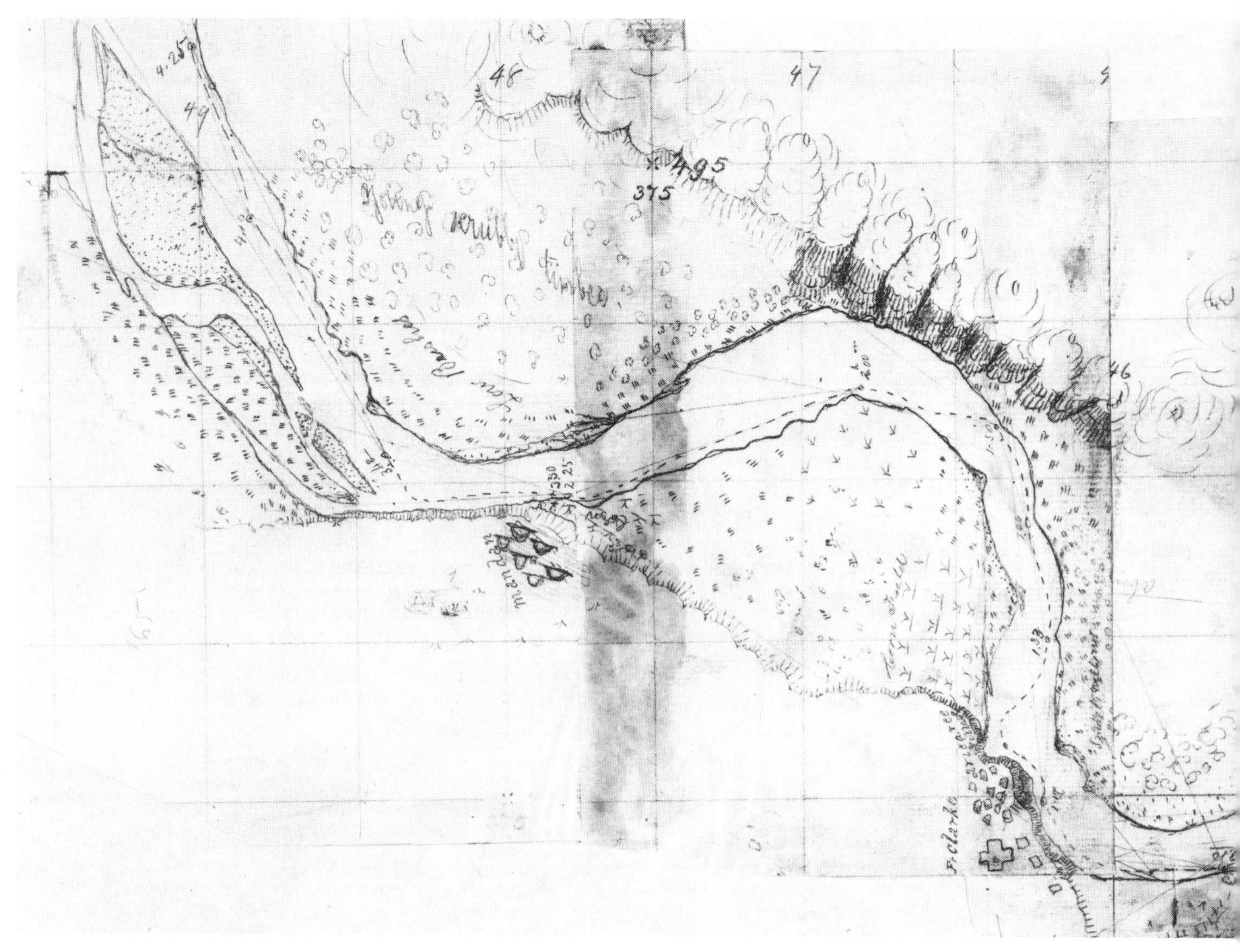

Fig. 2.7. Detail from the Warren 1856 manuscript map of the Missouri River, showing Fort Clark and its environs. (National Archives)

for its entire course. Later mapping made only minute improvements on these maps, principally in recording its ever-changing channel and in documenting the destruction of the river as a free-flowing natural stream in the post–World War II period. The Pick-Sloan construction program created six massive earthen dams across its valley that flooded nearly eight hundred miles of the river through the Great Plains states of North and South Dakota and Montana, transforming a once-proud river into what has been called a "federal canal."

This abbreviated account of the mapping of the Missouri River does little justice to the richness of the historical data available. Such a survey nevertheless permits us to envision the major steps by which the stream was gradually revealed to the world by the explorers and surveyors who charted this part of the American West.

Notes

The author thanks Robert E. Karrow, curator of maps at the Newberry Library, Chicago, Illinois, for his patience in providing access to the Karpinski map collection.

1. Phil E. Chappell, "A History of the Missouri River," *Transactions of the Kansas State Historical Society for 1905–1906* 9 (1906): 237–94; and Stanley Vestal, *The Missouri* (New York: Holt, Rinehart & Winston, 1945).
2. A. P. Nasatir, *Before Lewis and Clark*, 2 vols. (St. Louis: St. Louis Historical Documents Foundation, 1952).
3. Raphael N. Hamilton, "The Early Cartography of the Missouri Valley," *American Historical Review* 39 (1933–34): 645–62. See also Hamilton's Ph.D. dissertation, "A Cartography of the Missouri Valley to the Establishment of 'La Compagnie d'Occident,' 1717" (St. Louis University, St. Louis, 1931).
4. Basic sources for this history include John Logan Allen, *Passage through the Garden* (Urbana: University of Illinois Press, 1975); Aubrey Diller, "Maps of the Missouri River before Lewis and Clark," in *Studies and Essays in the History of Science*, ed. by Ashley Montague (New York: Henry Schuman, 1946), pp. 505–19; various studies by Father Jean Delanglez, cited below; Sara Jones Tucker, *Atlas: Indian Villages of the Illinois Country*, Illinois State Museum Scientific Papers, vol. 2, no. 1 (Springfield, 1942); Carl I. Wheat, *Mapping the Transmississippi West: 1540–1861*, 5 vols. (San Francisco: Institute of Historical Cartography, 1957–63); and citations following.
5. U.S. War Department, "Supplement to the Report of the Chief of Engineers," *Annual Report of the War Department for . . . 1902* (Washington, D.C.: GPO, 1902), pp. 175–83.
6. Bernard DeVoto, *Course of Empire* (Boston: Houghton Mifflin, 1952), p. 52.
7. Francis Borgia Steck, "The Jolliet-Marquette Expedition, 1673," *Catholic University of America Studies in American Church History* 6 (Chicago: Catholic University of America, 1928).
8. Father Jean Delanglez, "The Jolliet Lost Map of the Mississippi," *Mid-America*, n.s., 28 (January 1946): 67–144.
9. The Marquette "autograph map" is reproduced in Tucker, *Indian Villages*, pl. 5; the original is in the Archives de la Compagnie de Jésus (Quebec).
10. DeVoto, *Course of Empire*, pp. 121–22.
11. Father Jean Delanglez, *Hennepin's Description of Louisiana* (Chicago: Institute for Jesuit History, 1941), pp. 108–109.
12. Wayne C. Temple, *Supplement to Atlas: Indian Villages of the Illinois Country*, Illinois State Museum Scientific Papers, vol. 2, no. 1 (Springfield, 1975), pl. 60, contains a legible reproduction of this map.
13. Father Jean Delanglez, *Some La Salle Journeys* (Chicago: Institute for Jesuit History, 1938), p. 78.
14. Mildred Mott Wedel, "The Identity of La Salle's *Pana* Slave," *Plains Anthropologist* 18 (August 1973): 203–17.
15. Gilbert J. Garraghan, *Chapters in Frontier History* (Milwaukee: Bruce Publishing Co., 1934), pp. 56–57; Garraghan cites Pierre Margry, *Découvertes et Établissements des Français dans l'Ouest et dans le Sud de l'Amérique Septentrionale, 1614–1754*, 6 vols. (Paris, 1879–86), 2: 203, 325–26.
16. Father Jean Delanglez, "Franquelin, Mapmaker," *Mid-America*, n.s., 25 (January 1943): 34.
17. Temple, *Supplement to Atlas*, pl. 59.
18. John L. Champe and Franklin Fenenga, "Notes on the Pawnee," in *Pawnee and Kansa (Kaw) Indians*, ed. by David Agee Horr (New York: Garland Publishing, 1974), p. 49.
19. Tucker, *Indian Villages*, pl. 7.
20. Ibid., pl. 13.
21. Father Jean Delanglez, "The Sources of the Delisle Map of America, 1703," *Mid-America*, n.s., 25 (July 1943): 276–77.
22. Ibid., pp. 285–87; and Tucker, *Indian Villages*, pl. 14.
23. The activities of Le Sueur have been confused by a long and tangled publication record. For a review of his activities and a clarification of that record, see Mildred Mott Wedel, "Le Sueur and the Dakota Sioux," in *Aspects of Upper Great Lakes Anthropology*, ed. by Elden Johnson (St. Paul: Minnesota Historical Society, 1974), pp. 157–58.
24. Gilbert J. Garraghan, "The First Settlement on the Site of St. Louis," *Mid-America*, n.s., 20 (October 1927): 342–47; and Nasatir, *Before Lewis and Clark*, 1: 6.
25. The original map is in the Archives Service Hydrographiques, Paris, cataloged as 138BIS 3–2; a photocopy is in the Karpinski collection in the Newberry Library, Chicago.
26. The map was first recognized in 1979 by Mildred and Waldo Wedel in the Service Historique de la Marine, Chateau Vincennes, catalog number AM BSH 69–20. A description of the map by Elizabeth R. P. Henning may be found in W. Raymond Wood, *An Atlas of Early Maps of the Midwest*, Illinois State Museum Scientific Papers, vol. 18 (Springfield, 1983), p. 1, pl. 1.
27. The draft version was published by Tucker, *Indian Villages*, pl. 15; the published version is available in many sources, including Charles O. Paullin, *Atlas of the Historical Cartography of the United States*, ed. by John K. Wright (New York: Carnegie Institution of Washington and American Geographical Society of New York, 1932), pl. 24.
28. Hamilton, "Early Cartography," p. 655.
29. Ibid., p. 658; and Temple, *Supplement to Atlas*, pl. 70.
30. William E. Foley, *A History of Missouri* (Columbia: University of Missouri Press, 1971), 1: 14.
31. Nasatir, *Before Lewis and Clark*, 1: 84–91.
32. Ibid., p. 119.
33. Ibid., 2: 253: Wheat, *Transmississippi West*, 1: 157, and maps 234–35a.
34. Aubrey Diller, "A New Map of the Missouri River Drawn in 1795," *Imago Mundi* 12 (1955): 175–80.
35. Nasatir, *Before Lewis and Clark*, 1: 96; and Wheat, *Transmississippi West*, 1: 158, note 5.
36. The Lewis 1804 map may be found in reproduction in Paullin, *Historical Cartography*, pl. 28; and in Patrick Gass, *A Journal of the Voyages and Travels of a Corps of Discovery under the Command of Capt. Lewis and Capt. Clarke . . .* (Philadelphia: Matthew Carey, 1810); see also Wheat, *Transmississippi West*, 2: 14–15, and map 300.
37. Victor Collot's map of the upper Louisiana region was

not published until 1826 in *A Journey in North America,* 2 vols., one atlas (Paris: Arthur Bertrand, 1826), although the data on the map relate to the year 1796.

38. Although circumstantial evidence favors Mackay as the original author, the actual draftsman of the Indian Office map is unknown; see Diller, "Maps of the Missouri River," pp. 513–16; and W. Raymond Wood, "Notes on the Historical Cartography of the Upper Knife-Heart Region," Report prepared for the Midwest Archeological Center, Lincoln, Nebraska (Lincoln, 1978), pp. 26–31.

39. Gary E. Moulton, *Atlas of the Lewis and Clark Expedition,* vol. 1, *The Journals of the Lewis and Clark Expedition* (Lincoln: University of Nebraska Press, 1983), p. 6.

40. W. Raymond Wood, "The John Evans 1796–97 Map of the Missouri River," *Great Plains Quarterly* 1 (Winter 1981): 51, note 1.

41. Henry C. Pitot, *James Pitot (1761–1831): A Documentary Study* (New Orleans: Louisiana Landmarks Society, 1968), pp. 47–52, details the origins of this map. Various versions may be found in reproduction in the Pitot study, p. 47; in Nasatir, *Before Lewis and Clark,* vol. 1, opposite p. 110; and in Wheat, *Transmississippi West,* vol. 1, map 255.

42. Foley, *History of Missouri,* pp. 63–71.

43. The most recent summary is that of Moulton, ed., *Atlas of Lewis and Clark Expedition,* and the sources cited therein.

44. [Nicholas Biddle, ed.], *History of the Expedition under the Command of Captains Lewis and Clark . . . 1804–5–6,* prepared for the press by Paul Allen, 2 vols. (Philadelphia: J. Maxwell, 1814).

45. Wheat, *Transmississippi West,* 1: 46–48, map 286.

46. Edmund C. Bray and Martha Coleman Bray, *Joseph N. Nicollet on the Plains and Prairies* (St. Paul: Minnesota Historical Society, 1976), pp. 135–69; and the Joseph N. Nicollet Papers, vol. 2, part 2: sheets 354–418, in the Manuscripts Division, Library of Congress.

47. Hiram Martin Chittenden and Alfred Talbot Richardson, *Life, Letters and Travels of Father Pierre-Jean DeSmet, S. J., 1801–1873,* 4 vols. (New York: Francis P. Harper, 1905), 1: 179–90. The original maps are in the Jesuit Provincial Archives, St. Louis; they are cataloged as C-8, Atlas, item 11, in the microfilm copy in the Vatican Film Library, Pius XII Library, St. Louis University, St. Louis.

48. Wheat, *Transmississippi West,* 4: 71–72, map 865.

49. Lt. Gouverneur K. Warren, *Memoir to Accompany the Map of the Territory of the United States from the Mississippi River to the Pacific Ocean,* 33d Cong., 2d sess., H. Exec. Doc. 91, vol. 11 (1855), p. 67.

50. Wheat, *Transmississippi West,* 4: 84–91, map 936.

51. National Archives, Record Group 77, Q579/1–40.

52. Wheat, *Transmississippi West,* 4: 183–87, map 1012.

53. National Archives, Record Group 77, Q137.

54. U.S. War Department, *Annual Report of the Chief of Engineers,* part 4 (Washington, D.C.: GPO, 1890), p. 3398.

3. PATTERNS OF PROMISE

MAPPING THE PLAINS AND PRAIRIES, 1800–1860

John L. Allen

DURING the great drive of the American people to the Pacific, the vast area lying between the Mississippi River and the Rocky Mountains was, for the better part of the nineteenth century, a zone of passage rather than a region of settlement. "Crossing the plains" became an epithet for what, to many, was a tedious but necessary part of a long journey to the dramatic Rockies, the exotic Southwest, or the bucolic Pacific Coast. In the romanticism that gripped America during the years between the opening of the nineteenth century and the Civil War, the supposedly featureless plains were largely devoid of the symbols that lured Americans—both spiritually and physically—to other areas. Yet, unappealing as the plains may have been, they had to be crossed in the migration toward the Pacific. Among all the regions of western North America, the plains were unique in the extent to which they were traveled and mapped long before they were settled permanently by an American population.

It was even longer before the plains were understood; the fact that they were mapped and traveled does not suggest that they were known in any proper sense of the word. Although much progress was made in the mapping of the region between 1800 and 1860, there was little understanding of the plains environment at the time of settlement following the Civil War. The settlers themselves were uncertain about what the plains really were, perhaps because the great empty spaces presented too few visual images that were familiar or concrete. Where there is little to focus on, the mind's eye may behold a great deal and, not knowing what it sees, see only what it knows.

Such was the case with the many and varied groups who mapped the plains prior to the Civil War.[1] The visual images of the plains that were recorded on the maps tended to support the preconceptions of the cartographers as they focused on those things that were of critical interest to them. The groups who were eventually responsible for the mapping of the plains—trappers and travelers, merchants and missionaries, soldiers and surveyors—were charting not always what they found, but sometimes what they wanted to find. Their work was a mapping of the geography of hope and expectation rather than the geography of reality—a mapping of patterns of promise.

1880–1810: Early Exploration and Maps of the Plains

As the nineteenth century opened, maps of the plains were based almost entirely on imagination and conjecture. Knowledge of the area was sketchy, at best, and on most maps of the period the great vastness of the plains was exaggerated. This is particularly true of the products of commercial cartographers, the primary source of geographic information on the plains in the early 1800s. In a classic sample of this type of cartography the Philadelphia engraver Samuel Lewis shows the plains as

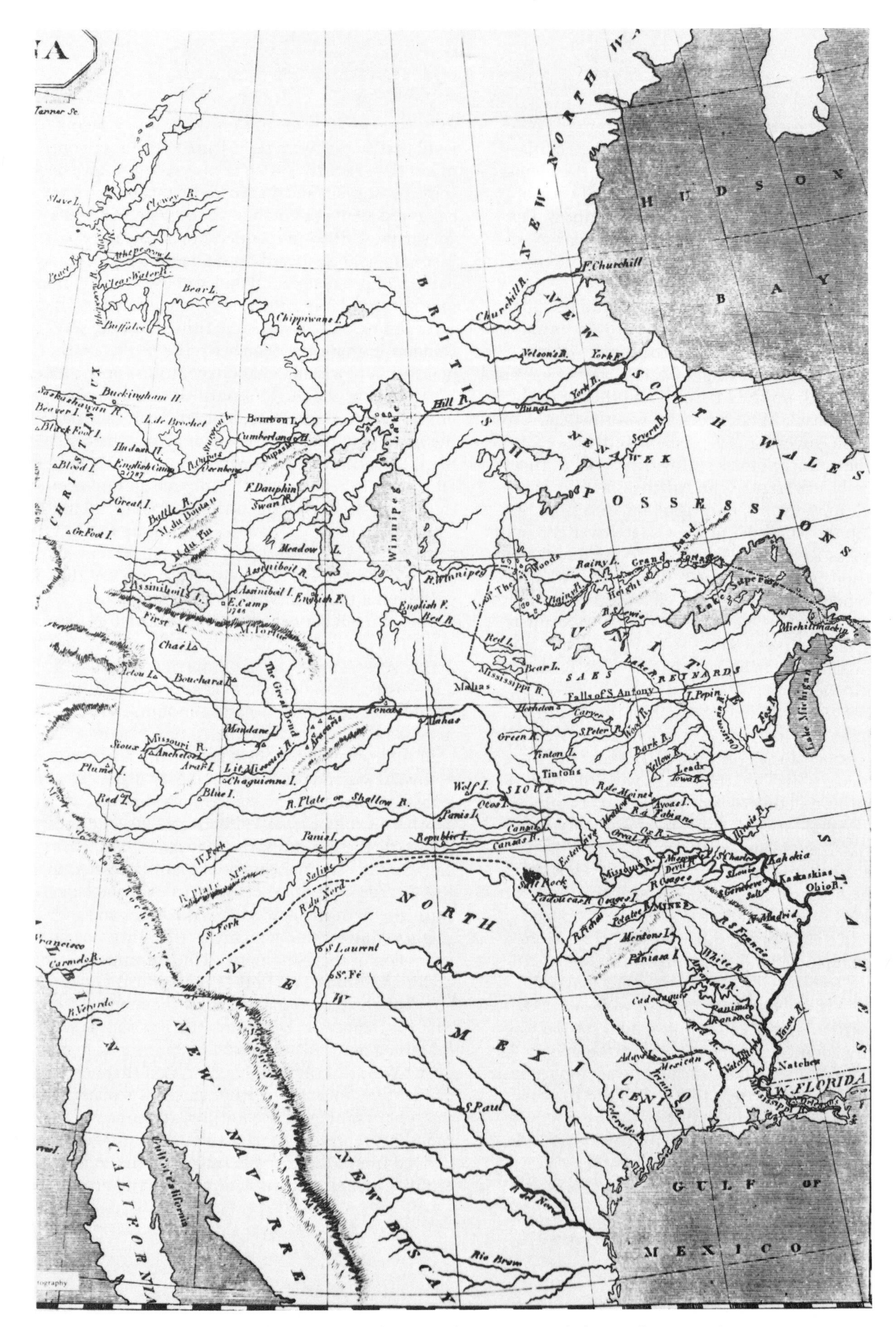

Fig. 3.1 "North America" by Samuel Lewis, 1804. (Library of Congress)

an immense open region stretching from the Mississippi to a range of mountains near the Pacific Coast (fig. 3.1). A few major streams of the plains are shown on this map, but they are sadly out of place and their courses are mostly imaginary. During the decade from 1800 to 1810, exploration was the primary agency by which some of these misconceptions began to be swept away.

Two major "official" explorations into the plains took place during the first decade of the century: Lt. Zebulon Pike's journey to the upper Arkansas and the expedition of Capts. Meriwether Lewis and William Clark up the Missouri and thence to the Pacific and back.[2] Pike was sent west by Gen. James Wilkinson, governor of Louisiana Territory, to explore the country of the plains tribes and to investigate the headwaters of the Arkansas and Red rivers. His trek in the summer of 1806 led him from St. Louis up the Missouri to the Osage River and on to the Kansas River, the Arkansas, and the Rockies. In a vain hope of locating the source of the Arkansas, Pike wandered into the tortured terrain of the southern Rockies, where he was captured by a Spanish patrol. Taken first to Santa Fe and then Chihuahua, he eventually made his way back across Texas to Natchitoches.

On his travels Pike saw much of the country Wilkinson had instructed him to map, as well as a good portion of the southern plains that he had not bargained for (fig. 3.2). His maps were the first of the southern plains to bear the stamp of "official surveys," and for their time, they were reasonably accurate.[3] Two major errors appeared on Pike's picture of the plains, however: one was his insistence that the headwaters of both the southern and northern plains rivers were in the same region of the Colorado Rockies; the other was his confusion of the Canadian River (a tributary of the Arkansas) with the upper Red River. Both errors plagued geographers and cartographers for well over a decade, particularly because Pike's information was hungrily adopted by commercial cartographers. In spite of his errors, Pike's mapping was important if for no other reason than that it reduced the east-west dimensions of the plains to their proper proportions, thus correcting an erroneous impression created by most previous maps of the region.

The mighty plains rivers on Pike's map, with their sources adjacent to those of waters flowing toward the Pacific, were represented with more optimism than accuracy. Although there may have been some pessimism and disillusionment in Pike's verbal assessments of much of the area he mapped as "desertlike," the interconnecting headwaters of eastward- and westward-flowing rivers became a pattern of promise that deluded those who studied maps of the plains for nearly forty more years.

At approximately the same time that Pike was wandering about the open stretches of the southern plains, the Lewis and Clark expedition was traversing the plains via the Missouri River, far to the north.[4] Sent forth by President Jefferson to locate the most practicable water route across the continent, Lewis and Clark believed, both before and after their expedition, in the inherent promise of the plains streams as the ultimate Passage to India. Their route in 1804–1805 took them up the Missouri to its head, across to Columbian waters, and down to the Pacific. On their return in 1806 they covered not only the Missouri but most of the course of its major tributary, the Yellowstone, as well. They did not locate a passage between the upper Missouri and the Columbian source region, but crossed the mountains via a tortuous four-hundred-mile stretch of rugged mountain terrain instead of a short portage (fig. 3.3).

In their mapping of the rivers of the northern plains, Lewis and Clark believed they had discovered a passage farther to the south. The Missouri's major southern tributary, the Yellowstone; its southern tributary, the Big Horn; the Platte; the Arkansas; and even the southern-most of plains streams, the Rio Grande—all had their sources near the headwaters of both the Lewis or Snake River and the mighty Multnomah, which was thought to be an extension of the Willamette. This information was obtained from the captains' geographical views after their explorations and from the maps of the first fur trappers in the plains and Rockies. It was presented on Clark's large 1810 manuscript map of the West and the published version that accompanied their journals.[5] Although Clark's maps were the best available on the northern plains for decades, his representation of the southern plains was derived from Pike and was subject to the same errors. Chief among those errors was the promising

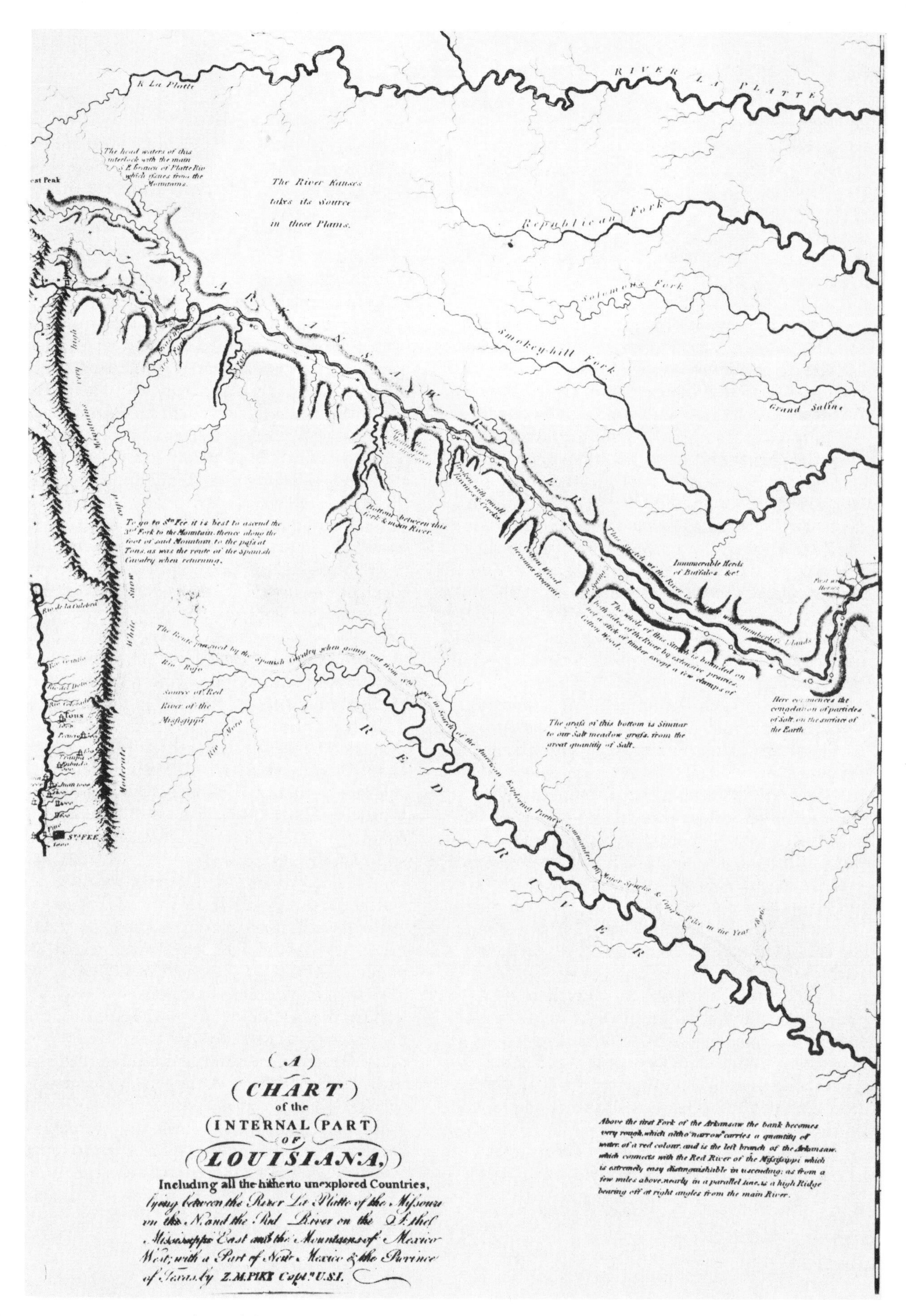

Fig. 3.2. "A chart of the internal part of Louisiana" by Zebulon Pike, 1810. (Library of Congress)

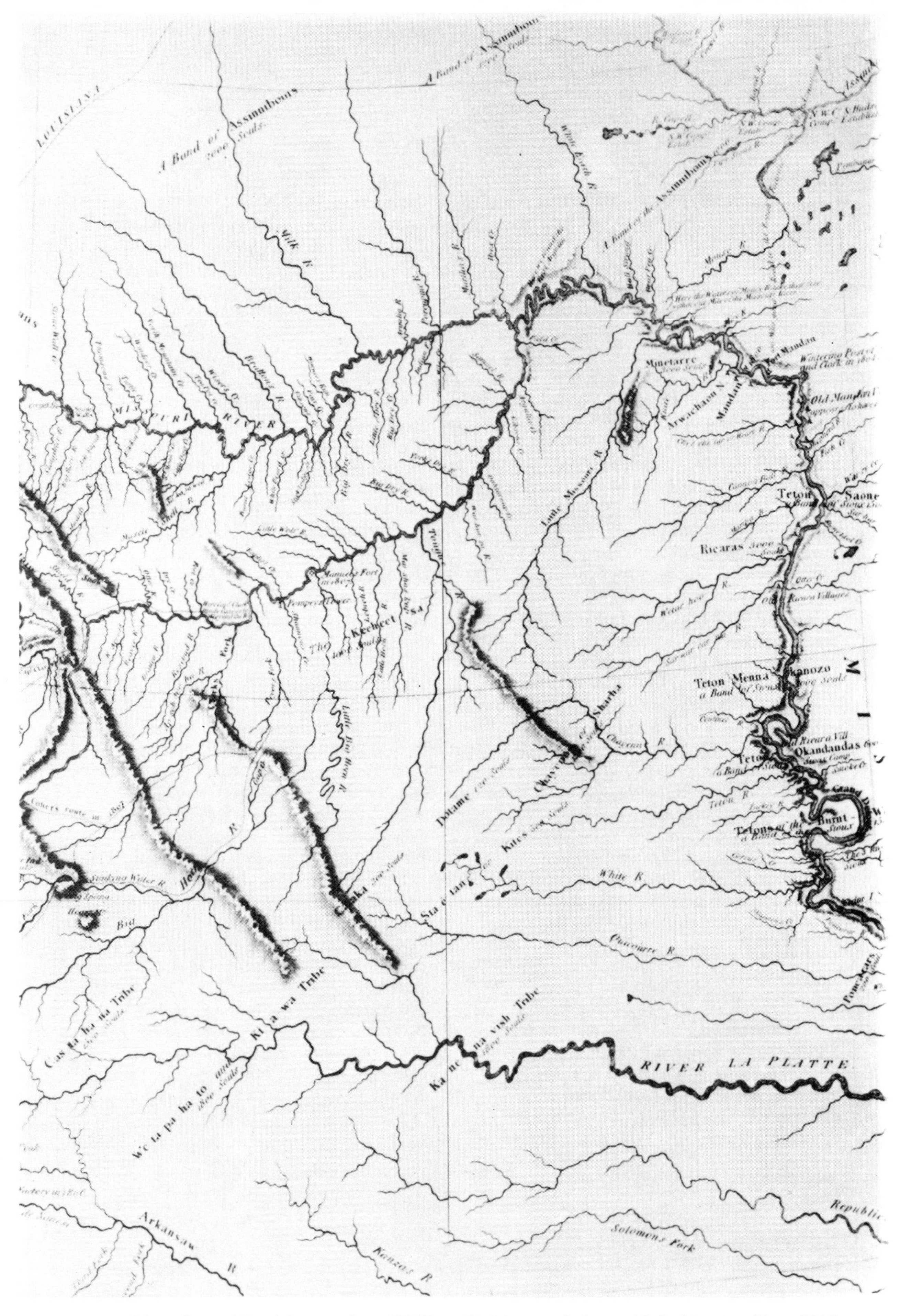

Fig. 3.3. Detail from Samuel Lewis's engraving of William Clark's map of 1810, published in 1814. (From Biddle, *History of the Expedition*, 1814)

but misleading pattern formed by the interlocking headwaters of plains streams and the rivers that drained to the Pacific.

1810–1820: Commercial Cartographers Discover the Plains

The great explorations of Lewis and Clark and of Pike continued to bear cartographic fruit into the second decade of the century. It was too early for fur trader lore to be incorporated into maps, and most of the significant maps of the decade were produced in the commercial ateliers of London and Paris.[6] But perhaps the greatest map of the decade was one produced not in the ateliers of Europe or even the eastern United States but in Natchez. This was the large map by Dr. John Hamilton Robinson, a member of Pike's party in 1806 (fig. 3.4). From the Platte north, Robinson followed the published version of the Lewis and Clark map. South of the Platte he relied on Pike's journals and map, on hearsay, and on hope. Near the western margins of the southern plains where commercial cartographers had shown areas of "light ash sand," a suggestion of aridity, Robinson found silver mines and added the notation "excellent wines made here." There were no pessimistic desert connotations for Robinson; the interstitial areas separating the plains rivers are clearly marked "prairies." Finally, via his colored boundaries, Robinson achieved an annexation that did not occur in reality for nearly thirty years; the whole of Texas and the region west and north through New Mexico is enclosed within a line bearing the legend "Limit of the United States." Robinson's map was an augury of the future rather than a reflection of the past, and among all the maps of the decade it most clearly depicted the patterns of promise.[7]

1820–1830: An Explorer, An Engraver, and An Entrepreneur

The decade of the twenties was important for the mapping of the plains, particularly by commercial cartographers. Of the three most significant maps of the decade, however, only one was a commercial production. A map produced by a government explorer was another influential cartographic contribution, not just during the decade but for the next generation. A third notable map of this period was not even a finished product, but a manuscript of field sketches.

The first map, Henry Tanner's "North America" from his 1822 atlas, had a great influence on later commercial cartography, particularly in its representation of the areas lying west of the crest of the Rockies.[8] Tanner's depiction of the plains was also important in that it fairly accurately indicated the state of geographical knowledge about that region in the early 1820s. The northern section of the plains on Tanner's map was drawn from Lewis and Clark data. Tanner made use of Pike's material as well, and depended on the work of Stephen Long for most of the detail for the central and southern plains.[9] Because of its relatively small scale, the Tanner map necessarily simplified much of the geography of the plains. Within those limitations, it was a reasonable, if not perfect, depiction of the region in its time.

The second map, compiled by Stephen Long in 1823 after his exploration of the central and southern plains in 1819–20, may have been one of the most important of all maps of the plains produced before the Civil War (fig. 3.5).[10] Long had traveled up the Platte River from Council Bluffs to the Rockies (via the South Platte) and then southward along the Front Range to the Arkansas River and across the plains to what he believed, on the basis of Pike's map, would be the source region of the Red River. Locating what he assumed was the upper Red, Long turned eastward and finally realized that he was on the Canadian Fork of the Arkansas, thus clearing up this misconception in plains cartography. Long's map, drawn as a result of these travels, was a "master map" in that it was followed by many later cartographers. This great map correctly depicted the course of the Platte to the forks and then the South Platte, although Long failed to get the North Platte with any degree of accuracy. Far to the south, the mistaken identification of the Canadian River with the Red River that had appeared on earlier maps was corrected. The country between the Platte and the Arkansas was shown with remarkable accuracy for the time and the nature of the geographical data. Far to the north, the Yellowstone and Big Horn were shown with their sources near their true location, correcting another major false impression from the

Fig. 3.4. "Map of Mexico, Louisiana, and the Missouri Territory" by John Robinson, 1819. (Library of Congress)

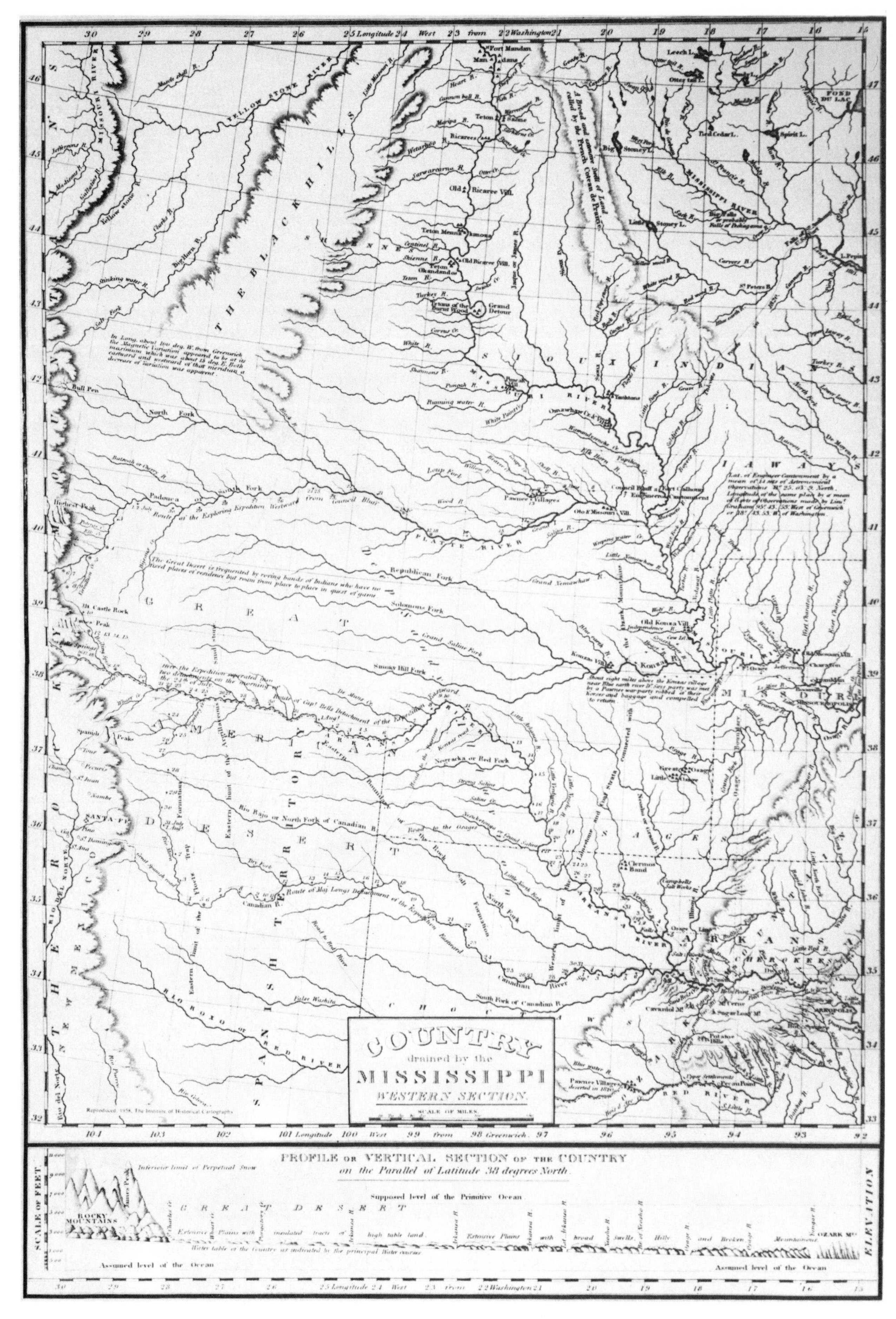

Fig. 3.5. "Country drained by the Mississippi, western section" by Stephen Long, 1823. (Library of Congress)

earlier Lewis and Clark and Pike maps. Like most maps of the plains, however, Long's map reflected the disjunction between mapping what was seen and seeing what was mapped, for across the western edge of the plains, in the area between the Platte and the Red, appear these words in large and bold type: "GREAT AMERICAN DESERT." Although Long was not the first to so label the area, his desert colored the view of the plains for many who read his map. Finally, Long made his contribution to the region's future by showing "the Great Spanish Road," a major portion of which soon became the Santa Fe Trail.

The third important map of the decade was, unlike the Long and Tanner maps, devoid of any features of the imagination in terms of rivers, mountains, or fuzzy estimations of land quality. Joseph Brown mapped and surveyed the Santa Fe Trail shortly after Long's prophetic rendering of the "Great Spanish Road," and although Brown's sketches were never published, they may have been accessible to later surveyors.[11] Closed to American traders during the period before American independence, Santa Fe and Taos began to welcome trade with Americans in 1821, and by 1824 enough trade was flowing across the southern plains to require an accurate survey of the road by the government of the United States. In 1826 Brown made the first survey from Fort Osage to Santa Fe, and his manuscript field sketches of the route might have been of great value both to traders and to the military in the southern plains. Although crudely drawn and demonstrating little or no care in topographic representation, Brown's sketches were relatively accurate because they were based on careful observation and instrument readings.[12] While other maps of the period may have been more important as base maps for printed government or commercial charts, Brown's use of instrumentation was significant. In this case, the pattern of promise was not just in the presaging of trade on the Santa Fe Trail but in the use of a method of obtaining data that became standard in later mapping of the plains.

1830–1840: Scientist, Soldier, and Priest

The decade of the thirties represents a watershed in the cartography of the plains. For nearly the first time, the geographical lore of the fur trade, which had been active in the West since the return of Lewis and Clark, began to work its way in small increments onto published maps. Naturally enough, most of the new information stemming from the fur trade concerned the Rockies and what lay beyond, and most of this material did not become available to cartographers for another decade or more. But the fur trappers and traders knew the plains intimately, and to fail to mention them in the annals of mapping the region would be to ignore the existence of one of the most brilliant bodies of geographical discovery in the nineteenth century.[13]

Three other mapping efforts of the decade had greater immediate significance, partly because they represented increments to earlier views of the plains that appeared immediately on maps, and partly because they signaled the entry of new groups into the cadre of those responsible for charting the plains before the Civil War. These new groups were military men (as opposed to army explorers and engineers), missionaries, and scientists.

The first of the three most important maps of the decade was drawn primarily to show the location of Indian tribes. Maps of tribal location in the plains had been made since 1809, when William Clark, then superintendent of Indians, made a manuscript map locating key tribal groups. As government strategies began to dictate the possible movement of Indian populations westward, other maps were drawn by government agents.[14] Finally, in the summer of 1835, a detachment of dragoons under the command of Col. Henry Dodge was sent westward across the plains to the Rockies with a mission of locating tribal patterns. Accompanying this expedition was Lt. Enoch Steen. His manuscript map of the dragoons' route shows both the state of geographical knowledge on the plains and tribal patterns on the frontier in the mid-thirties.

The second significant map of this period also shows Indian locations, but presumably for different reasons. In 1838 missionary Samuel Parker drew a large "map of Oregon Territory" that represents the northern plains with a fair degree of accuracy in terms of their key details, particularly the area between the Platte and the Missouri (fig. 3.6).[15] For one of the first times on a published map, the Yellowstone system is shown in its proper con-

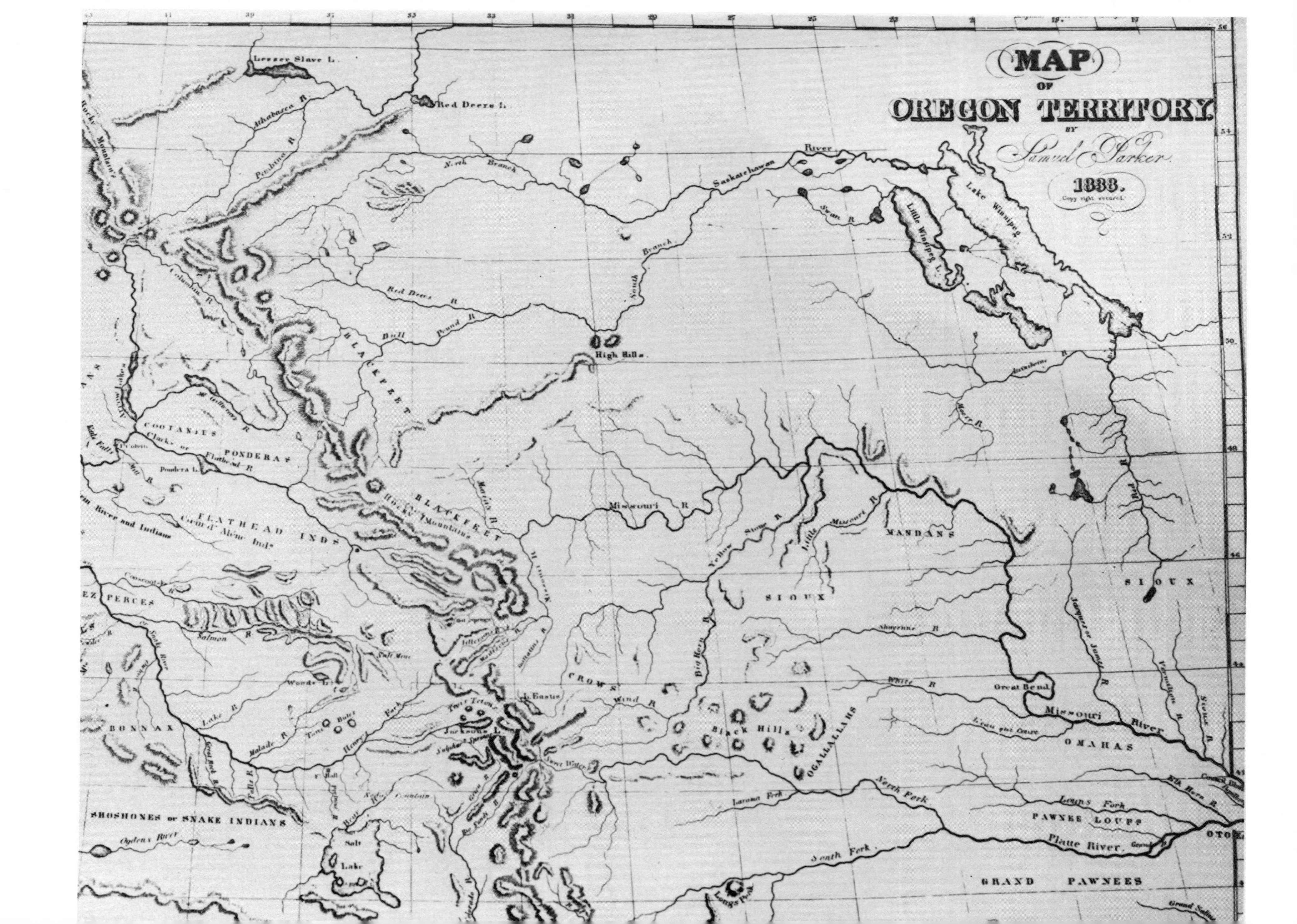

MAP
OF
OREGON TERRITORY
BY
Samuel Parker
1838.
Copy right secured.
Lesser Slave L.
Red Deers L.
Athabasca R.
Pembina R.
North Branch
Saskatchawan
River
Swan R.
Lake Winnipeg
Little Winnipeg L.
South Branch
Red Deers R.
Bull Pound R.
High Hills
BLACKFEET
COOTANIES
Clarks, or Flathead R.
PONDERAS
Pondera L.
FLATHEAD INDS
Coeur d' Alêne Inds
Moose R.
Missouri R.
Marias R.
Yellow Stone R.
Little Missouri R.
MANDANS
SIOUX
Rocky Mountains
Salmon R.
EZ PERCES
Salt Mine
Big Horn R.
Shayenne R.
White R.
Great Bend
Missouri River
Jaques or James R.
Vermillion R.
Sioux R.
CROWS
Wind R.
L. Eustis
Two Tetons
Jacksons L.
Sulphur Springs
Sweet Water
Black Hills
OGALLALLAHS
L'eau qui Coure
OMAHAS
Woods R.
Lake R.
BONNAX
Malade R.
Three Butes
Henrys Fork
Great Rock R.
Falls R.
SHOSHONES or SNAKE INDIANS
Ogdens River
Salt Lake
Green R.
Laramie Fork
North Fork
Loups Fork
Elk Horn R.
PAWNEE LOUPS
Platte River
OTO
South Fork
Longs Peak
GRAND PAWNEES
Grand Saline

figuration and the North Platte is accurate enough to suggest that Parker must have had access to fur trade lore. As on virtually all other maps of the plains until just before the Civil War, the Black Hills—a key feature of northern plains geography—wander about the country between the North Platte and the Missouri, stretched far out of proportion toward the south and west. Long's Peak is curiously detached from the main chain of the Front Range to stand as a single sentinel out in the plains to the east. These errors were significant in that Parker's map was widely read, particularly among potential migrants, many of whom were soon crossing the plains.[16]

The third key map of the thirties is less relevant in its representation of the plains than in its symbolism and promise of what was to come during the succeeding decade. This was the large map of the hydrographic basin of the Upper Mississippi, drawn by an immigrant French scientist, Joseph Nicollet, and his young American assistant, John Charles Frémont.[17] Nicollet possessed excellent scientific training and introduced to western cartography—not only through his own map but especially through his training of Frémont—a measure of scientific precision that had heretofore been lacking. Nicollet's map of the Mississippi's hydrographical basin was produced on the basis of nearly 100,000 instrument readings and 326 astronomical point observations. Although his contribution to plains cartography was restricted to the territory north and east of the Missouri, it was, nevertheless, the first really accurate map depicting any portion of the trans-Mississippi West. Nicollet's concept of the hydrographical basin, which was foreign to American science, bore fruit in Frémont's mapping efforts during the next decade.

1840–1850: The Great Decade in Plains Cartography

While the maps by the U.S. Army Corps of Topographical Engineers are inarguably the most important maps of the decade of the forties for the West in general, many other distinctive maps were produced during that period, particularly for the area of the plains. The trickle of maps in the beginning of the century had become a flood by the end of the 1840s, and during the decade the number of different groups engaging in mapping the plains increased greatly. Many of the plains cartographers of the forties represented vested interests; the notion that much of the critical work done in mapping the plains focused on promise rather than reality was at no time more true than during the period between 1840 and 1850.

Examples of such optimistic maps were the Santa Fe Trail charts of Josiah Gregg, important for their delineation of a vital trading link across the plains, for their description of the southern plains terrain, and for their wide circulation.[18] The most important Gregg map was the large map of "Indian Territory" published with his *Commerce of the Prairies* in 1844 (fig. 3.7). This map covers the entire southern half of the plains, from the Platte on the north to the Rio Brazos and Rio Colorado of Texas on the south. Showing a multitude of wagon roads, identified by their discoverers or primary users, and depicting "prairie" and wooded country, the 1844 map is also marked by a stippled pattern that may represent sandy areas adjacent to the watercourses. In the western plains, the Gregg map shows innumerable settlements, divided by symbols into forts and trading posts; Indian villages; and Spanish towns, villages, and "ranchos." Conveying the impression of a well-populated region, the map must have whetted the interest of prospective traders on the trail to New Mexico. Finally, in a concession to geographic reality. Gregg mapped—for the first time—the Llano Estacado, or Staked Plain, which he identified as "Arid Table Land nearly 2000 feet above the Streams." A blend of optimism and reality, Gregg's map was certainly one of the best of the southern plains before the Mexican War.

A work of nearly equal creditability for the northern plains, although carrying less detail than Gregg's map, was the first cartographic effort of the Belgian Jesuit missionary, Father Pierre Jean De Smet.[19] This map, published in France and compiled from several cartographic products of U.S. government mappers of the thirties and forties, is important in its illustration of the consensus view of the plains during the mid-forties and, more critically, in its demonstration of the variety of the mappers of the plains. De Smet's map faithfully renders that portion of the plains between the Platte on the south and the

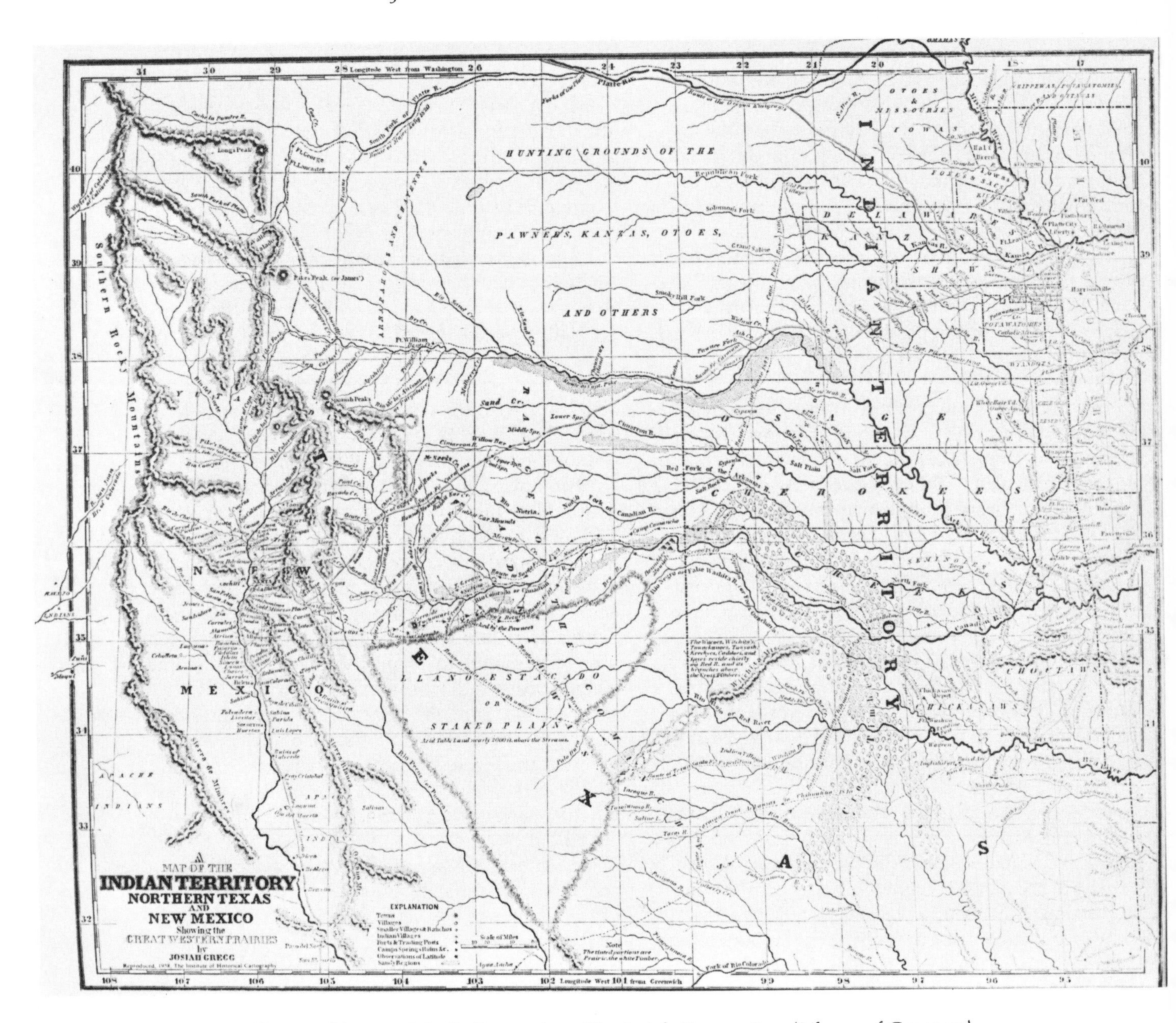

Fig. 3.7. "A map of the Indian territory" by Josiah Gregg, 1844. (Library of Congress)

Saskatchewan on the north, showing the North Platte, the Yellowstone, and the Missouri in essentially correct detail and depicting an almost unbelievable number of tributaries for each of those streams, many of them identified with the names by which they are known today. A dotted line shows what was fast becoming the Oregon Trail, and key landmarks along that vital folk migration route are identified. In a curious omission, there are no Black Hills. This migratory and mysterious upland region seems to have been as absent from De Smet's cosmography as from the images of earlier mappers of the plains. But De Smet missed little else, and his representation of the plains was one of the more precise efforts of the decade among those maps drawn by persons who were not trained surveyors or topographical engineers.

In addition to merchants such as Gregg and missionaries such as De Smet, other "lay" or "amateur" cartographers were active in drawing at least parts of the plains in the forties. Foremost among these were migrants or travelers along the Oregon, California, and Mormon trails across the plains. Most of the trail-related maps did not, however, dwell at length on the plains; the end of the journey was more alluring than the lands along the way. Those emigrant maps that are relevant in the context of plains cartography are those that were either done to illustrate emigrant guides or designed to accompany published journals of the transcontinental trek. Representative of the first group was a map of "Routes to California and Oregon," published with Oliver Steele's *Western Guide Book and Emigrants' Directory* in 1849, and limited in its portrayal of the plains to showing the major emigration routes, along with the plains rivers those routes paralleled.[20] Steele's map is probably derived from the Pike and Long maps of an earlier generation. The major rivers are shown with a fair degree of accuracy, but the mostly unnamed tributaries to those streams are laid out in a highly regular fashion, as if the cartographer were using calipers to measure precisely the distance between them so they would all enter into their parent streams at equal distances from one another. To the west, along the base of the Rockies, Steele resurrected Long's "Great American Desert," stretching from the Platte to the Red River. The map's rendition of the plains is striking in its simplicity—it is clear that the purpose of the map was to get travelers across the empty spaces as quickly as possible to the promise that lay beyond.

Among the second group of migrant and traveler maps, those accompanying journals of travels in the West, the cartographic efforts of Rufus B. Sage are both representative and among the finest examples of the genre. Sage's map of 1846, drawn to accompany his *Scenes in the Rocky Mountains,* is outstanding in its portrayal of the territory east of the Rocky Mountains (fig. 3.8).[21] From the Missouri on the north to the Canadian River on the south, Sage drew as accurate a map of the plains as any mid-nineteenth-century cartographer's, except for the maps of topographical engineers. His delineation of the courses of virtually all the major plains streams and their tributaries is nearly without fault; he identified both the Oregon and California trails with care and precision; and he located, as accurately as any, the territories of the major plains tribal groups. Like other cartographers of the period, Sage did not have the Black Hills correctly, showing them as a linear chain running northwest from the Sweetwater to the Missouri. In a concession to both the patterns of promise and the pessimism that were evident among mappers of the Plains, Sage's "Great American Desert" sprawls in flourishing letters across the plains south of the Arkansas, while in the heart of "proposed Ne-Bras-Ka Territory," straddling the Platte and identified in even more florid style, are the "Grand Prairies."

Another group whose work appeared in the forties was private citizens who were interested in mapping the plains for the purpose of locating what would become the ultimate nineteenth-century symbol of promise—a network of railroad lines linking the Mississippi Valley with the Pacific. Chief among these proponents of a new Passage to India was New Yorker Asa Whitney. Whitney's 1849 book, *A Project for a Railroad to the Pacific,* was graced with a small outline map limning the prospective rail routes across the plains.[22] It was not Whitney's purpose to show the geography of the plains with any degree of precision. The plains were an area to be crossed, and the central feature of his map was the depiction of the crossings themselves—or at least the proposals for such crossings. The plains are presented in the barest possible detail. Nothing adorns the map except a half-dozen or so place names and river names. But dominating the map, for all to see, are the straight lines of what Whitney hoped would become iron roads, linking forever the American territory east and west of the plains crossing. There is no clearer example of the patterns of promise than those bold lines etched across the face of the empty plains by Asa Whitney.

While hordes of emigrants were crossing the plains, sometimes drawing sketch maps as they went, and while other untrained but competent

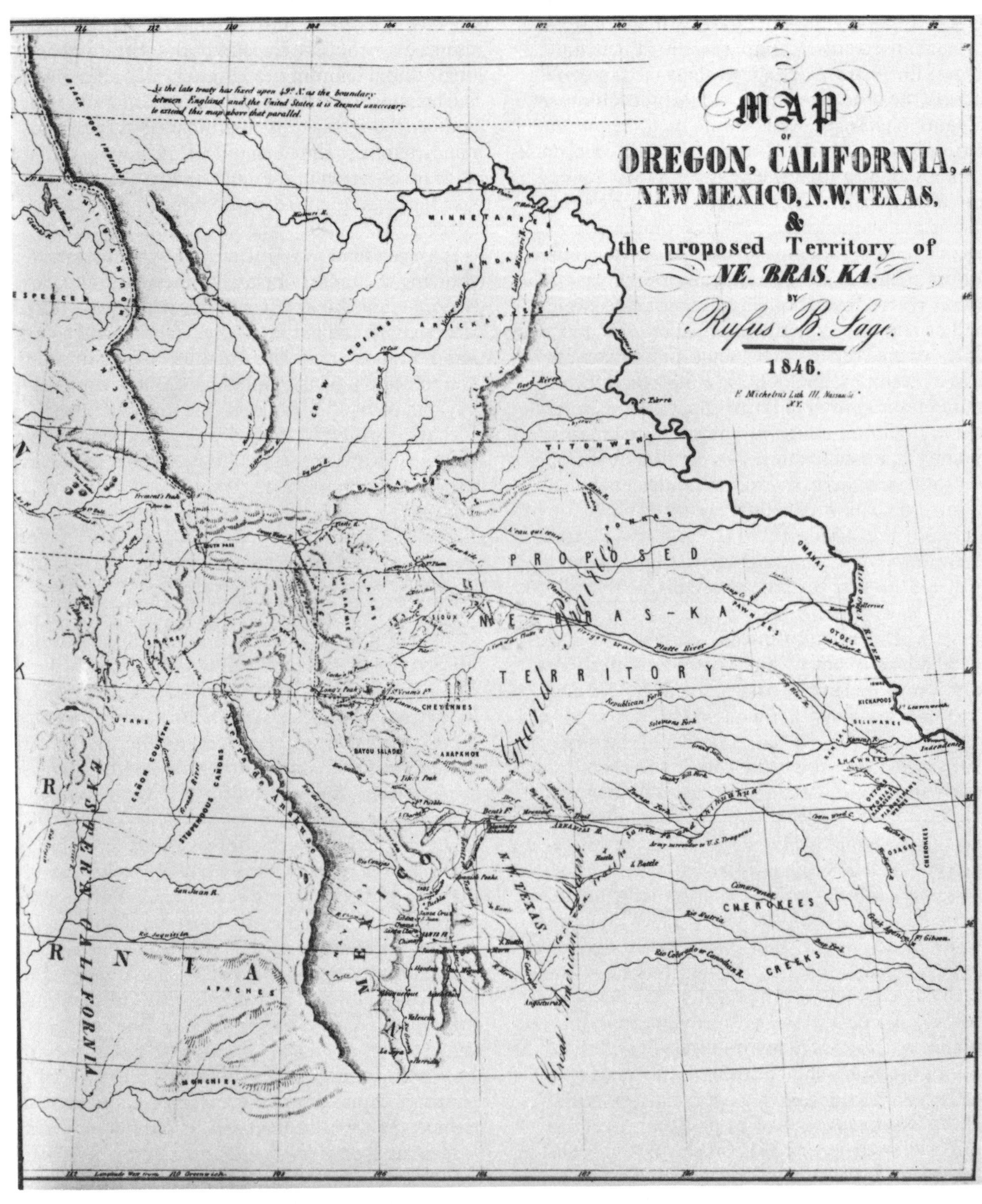

Fig. 3.8. "Map of Oregon, California, New Mexico, N.W. Texas, & the proposed Territory of Ne-bras-ka" by Rufus B. Sage, 1846. (Author's collection)

traveler-cartographers were drafting maps of western geography, a small group of well-trained and highly skilled cartographers was engaged in constructing the official maps of the West, including the plains, that must stand as the most important contributions to plains cartography in the decade of the forties and through the next ten years as well. These men were members of the U.S. Army Corps of Topographical Engineers.[23] Their outstanding maps, although more prosaic than the renditions of other draftsmen, were a great deal more precise. The work of the topographical engineers can be exemplified by the cartographic labors of two men in the middle years of the decade: Lts. John C. Frémont and W. B. Franklin. Each of these men produced or was directly responsible for the production of major maps. Each of them faithfully represented plains geography as it had been measured, not simply seen. It was not their task to amaze and enchant their audience. Nor were they trying to make a point about the suitability of the plains for agricultural settlement or as a prospective highway to the Pacific. Rather, in the spirit of science as it was emerging at West Point, where many of the engineers were trained, they viewed their job to be the true and accurate rendering of the topography in detail. In this, given the limitations of their time and place, they succeeded.

Typical in this regard was Frémont's large map of 1845, showing the West from the Missouri to the Pacific and illustrating his 1842 expedition up the North Platte to South Pass as well as his expedition of 1843–44 in which he crossed the Rockies, viewed the Great Salt Lake, circumscribed the Great Basin, and in general saw more of the west in a shorter period of time than almost anyone had.[24] Frémont's map was, in his own words, "strictly confined to what was seen and to what was necessary to show the face and character of the country." The portion of the map showing the "face and character" of the plains was therefore restricted to the lines of Frémont's crossings and delineates the Platte and Kansas systems, along with a short portion of the Arkansas River east of the mountains. There are vast areas of bare paper on the map—not suggesting that Frémont did not know anything of those regions but that, as a good scientist, he was confining his published statement to what he had been able to observe and measure. The courses of the rivers as laid down on Frémont's 1845 map are correctly represented for the first time in both latitudinal and longitudinal coordinates, as are key landmarks such as Chimney Rock and Scotts Bluff along the Oregon Trail. Although the real significance of Frémont's map lies in his depiction of the farther West, his accurate charting of the Platte and Kansas systems is in itself an outstanding contribution to plains cartography.

The 1845 map of W. B. Franklin was drawn to accompany the report of Col. Stephen Watts Kearny's expedition to the Rockies in the summer of 1845.[25] Although considerably smaller in scale and scope than Frémont's map, the Franklin map is similar in its reliance on measurement as well as observation. Franklin utilized Frémont's data for the Platte system and filled in some of the blank spaces on Frémont's map for the Arkansas River below Bent's Fort. Like Frémont's map, Franklin's production contains a lot of blank space—and for the same reason.

1850–1860: A Process Completed and a New Promise Begun

The last years of the forties were momentous for plains cartography. The new demands produced by the war with Mexico and by the discovery of gold in California meant that it was even more crucial that the best and fastest routes across the plains be located with some precision. Since these events brought people across the southern reaches of the plains, the area most frequently shown on maps derived from these twin imperatives was the region from the Arkansas to the Rio Grande. The most notable maps of the plains that were produced in the early 1850s were based primarily on source materials from either the war or the forty-niners' rush to California. As the fifties progressed, four new mapping incentives came to the fore: the surveys of the United States boundary in the north and south; the surveys for a railroad to the Pacific; the mapping of the routes to the newly discovered gold fields of the Rockies; and the surveys the U.S. Land Office made as the plains were prepared for the entry of agricultural settlers.

The first year of the decade saw the publication of a map that effectively distilled the knowledge gathered during the Mexican War and, at the same

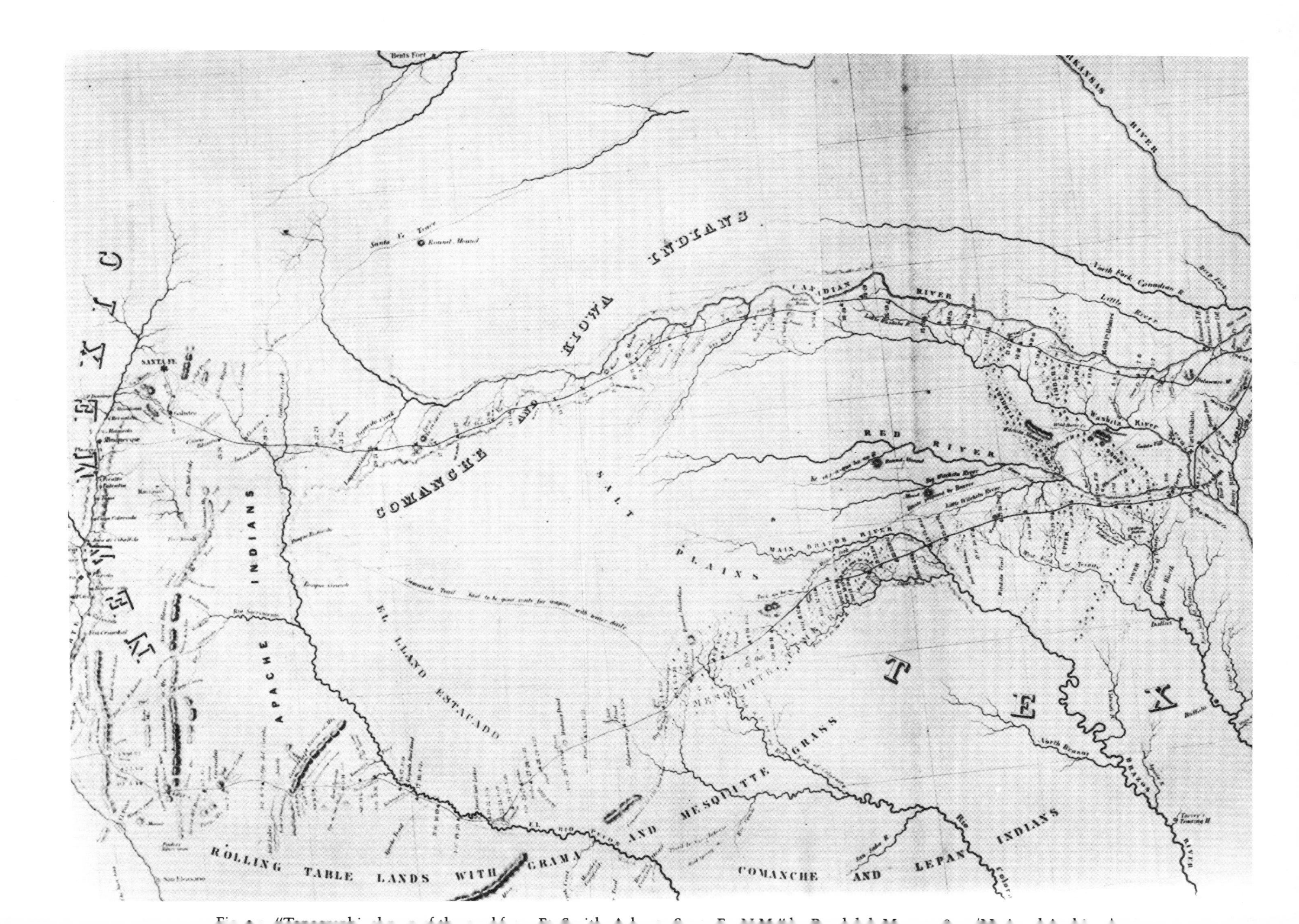

Bents Fort
ARKANSAS RIVER
North Fork Canadian R.
Deep Fork
Little River
Santa Fe Trace
Round Mound
KIOWA INDIANS
CANADIAN RIVER
Little Washita R.
Washita River
RED RIVER
Round Mound
Big Witchita River
Little Witchita River
MAIN BRAZOS RIVER
West Fork of Trinity
Witchita Trail
UPPER
LOWER
Dallas
SANTA FE
Galisteo
Pajarito Creek
Gallinas Creek
COMANCHE AND
SALT PLAINS
INDIANS
APACHE
Rio Sacramento
Bosque Redondo
Comanche Trail
EL LLANO ESTACADO
MESQUITTE GRASS
MESQUITTE AND
EL RIO PECOS
ROLLING TABLE LANDS WITH GRAMA
COMANCHE AND LEPAN INDIANS
North Brazos
BRAZOS RIVER
Buffalo
Torrey's Trading H
San Elzeario
Colorado
Rio
M E X I C
N
T E X

time, reflected the new demand of California migrants for accurate travel directions across the southern plains. This map was drawn to accompany the report of Capt. Randolph B. Marcy's 1849 expedition from Fort Smith, Arkansas, to Santa Fe.[26] Marcy's expedition had as its primary objective the discovery of a practical and reasonably easy trail to California via the southern route. Using data obtained on this journey by compass, chain, and viameter readings, and combining that data with War Department maps of the Mexican War campaigns, Marcy constructed a map of the southern plains that was both accurate and informative, carrying information on land quality as well as distance and direction (fig. 3.9). Marcy showed the Cross Timbers country, noted the presence of mesquite timber between the Brazos and the Pecos, and identified the salt plains that lie north of the Brazos, near the head of the Red River. He described the territory between the Pecos and the Rio Grande as "rolling table lands with grama and mesquite grass," and although he placed the legend "El Llano Estacado" north of the Pecos, he made no commentary on the character of that region. All in all, Marcy's map was an optimistic appraisal of both the land and the ease of passage across the southern plains. Indeed, in his journals, Marcy noted that over the greater portion of the area he covered, the terrain was so perfectly level that "it would appear to have been designed by the Great Architect of the Universe for a railroad"—a prophetic statement that was soon tested by the surveys for the Pacific railway.

Before those surveys began to yield results, however, another official government survey of the plains and other western territories was under way. The U.S. Boundary Survey was begun in the late 1840s to settle the dividing line between the United States and Mexico on the south and between the United States and Canada on the north.[27] The driving force in these surveys was the topographical engineer William H. Emory; his large map of the "United States and their Territories Between the Mississippi and the Pacific Ocean," which consolidates the results of the boundary surveys, and Lt. G. K. Warren's synthesis of the Pacific railroad surveys together represent the highwater mark of western cartography before the Civil War.[28]

The detail on Emory's magnificent map of 1857 is so great that it defies description. Suffice it to say that his map, like others produced by the topographical engineers, was highly accurate and presented data as obtained through scientific observation rather than speculation. From the Rio Grande to the Missouri, Emory outlined and named virtually every watercourse known in the plains, and on his map the future political divisions of the plains states begin to take shape. Only two areas of the plains are limited in detail: the Llano Estacado, although named, is shown as a blank region and labeled "unexplored"; and in the northern plains, the location of the Black Hills and their drainage system, along with the Badlands country, is inaccurate. The lettering in the legend "Black Hills" stretches from the Tongue River to the upper reaches of the Cheyenne, and the Badlands appear just north of the White River, not far from Fort Laramie. In the case of the first empty area, Emory was absolutely correct—the Staked Plains were unexplored. As for the Black Hills country, no one had ever come close to showing that region as it really is. To Emory's credit, he at least did not confuse the Black Hills with the Laramie Range extension of the Colorado Front Range, as so many others had done throughout the nineteenth century.

The other great set of maps produced in the 1850s emerged from the work of the Pacific railroad surveys, and once again, a single name, Lt. Gouverneur Kemble Warren of the Corps of Topographical Engineers, stands out among the rest.[29] The Pacific railroad surveys of 1853–55 were designed to locate, through a comprehensive and systematic exploration, the most practicable rail routes to the Pacific along the 47th, 41st 38th, and 32nd parallels. Some of the great names of western exploration were associated with the survey, but it fell to Warren to draw together the information gathered by the surveys into a "master map" of the American West. Before commencing the final stages of his great map of the West, however, Warren determined the need to reconnoiter further the route that had not been investigated by the survey. The so-called central route along the Platte Valley had been left unsurveyed by the official parties on the grounds that it was already very well known from the Frémont expeditions and the crossing of countless emigrant parties. In 1855, '56, and '57, while detached from his official duties with the Pacific railroad surveys, Warren

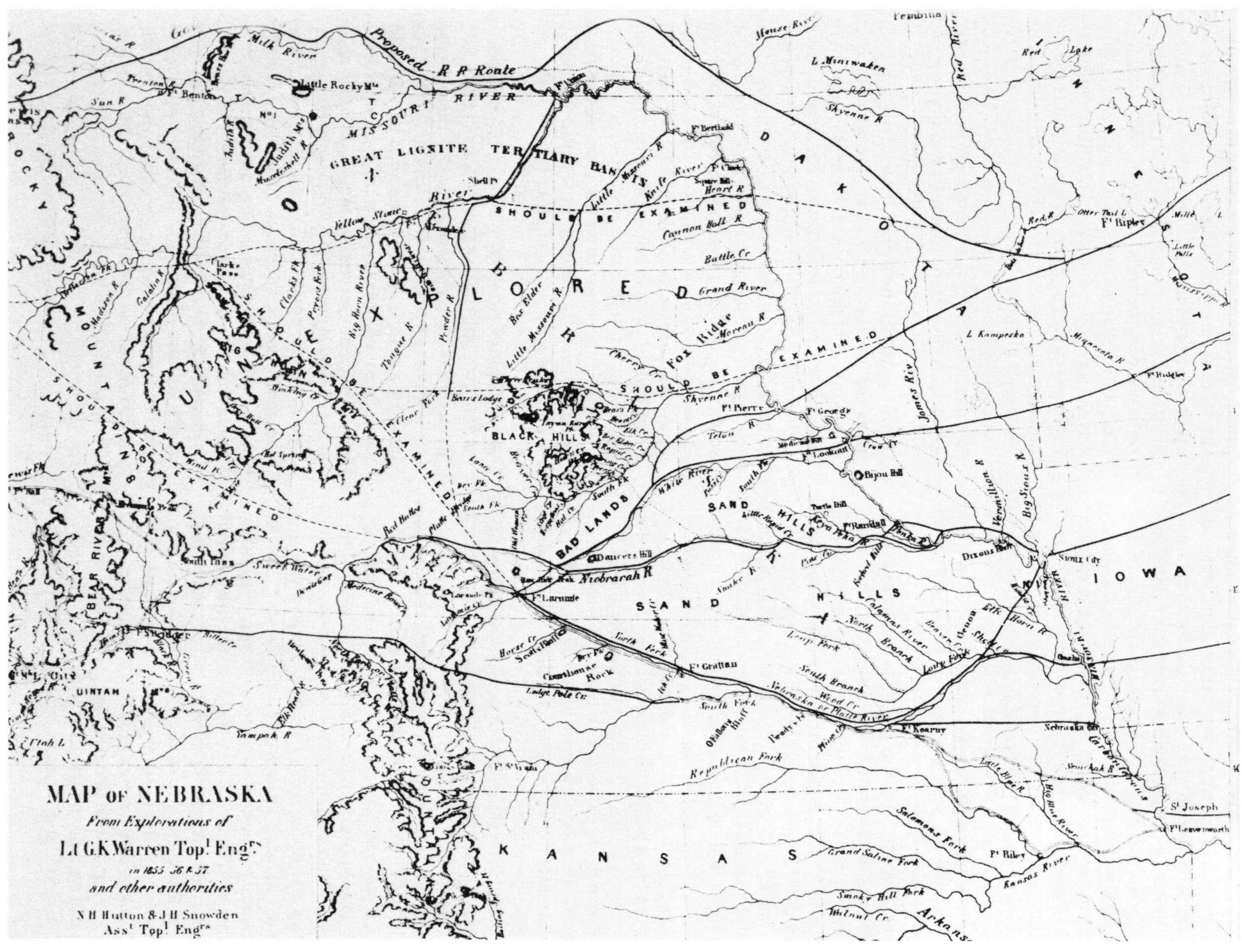

Fig. 3.10. "Map of Nebraska" by Gouverneur Warren, 1857. (National Archives)

explored the country north of the Platte; his "Map of Nebraska," based on that exploration, was produced in 1857 (fig. 3.10). A highly accurate map of the northern plains, Warren's first 1857 product was unique in two respects—it was the first map that accurately located and defined the Black Hills and the Badlands terrain on the southern and eastern flanks of that range, and it was "corrected" by reference to the largest collection of fur trapper maps that had ever been assembled.[30]

As remarkable as Warren's first 1857 map was, it was overshadowed by the completion of his "General Map of the Territory of the United States from the Mississippi to the Pacific Ocean," also completed in 1857.[31] In Warren's report, attached to the map, he noted that the map contained materials obtained through "official" exploration in the West since 1800. The list of authorities consulted was provided in the lower right corner of the map itself. It is a long list—forty-four entries in all, beginning with Lewis and Clark and ending with government surveys "to date" from the topographical engineers and the U.S. Land Office.

The Warren map of 1857 is the capstone of western cartography before the Civil War. William H. Goetzmann has called it "the first reasonably accurate map of the American West," and although that statement is a bit hyperbolic in reference to the plains alone, Warren's map certainly is the best map of the plains before 1860.[32] The "General Map" is equally precise for the northern and southern plains—a rare feat in itself. In the south, the area of the Llano Estacado is still largely a blank space; although Warren does not indicate that the area is unexplored, he clearly did not have information on it and, consequently, left the region empty of detail. The only comparable blank space in the northern plains is the area of the Powder River Basin and the upper courses of the Little Missouri and Knife rivers, marked "unexplored" on the map. To compensate for these two relatively small unknown regions, Warren added a geographical feature that had never been shown before on a major map. North of the Platte and south of the Niobrara River appears the legend "Great Sand Hills." Like the earlier 1857 map, the "General Map" shows the Black Hills, along with the Badlands. In addition, Warren's representation of the courses of the plains streams was not confined merely to drawing solid lines on blank paper. Adjacent to the streams he used hachuring and other techniques to show such features as the Missouri Breaks and the "coasts" of the Platte. Finally, in a recognition of their strategic importance, Warren shows the tribal territories of virtually all the Plains Indian nations, thus providing both a clear ethnographic picture of the plains prior to the Civil War and a prophetic comment on the significance of those locations for the coming conflicts between different cultures.

Although Warren's map signifies the culmination of American efforts to map the plains during the first six decades of the nineteenth century, it is not the end of the story. For the decade of the fifties ended as the first decade of the century had begun—with two types of maps that showed patterns of promise. The first of these was maps drawn to accompany the more than two dozen guide books published in 1859 to instruct the gold-hungry in the fastest and easiest routes to the newly discovered gold regions near Pike's Peak. Most of these maps show little of the plains other than the major streams along which the various routes to the Rockies are laid out. In this sense, they resemble many other maps of the period in demonstrating cartographically that the crossing of the plains is full of promise—primarily for what lies at the end of the journey rather than in the country traversed.[33]

The second set of maps from the sixties is that derived from the U.S. Land Office surveys in the eastern plains (fig. 3.11).[34] Like many others, they are maps of promise—but for the first time, the promise is in the plains region itself and not in what lies beyond. On these maps appear the tidy squares first envisaged by President Jefferson three-quarters of a century earlier: the squares of the range-township-section system of survey; the squares that promised good farmland and a familiar landscape; the squares that an agricultural population permanently stamped on the plains after the Civil War. As yet those neat geometric shapes were only symbols, but soon they were to become a reality for a settling rather than a traveling population—whether that reality was to be met in the form of poverty or prosperity. All of the mapping of the plains during the years preceding the great sectional conflict had led to a pattern of promise on, not beyond, the plains.

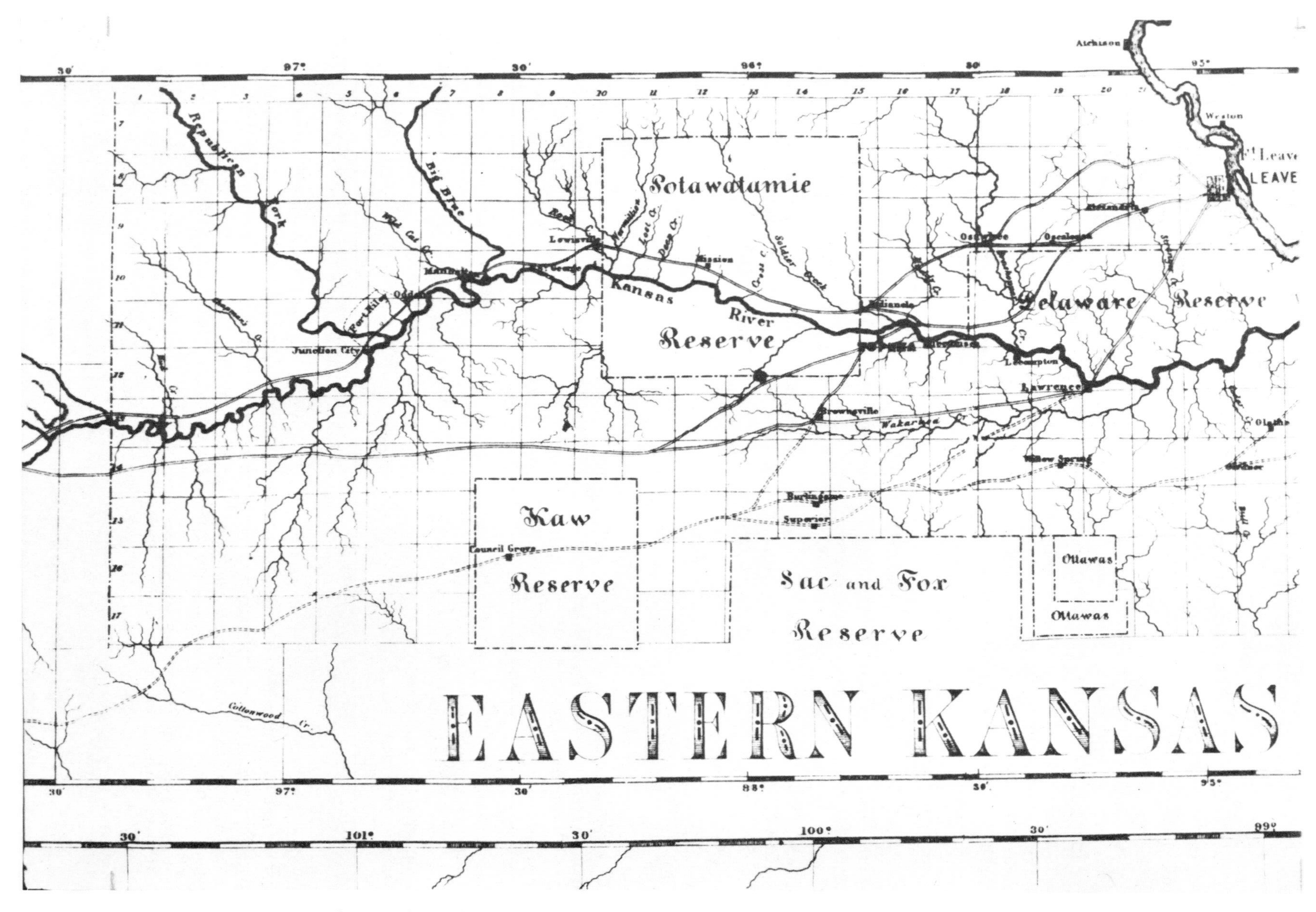

Fig. 3.11. "Central or Smoky Hill Route to the gold fields of western Kansas" by Tennison. (Denver Public Library)

Notes

1. The best source on the cartography of the American West in the pre–Civil War period is Carl Irving Wheat, *Mapping the Transmississippi West, 1540–1861,* 5 vols. (Menlo Park, Calif.: Institute of Historical Cartography, 1957–63). Wheat's work has served as a major reference for this article.

2. See Donald Jackson, ed., *Letters of the Lewis and Clark Expedition, with Related Documents, 1783–1854,* 2d ed., 2 vols. (Urbana: University of Illinois Press, 1978), and John L. Allen, *Passage through the Garden: Lewis and Clark and the Image of the American Northwest* (Urbana: University of Illinois Press, 1975) for the Lewis and Clark Expedition. The best source on Pike is Donald Jackson, ed., *The Journals of Zebulon Montgomery Pike with Letters and Related Documents,* 2 vols. (Norman: University of Oklahoma Press, 1966).

3. Pike's maps were published along with his account in 1810. See Zebulon Montgomery Pike, *An Account of Expeditions to the Sources of the Mississippi, and Through the Western Parts of Louisiana* (Philadelphia: J. A. Conrad & Sons, 1810).

4. On the expedition of Lewis and Clark, see the works by Allen and Jackson above, along with the explorers' original journals, in Reuben Gold Thwaites, ed., *Original Journals of the Lewis and Clark Expedition, 1804–06,* 8 vols. (New York: Dodd and Mead, 1904–1905).

5. Clark's manuscript map has been reproduced and published by Yale University (1950); the published version of the map first appeared in the official history of the expedition, the so-called Biddle edition, edited by Nicholas Biddle of Philadelphia but with Paul Allen's name appearing in the work as the editor: Paul Allen, ed., *History of the Expedition under the Command of Captains Lewis and Clark to the Sources of the Missouri, Thence across the Rocky Mountains and Down the River Columbia to the Pacific Ocean,* 2 vols. (Philadelphia: Bradford and Innskeep, 1814).

6. Wheat (*Mapping the Transmississippi West,* vol. 2, ch. 3) gives a fine account of commercial mapping of the West during this period.

7. See Jackson, *Journals of Pike,* vol. 2, appendix 3; an obituary of Robinson was published in the St. Louis *Missouri Gazette* on 24 November 1819.

8. Henry S. Tanner, *A New American Atlas* (Philadelphia: H. S. Tanner, engraver, 1823).

9. The Tanner map became a progenitor of many other commercial productions during the 1820s and 1830s; in this way Pike and Long material was passed along to those who had not read their official exploratory accounts.

10. Stephen H. Long, *An Account of an Expedition from Pittsburg to the Rocky Mountains* (Philadelphia: H. C. Carey and Lea, 1823); see also Roger L. Nichols and Patrick L. Halley, *Stephen Long and American Frontier Exploration* (Cranbury, N.J.: Associated University Presses, 1980).

11. Brown's field sketches and notes are in the collection of the National Archives, Washington, D.C. Wheat gives a good summary of the Brown maps.

12. See Gouverneur Kemble Warren, *Memoir to Accompany the Map of the Territory of the United States from the Mississippi to the Pacific Ocean . . . ,* 33d Cong., 2d sess., Sen. Exec. Doc. 78 (1859), p. 19.

13. The greatest of the fur trade cartographers was Jedediah Strong Smith, whose master manuscript map of the American West, drawn in the winter of 1830–31, has been lost for nearly a century and a half. Evidence of this map exists in the form of annotations made by a George Gibbs on a Frémont 1845 map. The discovery of this "Gibbs-Frémont-Smith" map was made by Carl Wheat and is discussed in his *Mapping the Transmississippi West,* 2: 120–29. Evidence that Gibbs had seen Smith's manuscript map is found in an entry by geologist F. H. Bradley in Ferdinand V. Hayden, *Sixth Annual Report of the U.S. Geological Survey of Territories* (Washington, D.C.: GPO, 1873), p. 233. Information on smith and his mapping may be found in Carl I. Wheat and Dale L. Morgan, *Jedediah Smith and his Maps of the American West* (San Francisco: California Historical Society, 1954). See also LeRoy R. Hafen, ed., *The Mountain Men and the Fur Trade,* 10 vols. (Glendale, Calif.: A. H. Clark, 1972), 9: 224.

14. See the *Report* of Isaac McCoy, 20th Cong., 2d sess., H.R. Report 87, ser. 177 (Washington, D.C., 1832); also Lt. J. P. Kingsbury, *Journal of an Expedition of Dragoons . . . to the Rocky Mountains,* 24th Cong., 1st sess., H. Exec. Doc. 181, ser. 289 (Washington, D.C., 1836).

15. See Samuel Parker, *Journal of an Exploring Tour beyond the Rocky Mountains* (Ithaca, N.Y.: Andrus, Woodruff, and Gauntlett, 1838).

16. This is particularly true for New Englanders; judging from the number of Parker maps still available in New England archives and "family" historical collections, his work must have been very popular indeed.

17. See Martha Coleman Bray, *Joseph Nicollet and His Map* (Philadelphia: American Philosophical Society, 1980); several Nicollet maps, including the 1838 "Hydrographical Basin . . ." are reproduced in Bray's volume.

18. See Josiah Gregg, *Commerce of the Prairies* (New York: Henry E. Langley, 1844). Some Gregg manuscript maps are in the collection of the National Archives, but it is not known whether Gregg was the actual author of the map appearing in his published work or whether the final map was simply completed under his supervision.

19. The De Smet map was published in *Voyages aux Montagnes Rocheuses* (Paris: Malines, P. G. Hanicq, 1844).

20. See Oliver G. Steele, *Steele's Western Guide Book and Emigrants' Directory* (Buffalo, N.Y.: Oliver G. Steele, 1849). A similar work, also with important maps of the emigrant trails, is Joseph E. Ware, *The Emigrant's Guide to California* (St. Louis: J. Halsall, 1849).

21. Rufus B. Sage, *Scenes in the Rocky Mountains* (Philadelphia: Carey and Hart, 1846). Whether Sage drew the map himself or had it drawn for publication is not known.

22. Wheat gives an excellent description of the railroad promoters, including Whitney, in vol. 3 of *Mapping the Transmississippi West,* The Whitney manuscript map is in the collection of the National Archives. See also Asa Whitney, *A Project for a Railroad to the Pacific* (New York, privately printed, 1849).

23. An excellent study of the U.S. Topographical Engineers is William H. Goetzmann, *Army Exploration in the American West, 1803–1863* (New Haven, Conn.: Yale University Press, 1959). A later and more abbreviated study is Frank N. Schubert, *Vanguard of Expansion: Army Engineers in the Trans-Mississippi West, 1819–1879* (Washington, D.C.: GPO, 1980).

24. Frémont's work is examined in detail in Wheat, *Map-*

ping the Transmississippi West, vol. 2. See also Donald Jackson and Mary Spence, eds., *The Expeditions of John Charles Frémont,* 3 vols. and map portfolio (Urbana: University of Illinois Press, 1970–84). Frémont's most important map was the large 1845 map, published in the *Report of the Exploring Expedition to the Rocky Mountains in the Year 1842 and to Oregon and North California in the Years 1843–44,* 28th Cong., 2d sess., Sen. Exec. Doc. 174, ser. 461 (Washington, D.C., 1845). While it has become conventional to refer to this map as "Frémont 1845," the cartographer was actually George Carl Ludwig Preuss, a German draftsman who accompanied Frémont and who had emigrated to the United States in 1834. For details on Preuss, see E. C. and E. G. Gudde, *Exploring with Frémont* (Norman: University of Oklahoma Press, 1958).

25. This map was printed as part of W. B. Franklin, *Report of a Summer Campaign in the Rocky Mountains,* 29th Cong., 1st sess., Sen. Exec. Doc. 1, ser. 470 (Washington, D.C., 1846).

26. Randolph B. Marcy, *Route from Fort Smith to Santa Fe,* 31st Cong., 1st sess., H. Exec. Doc. 45, ser. 577 (Washington, D.C., 1849).

27. William H. Emory, *Report of the United States and Mexican Boundary Survey* (Washington, D.C.: Cornelius Wendell, 1857).

28. The reports of the Pacific railroad surveys were published in twelve volumes by the government as *Pacific Railroad Reports,* 33d Cong., 2d sess., Sen. Exec. Doc. 78 (Washington, D.C., 1855).

29. There is no biography of Warren, a curious gap in the scholarly literature in light of the significance of Warren's role in the pre–Civil War West. Some information may be found in William H. Goetzmann's *Exploration and Empire: The Explorer and the Scientist in the Winning of the American West* (New York: Alfred A. Knopf, 1967), and his *Army Exploration in the American West;* the primary manuscript source is the Warren Papers in the New York State Library in Albany.

30. These fur trapper maps, seven in all, are in the Warren Papers, New York State Library, Albany, New York.

31. This map was published by the government as part of the reports on the Pacific railroad surveys; 33d Cong., 2d sess., Sen. Exec. Doc. 78.

32. See Goetzmann, *Exploration and Empire,* pp. 314–16.

33. Cf. T. Parker, *The Illustrated Miner's Handbook and Guide to Pike's Peak* (St. Louis: Parker and Huyett, 1859).

34. The map illustrated was drawn by a cartographer named Tennison and lithographed by Middleton, Strowbridge and Co. of Cincinnati, Ohio, in 1859. A photostat copy is in the Denver Public Library.

4. INDIAN MAPS

THEIR PLACE IN THE HISTORY OF PLAINS CARTOGRAPHY

G. Malcolm Lewis

REFERENCES to the maps and mapping activities of North American Indians have appeared in scholarly writings for approximately two hundred years and in contemporary accounts of discovery and exploration for more than four hundred years.[1] The topic has received relatively little attention, however, from modern scholars.[2] In view of the recent expansion of Indian studies in both Canada and the United States, this lack may at first seem surprising. In part it reflects the fact that there are relatively few extant examples of Indian maps because Indians and most whites have tended to treat them as ephemera, not for the most part worthy of preservation. In part it also reflects both the geographical scatter of extant examples through the libraries, museums, archives, and private collections of Europe as well as North America and the problems of searching through the vast literature and the large number of dispersed archival collections in which accounts of the mapping activities of Indians are occasionally to be found.

It would be possible to write on this topic with reference to any major region of North America, but areas within and immediately adjacent to the Great Plains are particularly frequently represented on extant examples of Indian maps. While most of these date from the nineteenth century, some are of eighteenth-century origin, and the earliest (1602) of all the extant examples of Indian maps from within North America covers part of the southern Great Plains.

North American Indians were in no sense unusual among the world's historic nonliterate peoples in making things which we call "maps" and which undoubtedly had many of the functions we associate with maps. Mapmaking is a universal trait—an aspect of pictographic communication that, in Europe at least, probably originated in the Upper Paleolithic, perhaps as early as 20,000 B.P. Within nonliterate societies it probably did not have a clearly differentiated status until after contacts with whites in the historic period. Throughout the world, white aliens solicited from native peoples geographical information about their terrae incognitae, and the response was frequently in map form. This experience was particularly common in North America, where Inuits and Aleuts as well as Indians conveyed information to whites in this way.

The "maps" of nonliterate peoples deserve far more attention than they have received to date in at least five contexts. (1) They are important cognitively for what they reveal about a people's spatial structuring and evaluation of the earth's surface. (2) For archeological purposes, they provide evidence upon which to base searches for settlements and other prehistoric sites. (3) They have ethnological significance as well, particularly regarding their roles within the religious, social, and information systems of indigenous peoples. (4) As historical documents, they should be studied for their roles in communications and negotiations between nonliterate native populations and alien whites. Finally

(5), their cartographic importance lies in their influence on maps made by whites.

This article examines Indian maps in historical and, more specifically, cartographic contexts. It draws its evidence almost exclusively from within the American Great Plains and Canadian prairies, even though in some cases better examples are available from other North American regions.

Possible Maps from Prehistoric Times

Peoples on the plains may have made maps in prehistoric times or incorporated simple cartographic principles in some of their graphic art. According to a recent comprehensive survey of North American Indian rock art:

> Here and there [in the Great Plains region], meandering lines and other abstract elements on certain [rock art] panels have suggested maps to experienced observers since the designs appeared to correspond rather closely with the features of nearby natural formations such as the contour of a mountain range or the course of a river. In various parts of the country, including the Plains, Indians of the historic period not infrequently drew maps on material other than rocks. . . . It is possible [therefore] that on occasion rocks were used for the same purposes, but reasonably convincing examples of this type are very infrequently encountered.[3]

No completely convincing examples of maps in prehistoric Plains rock art are known, however, and it seems unlikely that any will be discovered. The reasons behind the ambiguity of the evidence are threefold. Of these, the first two are universal and arise from the properties of maps drawn by nonliterate peoples. Such maps are always structured topologically; that is, they do not conserve true distance or true direction. Likewise, they never consistently represent the relative physical magnitudes of topographical features; culturally significant but topographically inconspicuous features are prominently represented and vice versa. In the absence of other evidence it is therefore virtually impossible to match maplike patterns in rock art with patterns of features on the earth's surface. "Other evidence," in the prehistoric context, takes the form of unambiguous pictographic representations of locally or regionally distinctive topographic features, such as lakes with distinctive shapes, mountains with distinctive profiles, bold escarpments, or sharp breaks in vegetation. Unfortunately, plains rock art, as indeed the region's early historic Indian art, is deficient in such elements.[4] For these reasons, the case for cartographic elements in prehistoric plains rock art is, and seems likely to remain, tantalizingly unprovable, yet at the same time irrefutable.

The northwestern Great Plains contain a number of pre- or protohistoric structures that are generally known as medicine wheels. Typically they consist of a central cairn (or small circle) of stones, from which radiate at unequal angular intervals stone lines of unequal length. At the distal ends of some of these are smaller stone cairns. The age and function(s) of these structures are matters of debate, but according to at least one twentieth-century Indian informant they incorporated cartographic principles. Supposedly memorials commemorating the war exploits of great chiefs, the stone lines show the directions of each expedition, their lengths the relative distances covered, and the presence or absence of distal cairns whether or not any of the enemy were killed.[5] This technique incorporates several of the cartographic principles employed by Chickasaw Indians in Mississippi in 1737 in conveying information to the French about intertribal relationships.[6]

The Skidi band of Pawnees provides a more certain example of the spatial arrangement of structures according to cartographic principles. Before these people were removed to Oklahoma from the middle Platte valley, the band was divided into several villages, each with a different shrine, the contents of which were determined by a specific star. The star gave its name to the shrine, and the name of the shrine became the name of the village. Five villages formed a central group, and their relative positions were fixed by the relative positions of the stars that had given them their shrines. Around these were approximately seventeen other villages, each likewise located according to the position of its star relative to the others. In this way, the villages of the Skidi people on the earth were located as a reflected picture of their stars in the heavens.[7] Each Skidi band possessed a sacred bundle, and within some of these were star charts painted on skin. One in the Field Museum, Chicago, is on tanned elk skin (fig. 4.1).[8] It represents the positions of stars by four-pointed symbols, drawn according to

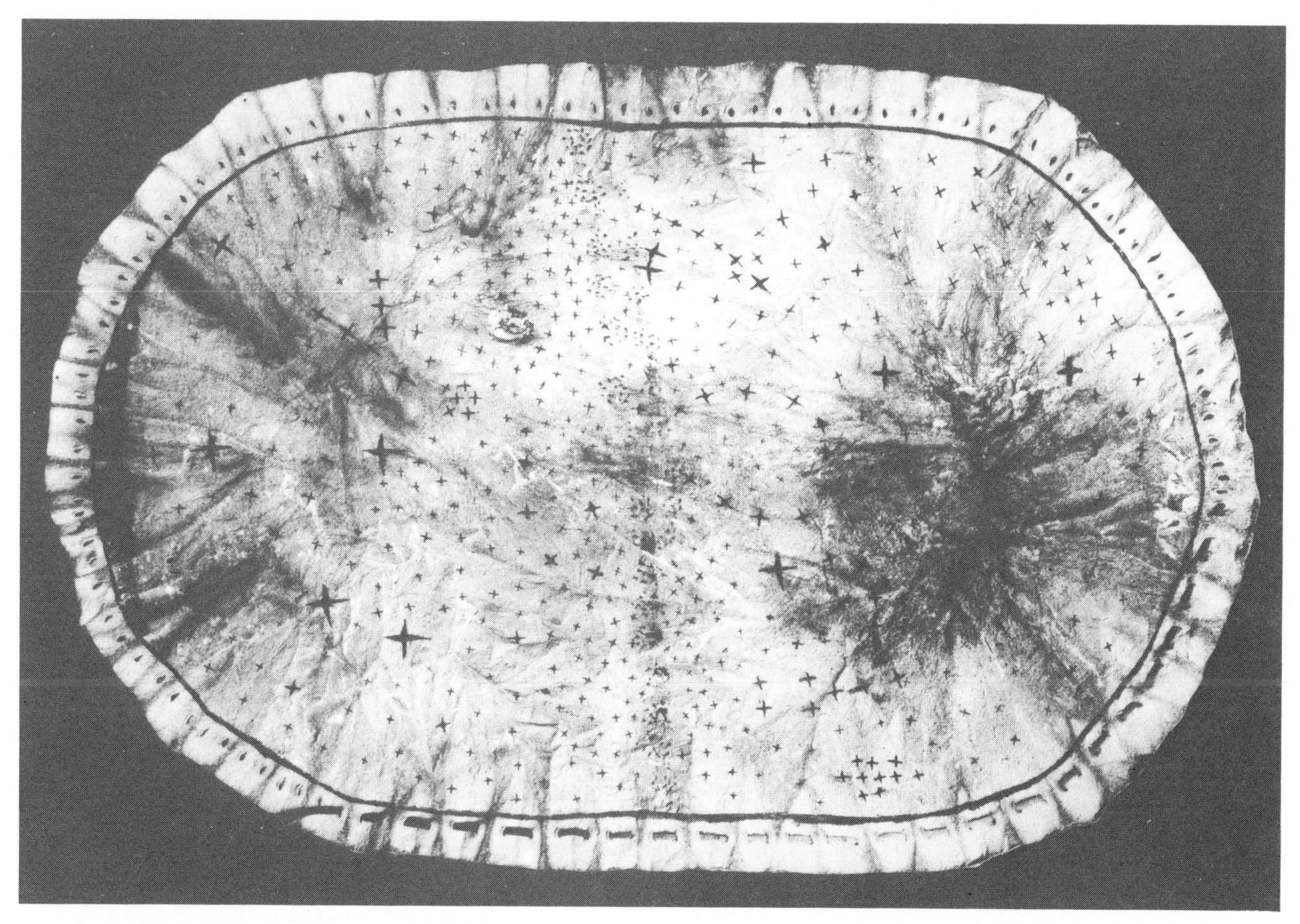

Fig. 4.1. Pawnee (Skidi band) celestial chart on elk skin, pre–1906 and supposedly much older. Original 65 × 43 cm. (Field Museum of Natural History, Chicago)

five different magnitudes: eleven of the first order; nine second order; forty-four third order; and many more fourth and fifth order, which, unlike the others, are apparently distributed at random. Across the center of the chart is a band of many small symbols representing the Milky Way. According to one analysis, the chart, which is supposed to predate any white influence, recognizes the same constellations that we do, shows seasonal changes, and records some double stars.[9]

Indigenous Maps of Historic Times

The Southern Ojibway Indians of northern Minnesota and western Ontario kept in their medicine bags birchbark scrolls, some of which have been fairly convincingly interpreted to be migration charts, representing in cartographic form the route via which the Ojibways believed they received the Midé religion—that is, from beyond the Great Salt Water (Atlantic Ocean), up the St. Lawrence River, through the Great Lakes, and from the head of Lake Superior via the St. Louis River to Leech Lake in central Minnesota. Grossly distorted, highly schematic, and with many mythical elements, these charts can be related to the Great Lakes–St. Lawrence drainage system as we now know it, but more convincingly so to the west than to the east. They are of course topological and would probably not have been recognized as maps but for the oral evi-

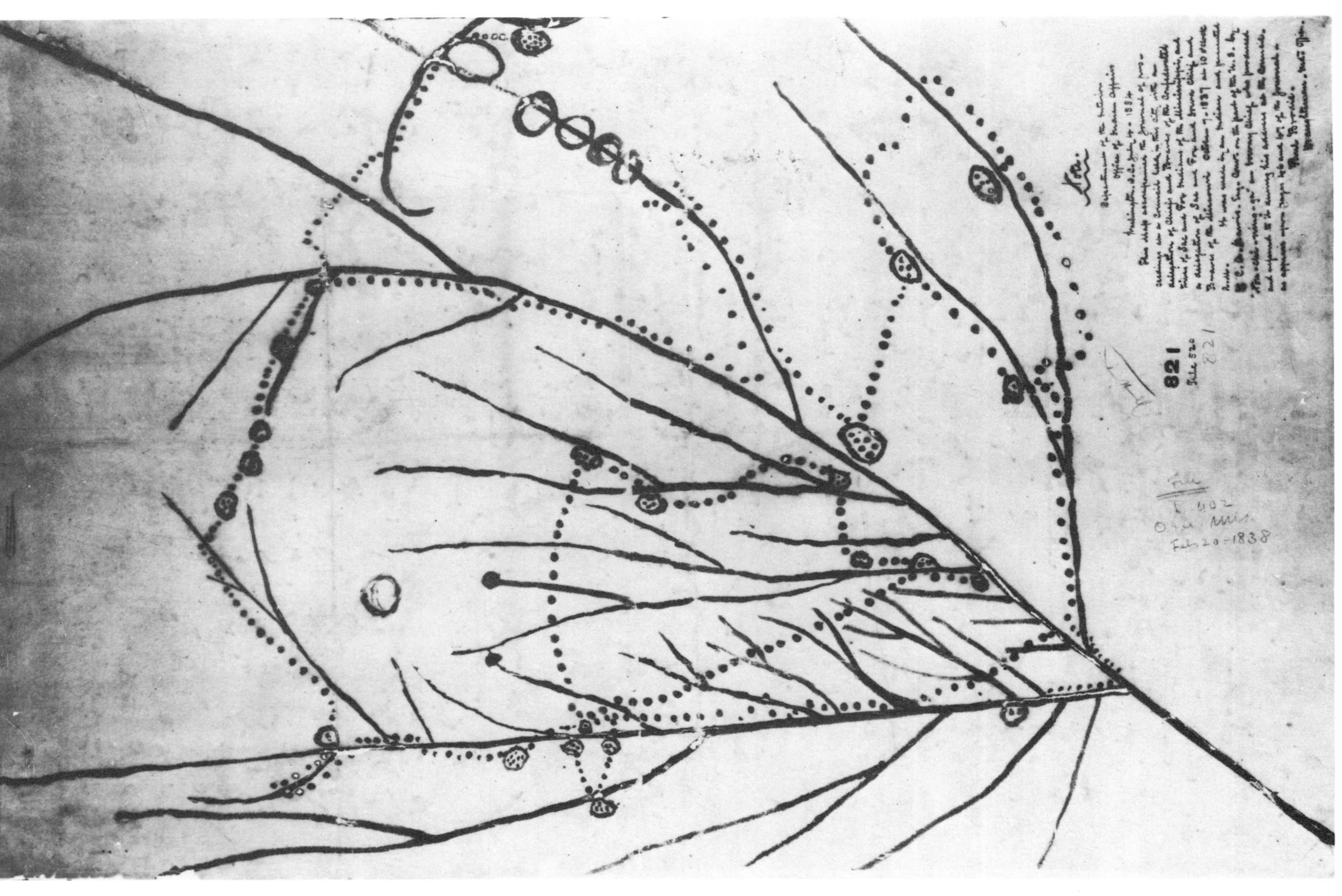

Fig. 4.2. Non-Chi-Ning-Ga's (Iowa) map of tribal migrations in the upper Mississippi and Missouri drainage basins, 1837. MS original, 104 × 69 cm. (Cartographic and Architectural Branch, National Archives)

dence collected from twentieth-century members of the tribe.[10]

Given the evidence of indigenous mapping for religious purposes by the Skidis and the Southern Ojibways, it is reasonable to speculate that historians of cartography may still have much to learn from ethnologists and that, conversely, ethnologists might discover more about the functions of certain art and artifacts if they had a greater awareness of the ability of Indians to make maps and of the characteristics and functions of the maps that they are known to have made. Evidence presented by Iowa Indians in the course of nineteenth-century treaty negotiations suggests that maps (both mental and artifact) were used as a means of preserving important information about the history of the tribe's migrations, although such maps are not known to have been preserved in medicine bundles (figs. 4.2 and 4.3).[11] At a more ephemeral level, Plains Indians certainly made and used maps for purely indigenous purposes. In the early nineteenth century, Comanche braves on the Texas plains were briefed before raids by older members of the tribe using maps. Routes were planned in units of a day, and the days were each recorded by notched sticks. A crude map was

> drawn on the ground with a finger or piece of wood illustrating the journey of the day represented by the notched stick. The larger rivers and streams are indicated, the hills, valleys, ravines, hidden water holes and dry countries, every natural object, peculiar or striking. When this was understood, the stick representing the next day's march was illustrated in the same way, and so on to the end.

Briefed in this way, a group of teenage braves, none of whom had previously undertaken the journey, memorized and successfully undertook a return journey of approximately one thousand miles from Brady's Creek, Texas, to Monterey, Mexico.[12] John Hunter, who as a captive had traveled between the Illinois Country and the Rocky Mountains with several tribes in the late eighteenth and early nineteenth centuries, later recorded that Indians in general could delineate maps of countries with considerable accuracy."[13] After ascending the Missouri River as far as the confluence of the Cheyenne River in what is now central South Dakota, Jean Baptiste Truteau reported that the Indians made

> delineations upon skins, as correctly as can be, of the countries with which they are acquainted. Nothing is wanting but the degrees of latitude and longitude. They mark the northern direction according to the polar star, and conformably to that, mark out the windings and turnings of the rivers, the lakes, marshes, mountains, woods, prairies and paths; . . . they compute distances by day's and half day's journeys.[14]

Throughout the nineteenth century some Plains Indians continued to make maps for indigenous purposes, although acculturation increased as the century progressed. Sometime between the founding of Rapid City in 1876 and his death in 1913, Amos Bad Heart Bull, an Oglala Sioux, drew a map of the Black Hills and surrounding plains on which the former are represented pictographically and the latter cartographically (fig. 4.4).[15] The almost circular Black Hills, with the traditional "Race Track" in yellow, are represented as an assemblage of symbolized features; for example, a bear's head for Bear Butte, a horned head for Ghost Butte, a barren rounded profile for Old Baldy, and a test-tube-like shape inverted on the top of a sketch of a bear's head for the towering cylindrical column of Bear Lodge Butte. In sharp contrast, the drainage networks on approximately thirty thousand square miles of the surrounding plains are represented in plan and with remarkable accuracy. Additional information includes the meridians 103° and 104° west. It would appear, therefore, that Amos Bad Heart Bull, who had been enlisted as a scout in the U.S. Army at Fort Robinson in 1890–91, when he must almost certainly have been exposed to official maps and surveys, placed an Indian-style pictographic map in the context of a Euro-American-style topographic map. In most examples of maps drawn by Plains Indians at this period, however, the mix is far less obvious.

Cartographic Elements in Plains Indians' Art

Throughout the nineteenth century, cartographic principles were also incorporated in much of the two-dimensional representational art of the Plains Indians. That done by women tended to be abstract, but men painted more naturalistically, frequently representing hills, buttes, men, horses, trees, and wild animals in profile against a planimetric representation of river and route networks. Many of the

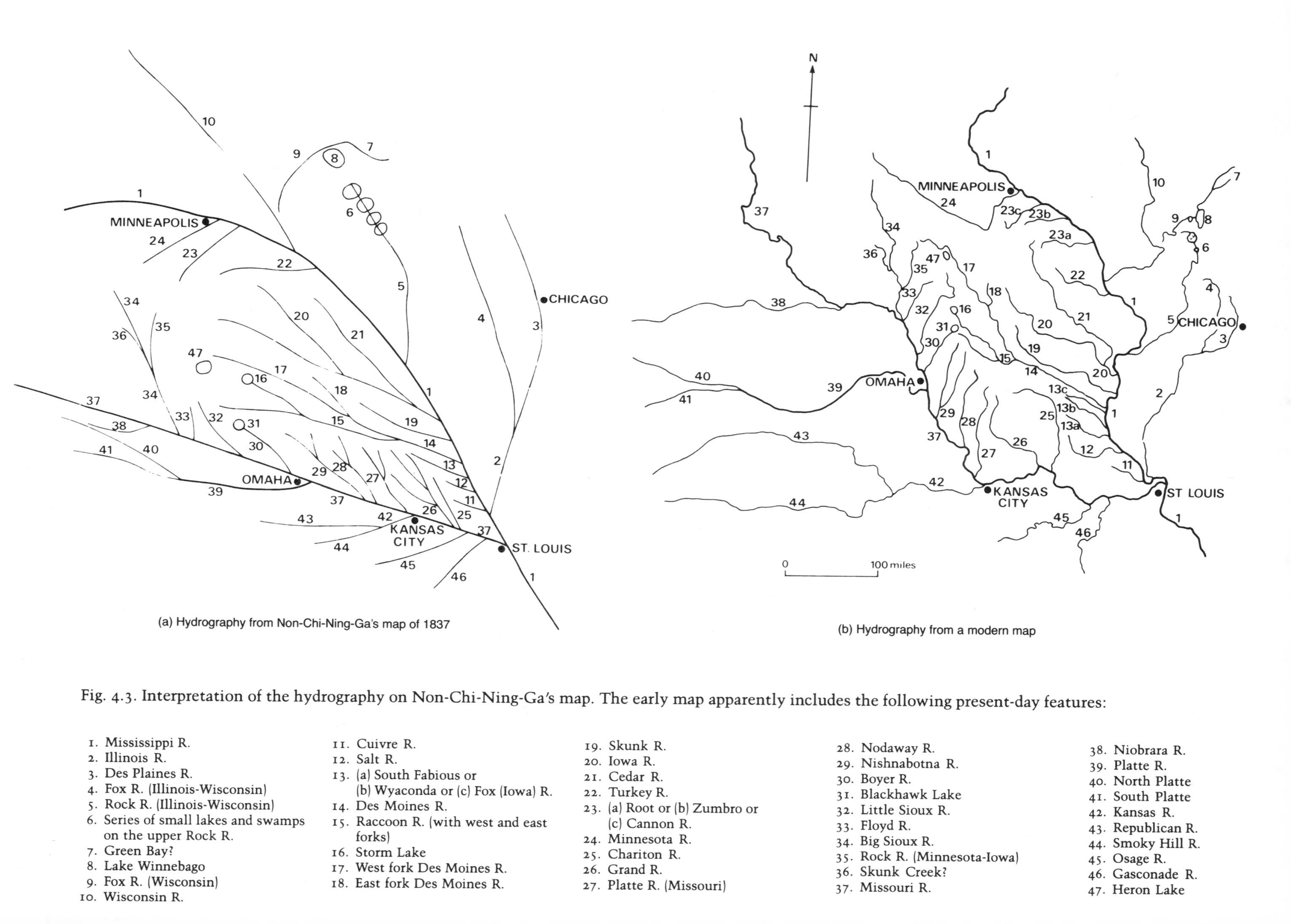

(a) Hydrography from Non-Chi-Ning-Ga's map of 1837

(b) Hydrography from a modern map

Fig. 4.3. Interpretation of the hydrography on Non-Chi-Ning-Ga's map. The early map apparently includes the following present-day features:

1. Mississippi R.
2. Illinois R.
3. Des Plaines R.
4. Fox R. (Illinois-Wisconsin)
5. Rock R. (Illinois-Wisconsin)
6. Series of small lakes and swamps on the upper Rock R.
7. Green Bay?
8. Lake Winnebago
9. Fox R. (Wisconsin)
10. Wisconsin R.
11. Cuivre R.
12. Salt R.
13. (a) South Fabious or (b) Wyaconda or (c) Fox (Iowa) R.
14. Des Moines R.
15. Raccoon R. (with west and east forks)
16. Storm Lake
17. West fork Des Moines R.
18. East fork Des Moines R.
19. Skunk R.
20. Iowa R.
21. Cedar R.
22. Turkey R.
23. (a) Root or (b) Zumbro or (c) Cannon R.
24. Minnesota R.
25. Chariton R.
26. Grand R.
27. Platte R. (Missouri)
28. Nodaway R.
29. Nishnabotna R.
30. Boyer R.
31. Blackhawk Lake
32. Little Sioux R.
33. Floyd R.
34. Big Sioux R.
35. Rock R. (Minnesota-Iowa)
36. Skunk Creek?
37. Missouri R.
38. Niobrara R.
39. Platte R.
40. North Platte
41. South Platte
42. Kansas R.
43. Republican R.
44. Smoky Hill R.
45. Osage R.
46. Gasconade R.
47. Heron Lake

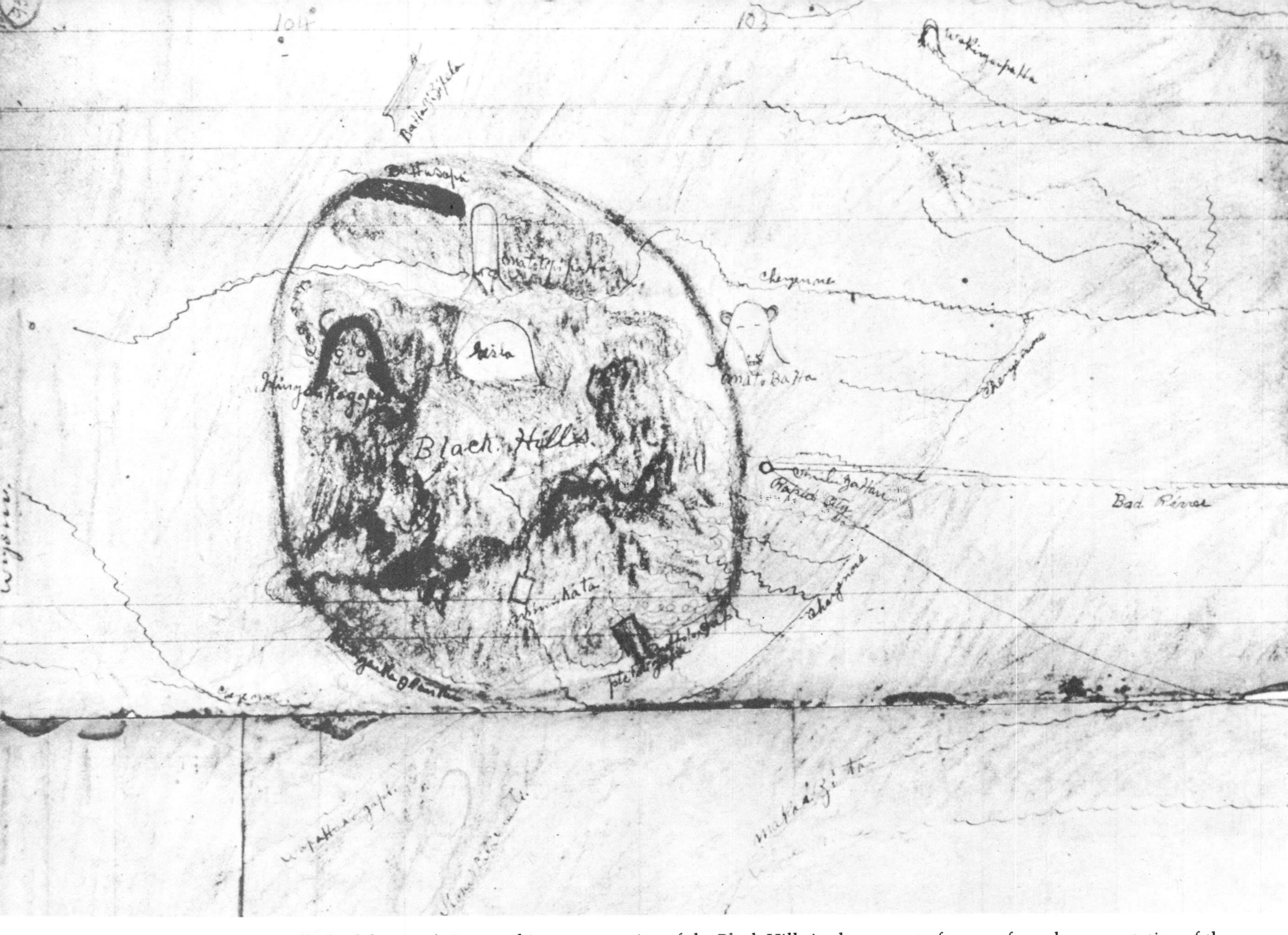

Fig. 4.4 Part of Amos Bad Heart Bull's (Oglala Sioux) pictographic representation of the Black Hills in the context of a more formal representation of the surrounding plains, done between 1876 and 1913. MS original, approximately 17.8 × 30.5 cm. (Reprinted from *A Pictographic History of the Oglala Sioux* by Amos Bad Heart Bull, text by Helen Blish, by permission of the University of Nebrasks Press, copyright © 1967 by the University of Nebraska Press)

Fig. 4.5. White Bird's (Northern Cheyenne) painting of the Battle of the Little Big Horn, painted in 1894–95. Original on muslin, 171 × 249 cm. (West Point Museum Collections, United States Military Academy)

scenes were of important events in tribal history. At first they were always painted on skin, though this practice may have been a continuation of an earlier rock-art tradition.[16] Later, the Plains Indians began to use other materials. Almost twenty years after the event, White Bird, a Northern Cheyenne, painted on muslin a representation of the Battle of the Little Big Horn, in which tepees, Indians, horses, and soldiers of the Seventh United States Regiment were represented in profile but placed in relation to a plan of the Little Big Horn and its tributaries and the prebattle routes of Maj. Gen. George Custer and Maj. Marcus Reno (fig. 4.5).[17]

Between 1878 and 1881, Howling Wolf, a Cheyenne, painted twelve colorful scenes in a sketchbook, reconstructing according to tribal lore significant events in his people's history. Two of these are presented against a cartographic background of rivers, wooded valley floors, buffalo trails, human tracks, and other features.[18] One scene shows the first white men seen by the Cheyennes, supposedly more than one hundred years earlier on the Missouri River at a camp above the Cheyenne River. The other reconstructs the first trading for horses by the Cheyennes with the Kiowas on the Arkansas River. A few years before painting these scenes, Howling Wolf had been transported to Fort Marion in Florida. During his exile he was sent by sea from St. Augustine for medical treatment in Boston. Enroute he mailed to his father a prepaid postal card on which he represented his route from St. Augustine to a point off the Atlantic Coast somewhere to the north of Savannah, Georgia. Buildings, the steamer, and persons with their totems were represented in profile but positioned in relation to a map of the coastline and estuaries. The shape of these geographic features had been grossly distorted, partly, at least, because of the restrictive rectangular format of the small (3 in. × 5.3 in.) card.[19]

Of the 414 posthumously published drawings made by Amos Bad Heart Bull, the Oglala Sioux, between 1891 and 1913, nine are eminently cartographic.[20] One of the most interesting is in black ink and crayons and shows the setting of the Black Hills conference of 1876. The actual talks, seven Indian camps, and Fort Robinson are depicted in oblique perspective, as are White and crow buttes. Each is located in relation to the line of Pine Ridge, shown by means of a belt of pine symbols, and the valley system of the White River and its tributaries, shown not by the courses of the rivers but by deciduous tree symbols representing the valley-floor galeria forests of cottonwoods.[21]

In 1906, the secretary of the State Historical Society of North Dakota commissioned what must have been one of the last historical-record maps to have been made by a Plains Indian. Painted on canvas in eleven segments by a Mandan, Sitting Rabbit, it depicts various natural and past and present cultural features along a stretch of the Missouri River more than four hundred miles long, from about the mouth of the Yellowstone River to near the North Dakota–South Dakota boundary.[22] The representation of the course of the Missouri is not original, but was derived somewhat freely from the sectional chart published by the Missouri River Commission between 1892 and 1895, with which Sitting Rabbit had been supplied. Thomas Thiessen and his coauthors have recently described this map and stressed its significance in providing leads for later archeological surveys.[23] Apparently, however, they failed to recognize the unexplained use of color: blue for the channels of the Missouri and Heart rivers; orange for smaller tributaries; dark black riverine borders of variable width (floodplains?), which are sometimes separated from the channel by narrow bands of brown (bare sand and silt?); and brown (treeless?) islands. Most cultural features are shown in profile, but American settlements are depicted, albeit schematically, according to their grid plan.

One of the earliest surviving examples of a map depicting events dates from 1825 (fig. 4.6).[24] Sketched by an Oto Indian, Gero-Schunu-Wy-Ha, it shows events in that year on the middle Missouri River between Council Bluffs and the Little Missouri River and also traces the route of an Oto war party that attacked the Arapahos in the area between the upper Arkansas and upper Cimarron rivers.[25] The events are depicted in typical Indian pictographic style against a network of rivers. The gross distortion of the network reflects the constraints imposed on the Indian by the rectangular sheet of paper (fig. 4.7a). Even so, it is a remarkable map, covering about a third of a million square miles of the northern and central American plains.

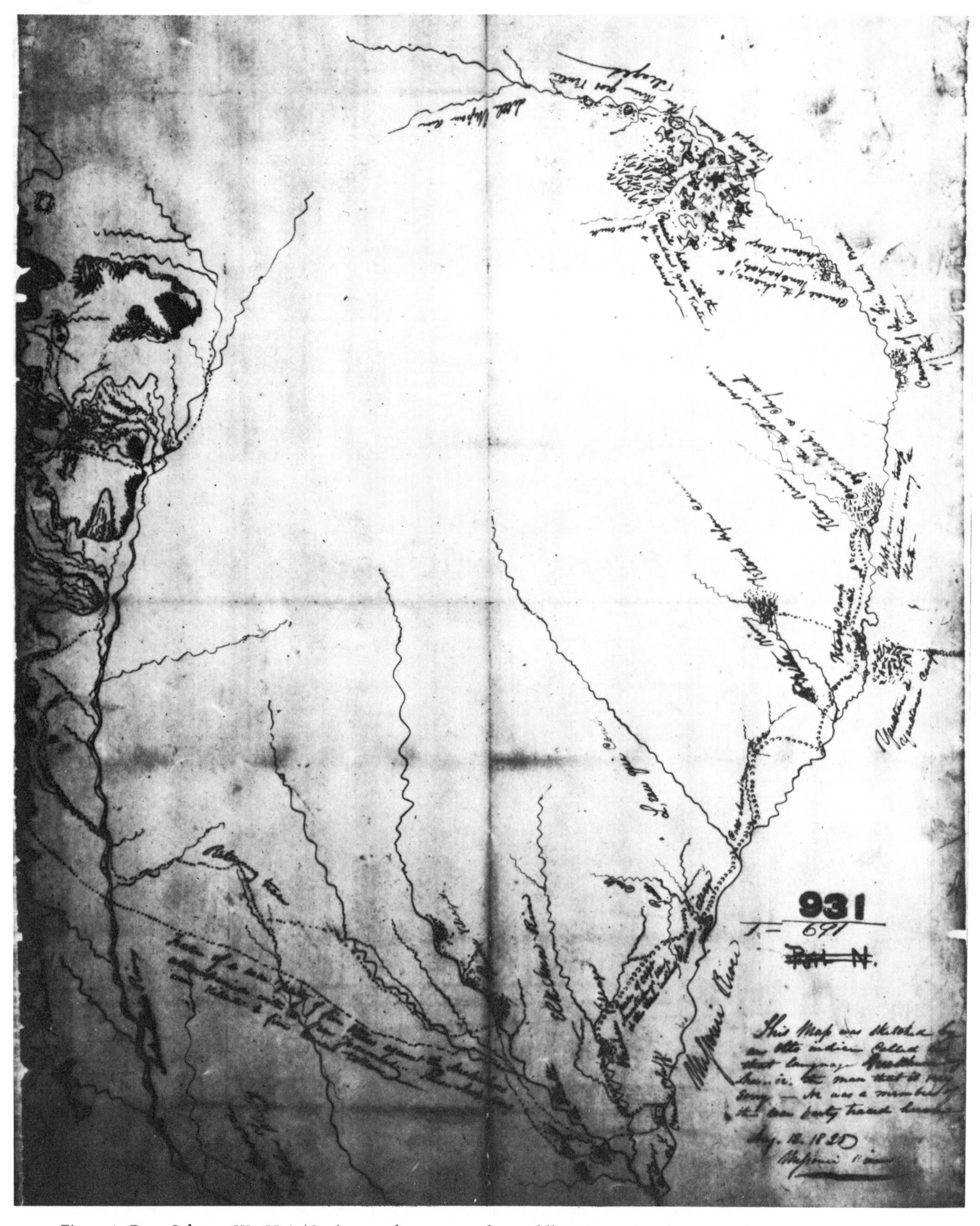

Fig. 4.6. Gero-Schunu-Wy-Ha's (Oto) map of events on the middle Missouri and upper Arkansas rivers, 1825. MS original with annotations, 53 × 42 cm. (Cartographic and Architectural Branch, National Archives)

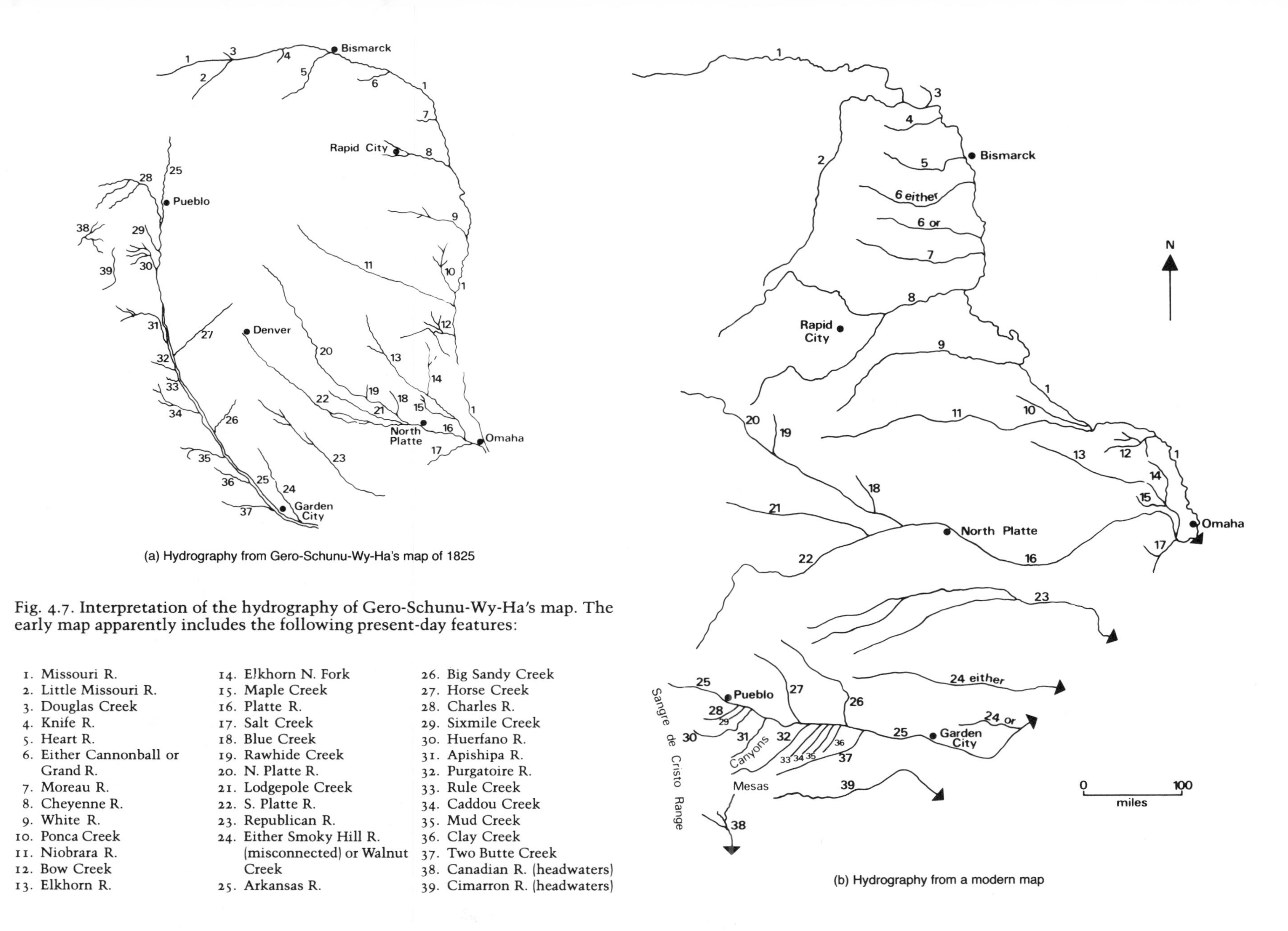

(a) Hydrography from Gero-Schunu-Wy-Ha's map of 1825

(b) Hydrography from a modern map

Fig. 4.7. Interpretation of the hydrography of Gero-Schunu-Wy-Ha's map. The early map apparently includes the following present-day features:

1. Missouri R.
2. Little Missouri R.
3. Douglas Creek
4. Knife R.
5. Heart R.
6. Either Cannonball or Grand R.
7. Moreau R.
8. Cheyenne R.
9. White R.
10. Ponca Creek
11. Niobrara R.
12. Bow Creek
13. Elkhorn R.
14. Elkhorn N. Fork
15. Maple Creek
16. Platte R.
17. Salt Creek
18. Blue Creek
19. Rawhide Creek
20. N. Platte R.
21. Lodgepole Creek
22. S. Platte R.
23. Republican R.
24. Either Smoky Hill R. (misconnected) or Walnut Creek
25. Arkansas R.
26. Big Sandy Creek
27. Horse Creek
28. Charles R.
29. Sixmile Creek
30. Huerfano R.
31. Apishipa R.
32. Purgatoire R.
33. Rule Creek
34. Caddou Creek
35. Mud Creek
36. Clay Creek
37. Two Butte Creek
38. Canadian R. (headwaters)
39. Cimarron R. (headwaters)

Maps Drawn for Whites

Given the map-drawing ability of Indians in the already settled parts of North America, it is not surprising that from the outset explorers approaching and entering the Great Plains obtained information about its geography in map form from indigenous peoples.

The earliest of all extant examples (albeit in contemporary transcript form) of a map made on paper by a North American Indian at the request of whites was made in Mexico City on 29 April 1602 (fig. 4.8).[26] In the course of an official inquiry concerning the expedition onto the Great Plains made by Don Juan Oñate, governor of New Mexico, in the previous year, a captured Indian named Miguel was asked to

> mark with pen and ink on a sheet of paper . . . the pueblos of his land. Miguel proceeded to mark on the paper some circles resembling the letter "O," some larger than others; in a way easily understood he explained what each circle represented. . . . Then he drew lines, some snakelike and others straight, and indicated by signs that they were rivers and roads.[27]

Miguel was probably a native of southern Texas, from where he had been taken north as a captive by other Indians. The upper right-hand part of the map may well show Indian villages on the west Texas coastal plain, but the Aces Rio certainly indicates the river far to the north where he was taken by the Spaniards. The most authoritative interpretation published to date suggests that the Aces Rio was one of the east-bank tributaries of the Arkansas River in north-eastern Oklahoma.[28] I tend to favor the Trinity River some sixty miles or so downstream from Dallas, but in any case a careful interpretation needs to be undertaken with reference to detailed archeological, linguistic, and environmental evidence.

French, English, and American approaches to the Great Plains did not commence until approximately one to two hundred years later, but they also involved the use of information collected in map form from Indians. Indeed, one of the earliest pieces of non-Spanish evidence for the region's existence, as distinct from its environmental characteristics, may have been received in map form. In 1694, Lawrence van den Bosh sent a map of the lower Mississippi valley to Governor Francis Nicholson of Maryland. According to the accompanying letter, information on all the country "on the left side of the Messacippi River" on the map had been obtained "from a French Indian."[29] The Red and Sabine rivers are shown for an indeterminable distance upstream, but despite two legends indicating that they are only sixty-five leagues beyond the Sabine River, the "Mountains of Silver and Gold Mines" to the far left of the map must, on geological grounds, be beyond the southern plains in western Texas and southern New Mexico.

Approximately thirty years later, Nicholson, who was then the governor of South Carolina, received from an Indian chief a map of the whole of southeastern North America painted on a deer skin.[30] Showing rivers, coastlines, tribal locations, and Indian trails, it appears on first examination to terminate to the west at the Mississippi. However, squeezed between that river and the hind end of the skin are grossly distorted representations of the Red and Arkansas rivers, as well as named sites of Indian villages associated with the rivers. Of these villages, the following can be recognized with reasonable certainty: on the Red, the Natchitoches (Notaukw), Kadohadacho (Katutaucejo), and Kichai (Kejoo); on the Arkansas, the Tawakoni (Tovocolau), Kansa (Causau), Pawnee (Pauncasau), and Comanche (Commaucerlau).

Neither of the maps received by Governor Nicholson indicates the existence of the grassland environment far up the Red and Arkansas rivers, but a map drawn in chalk in 1743 by Joseph la France, a "French-Canadese Indian," on the floor of a dining room at the Golden Fleece in New Bond Street, London, was among the first to indicate grasslands. La France, an ignorant metis who "knew nothing of figures," had spent the previous three years traveling, fishing, and hunting with Indians at and beyond the northeastern edge of the plains, and in culture he was almost certainly more Indian than French.[31] His chalked map of the Lake Winnipeg–Nelson River drainage system was incorporated in an important printed map of 1744. Grossly distorted, and hence difficult to interpret, it places the "Assinibouels of the Meadows" to the west of what would appear to be Lake Manitoba—that is, in the grassy openings and wheatgrass prairies of southwestern Manitoba.[32] Beyond, and to the southeast, were the

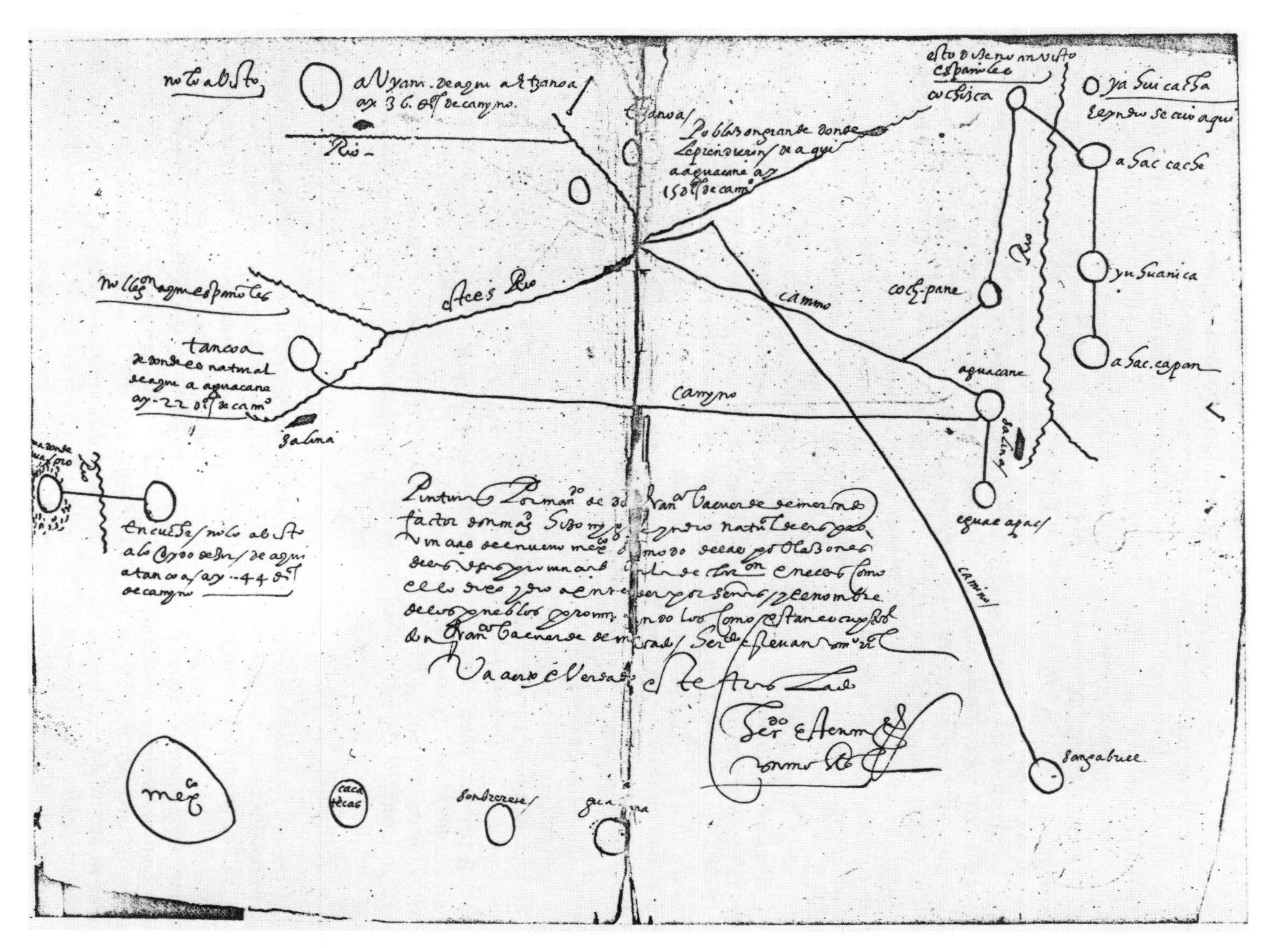

Fig. 4.8. Miguel's (Southern Plains or West Gulf Coast Indian) map of an area that includes San Gabriel, New Mexico, either the middle Arkansas valley or the Texas coastal plain, and probably parts of Mexico, 1602. MS, contemporary transcript, 31 × 43 cm. (Archivo General de las Indias, Seville, Spain)

Nation de Beaux Hommes (generally supposed to be the Blackfeet).

The French approach to the northern plains was also preceded by the solicitation of geographical information from Indians, and some of this was likewise in map form. It differed, however, from that supplied to the English in that some of it became incorporated in important printed maps, with unfortunate consequences for the growth of knowledge about the area. In a report to the governor of New France dated 10 October 1730, the fur trader Pierre Gaultier de Varennes de la Vérendrye mentioned obtaining information from various groups of Cree Indians.[33] Much of this information, concerning lands to the west and northwest of Lake Superior that were then virtually unknown, was in map form. Some of these maps have survived, though none, apparently, in their original form. The closest to an original is a crude compilation from several Indian sources showing the drainage systems between the head of Lake Superior and the northern end of Lake Winnipeg.[34] The map indicates that the information for the area near the latter lake is based on a map drawn by Cree Indians. It shows prairies occupied by the Assiniboeles and Sioux to the west and southwest of what we may now infer to be Lake Winnipeg. This was probably the first map to indicate the existence of the wheatgrass prairies of the Canadian and Dakotan plains. Far less boldly the map shows a "River of [or from] the West," (*Flouve de L'ouest*), passing prairies on one side and a "Mountain of Bright Stone" (*Montagne de pierre Brillante*) on the other. This river terminates at the edge of the map, where "the waters begin to ebb and flow" (*Commencemens de flux et reflux*).

This map, or one or more transcripts of it, evidently came into the hands of three eminent French cartographers. In 1754 Philippe Buache merely reproduced it as an inset to his elegant map of western Canada.[35] In 1730 or soon after, Guillaume Delisle rescaled the map and pasted it as an insert on a manuscript map of North America.[36] In doing so, he placed Lake Winnipeg and the source of the Mississippi in the same longitude as the Gulf of California, and both the Mountain of Bright Stone and the source of the River of the West were located almost on the Pacific Coast. This false geography was widely disseminated in 1743, when it was incorporated in Nicolas Bellin's "Carte de l'Amerique Septentrionale."[37]

The supporting evidence is far from conclusive, but it would appear that the Crees' information was correct according to their principles of mapping and that it was the French cartographers' interpretations of topologically arranged information which were wrong. The River of the West as conveyed by the Cree Indians was not what was later to become known as the Saskatchewan River (i.e., a great river flowing from the west) but the thirty- to forty-mile-long Echimamish River, a small, east-bank tributary of the Nelson River by which Cree Indians, using the Hayes River, traveled west on an important route inland from Hudson Bay. I further believe that the "Mountain of Bright Stone" was not a prominent peak within the Rocky Mountains but an important painted stone very near to the portage between the headwaters of the Hayes and Echimamish rivers; that the prairies near the Echimamish were not grassy openings in the forest, of the type already familiar to the French, but swampy openings, of which there are many to the east and northeast of Lake Winnipeg; and that "the place where the waters begin to ebb and flow" was not a tidal zone near the Pacific Ocean (Bellin's River of the West was actually represented as flowing toward the west) but the area around the twenty-pace-long portage between the Echimamish and Hayes rivers, which, to quote an early nineteenth-century explorer, was "remarkable for the marshy streams which rise on each side of it, taking different courses."[38]

It is not my purpose to review in any detail the collection and use of Indian maps by the staff of the Hudson's Bay Company as they moved southwestward through the forests into the grasslands during the last few years of the eighteenth century; by the Spaniards during their brief period of exploration on the middle Missouri in the late eighteenth century; or by American explorers from Lewis and Clark onward.[39] Five maps embracing all or parts of the plains of what are now Alberta and Montana were collected from Blackfeet Indians by the Hudson's Bay Company surveyor Peter Fidler in 1801–1802, and at the Mandan villages in the late winter or early spring of 1805, William Clark collected two similar maps of the upper Missouri and Yellowstone basins.[40] These are relatively well known, but an even more remarkable map collected at approximately the same period has virtually escaped attention. Drawn on paper with a watermark dated 1801, it has names that appear to be in the hand of the fur trader David Thompson, as does the endorsement on the verso: "Indian Chart Rocky Mountains."[41] Although there is no reason to doubt the source of information, the latter part of the endorsement is a misnomer, as the map shows the rivers and lakes from Hudson Bay westward to the Canadian Rockies and from Lake Athabaska southward to the Great Plains in what are now Alberta, Saskatchewan, and the Dakotas—an area of more than one million square miles.

Indian Influence on Printed Maps

Only rarely can one demonstrate the influence of Indian maps on the maps of explorers and on the printed maps on which they in turn had an influence. Eighteenth- and early-nineteenth-century cartographers seldom cited the sources of their information or explained the ways in which they selected from and modified it. For the most part one can only infer the Indian contributions. One line of inference is that maps made by whites often represented with some degree of truth the geography of areas well beyond the frontiers of exploration. For example, Peter Pond's map of 1785 shows the South Saskatchewan, Assiniboine, Qu'Appelle, Souris, and upper Missouri rivers and depicts, albeit schematically, the range of mountains, "called by the natives Stony Mount," in which the first and last of these rivers have their source.[42] Having traded in what are now western Minnesota, southern Manitoba, and southern Saskatchewan between 1774 and 1777, Pond probably obtained this information from Indians.

Other lines of inference can be deduced from the general characteristics of maps made by Indians. First, Indians frequently failed to distinguish on their maps between river courses, portages, and trails, which to them were merely parts of a single communication network. Second, they would represent small rivers of strategic route significance as equal to or greater in importance than hydrologically major rivers. Third, arrows indicated the direction of travel, which, as often as not, was opposite to the direction of the river flow. Fourth, in the absence of any sense of or need to use absolute scale or direction, shapes and patterns were grossly distorted. When copied by explorers who had little knowledge of these characteristics and incorporated by cartographers who had even less, such patterns often appear to us, in the light of our knowledge of the geography, as so erroneous as to have been based on little more than myth. Had the plains not had an aboriginal population, this may well have been the case, but in fact, many of these patterns were derived from Indian sources. We need to reexamine early white maps of the region in order to recognize the seams where the *terrae semicognitae*, as communicated by the Indians, are welded to the *terrae cognitae* of the Europeans, Canadians, and Americans—for example, where rivers *appear* to cross, to bifurcate downstream, or to flow uphill.

Indian maps often incorporated pictographic conventions, especially those relating to quantity. Camps, villages, and tribes, for example, were frequently represented on their maps by clusters of dots (or of symbols) approximately proportional in number to their respective importance, or by symbols (such as circles) the sizes of which denoted relative importance. Such techniques are found on several of William Clark's transcripts of maps made for him by Indians, though we do not know whether he was faithful in the manner in which he copied them. It is interesting that Collot's "Map of the Missouri," which is generally supposed to have incorporated the results of Truteau's expedition up the Missouri in 1794–96, uses clusters of between two and five tepee symbols to denote the locations of Plains Indians tribes far to the west and northwest of the limit of Collot's explorations.[43] This device, together with the strange representation of the courses of the rivers of the central and northern plains, the naming of the "Yellow [Rocky] Mountains according to the Indians living on the Missouri," and the use of a different proportional symbol to represent the tribes of earth-lodge dwellers of the east-central plains and domed-house dwellers of the Great Lakes region, suggests that much of Truteau's information may well have been obtained directly or indirectly from Indians in map form—perhaps from those whose "delineations on skin" he praised so highly.[44]

The Way Ahead

Given that the outer limit of the mapping of the plains by whites in the eighteenth and early nineteenth centuries was frequently several hundred miles in advance of their exploration, it is reasonable to suppose that Indians supplied much of the information about these fringe zones. The more we understand about the maps known to have been made by Indians, the more we will understand about the whites' representations of these *terrae semicognitae* on their maps, about the false geographies that were thus disseminated, and about the consequences thereof for those who were then beginning to push forward the limits of the *terrae cognitae.* These relationships are an important source of Indian-white misunderstanding that, to date, has barely been recognized by scholars or, indeed, by those involved in litigations regarding Indian land claims. Historians of cartography ought to be contributing far more to the elucidation of these former misunderstandings, not only within the context of the Great Plains but for the continent as a whole. To do so they will need to increase their knowledge of both the real world and the Indians' worlds in the past. They will also need to apply more sophisticated conceptual approaches to problems of cartographic interpretation.

Notes

The author gratefully acknowledges financial assistance received from the Canadian High Commission, London; the Newberry Library, Chicago; The Social Science Research Council (U.K.); and the University of Sheffield Research Fund.

1. William Bray, "Observations on the Indian Method of Picture Writing," *Archaeologia* 6 (1782): 159–62; "The relation of the navigation and discovery which Captaine Fernando Alarchon made by order of the right honourable Lord Don Antonio de Mendoca Vizeroy of New Spaine . . ." [in 1540], in Richard Hakluyt, *The Third and Last Volume of the Voyages, Navigations, Traffiques and Discoveries of the English Nation* . . . (London: George Bishop et al., 1600), p. 438. Bray explains and reproduces marks made by a Delaware Indian on a tree by the Muskingum River, which show several events of war in relation to the sites of Fort Pitt and Fort Detroit and the courses of the Allegheny, Monongahela, and upper Ohio rivers. Alarchon describes how, on request, an Indian on the lower Colorado River drew a map of the river and settlements upstream.

2. Early modern writings on Indian, as distinct from Inuit, maps and mapping include: Cottie A. Burland, "American Indian Map Makers," *Geographical Magazine* 20 (1947–48): 285–92; Delf Norona, "Maps Drawn by Indians in the Virginias," *West Virginia Archeologist* 2 (1950): 12–19, and "Maps Drawn by North American Indians," *Bulletin of the Eastern States Archeological Federation* 10 (1951): 6; and Robert F. Heizer, "Aboriginal California and Great Basin Cartography," *Report of the California Archeological Survey* 41 (1958): 1–9. Recent and in general more interpretative contributions include: John Warkentin and Richard I. Ruggles, *Historical Atlas of Manitoba* (Winnipeg: Historical and Scientific Society of Manitoba, 1970), pp. 72–73, 86–91; David H. Pentland, "Cartographic Concepts of the Northern Algonquians," *Canadian Cartographer* 12 (1975): 149–60; Wayne Moodie and Barry Kaye, "The Ac Ko Mok Ki Map," *The Beaver,* Outfit 307, no. 4 (Spring 1977): 4–15; Louis De Vorsey, "Amerindian Contributions to the Mapping of North America: A Preliminary View," *Imago Mundi* 30 (1978): 71–78; Thomas D. Thiessen, W. Raymond Wood, and A. Wesley Jones, "The Sitting Rabbit 1907 Map of the Missouri River in North Dakota," *Plains Anthropologist* 24, pt. 1 (1979): 145–67; and G. Malcolm Lewis, "The Indigenous Maps and Mapping of North American Indians," *Map Collector* 9 (1979): 25–32, and "Indian Maps," in Carol M. Judd and Arthur J. Ray, eds., *Old Trails and New Directions: Papers of the Third North American Fur Trade Conference* (Toronto: University of Toronto Press, 1980), pp. 9–23. To date there has been only one monograph on the subject: Rainer Vollmar, *Indianische Karten Nordamerikas: Beiträge zur Historischen Kartographie* (Berlin: Dietrich Reimer, 1981).

3. Klaus F. Wellmann, *A Survey of North American Indian Rock Art* (Graz, Austria: Akademische Druck- und Verlagsanstalt, 1979), p. 131.

4. David S. Gebhard, *Indian Art of the Northern Plains* (Santa Barbara: The Art Galleries, University of California, 1974), p. 13.

5. Thomas F. Kehoe, "Stone 'Medicine Wheel' Monuments in the Northern Plains of North America," *Proceedings of the 40th International Congress of Americanists, Rome, 1972,* II: 184.

6. [A Chickasaw Indian], "Nations Amies et Ennemies des Tchikachas . . . le Sept Septembre 1737," transcribed by Alexander de Batz, MS, General Correspondence of Louisiana C.13, V.22, Archives des Colonies, Paris.

7. Alice C. Fletcher, "Star Cult among the Pawnee: A Preliminary Report," *American Anthropologist,* n.s. 4 (1902): 730–36.

8. Pawnee (Skidi band) celestial chart on an oval elk skin, undated but collected at Pawnee, Oklahoma, in 1906 from a Skidi named Big Black Meteor. Artifact, 65 × 43 cm., no. 71, 898–10, Department of Anthropology, Field Museum of Natural History, Chicago.

9. Ralph N. Buckstaff, "Stars and Constellations of a Pawnee Sky Map," *American Anthropologist*, n.s. 29 (1927): 279–85.

10. Selwyn Dewdney, *The Sacred Scrolls of the Southern Ojibway* (Toronto and Buffalo: University of Toronto for the Glenbow-Alberta Institute, Calgary, Alberta, 1975), chap. 5.

11. Non-Chi-Ning-Ga (Iowa), untitled map of the Upper Mississippi and Missouri drainage systems between Lake Michigan and the Great Plains showing "the route of my (Ioway) forefathers—the land that we have always claimed," presented at a council between Indians of the Mississippi and Missouri in Washington, D.C., on 7 October 1837, MS, 104 × 69 cm., Record Group 75, Map 821, Tube 520, Cartographic Branch, National Archives, Washington, D.C.; "Map of the Country formerly occupied by the Ioway Tribe of Indians from a map made by Waw-Non-Que-Skoon-A An Ioway Brave," in Henry R. Schoolcraft, *Information Respecting the History, Condition, and Prospects of the Indian Tribes of the United States* (Philadelphia: Lippincott, Grambo, 1853), vol. 3, plate 30, facing p. 256.

12. Richard I. Dodge, *The Hunting Grounds of the Great West* (London: Chatto and Windus, 1877), p. 414.

13. John D. Hunter, *Manners and Customs of Several Indian Tribes Located West of the Mississippi, . . .* (Philadelphia: J. Maxwell, 1823), p. 209.

14. Jean Baptiste Truteau, "Remarks on the Manners of the Indians living high up on the Missouri," *Medical Repository*, 2d hexade, 6 (1809): 56.

15. Amos Bad Heart Bull and Helen Blish, *a Pictographic History of the Oglala Sioux* (Lincoln: University of Nebraska Press, 1967), no. 198.

16. Jacob J. Brody, *Indian Painters and White Patrons* (Albuquerque: University of New Mexico Press, 1971), p. 25.

17. White Bird (Northern Cheyenne), untitled painting of the Battle of the Little Big Horn done on muslin in 1894–95, 171 × 249 cm. no. 407, West Point Museum, United States Military Academy, West Point, New York.

18. Howling Wolf (Cheyenne), nos. 1 and 2 of a set of twelve paintings made in a commercial sketchbook between 1878 and 1881, owned by Mrs. A. H. Richardson and on loan to the Joslyn Art Museum, Omaha; reproduced in Karen D. Petersen, *Howling Wolf: A Cheyenne Warrior's Graphic Interpretation of His People* (Palo Alto, Calif.: American West, 1968), plates 1 and 2.

19. Howling Wolf (Cheyenne), postal card addressed to his father Minimic at Fort Marion with a pictographic map done at sea in July 1877 after leaving Savannah, Georgia, showing coastline and major settlements along the route from St. Augustine, Florida, 7.6 × 13.0 cm, Francis Parkman Papers, Massachusetts Historical Society, Boston, Mass.; reproduced in Karen D. Petersen, *Plains Indian Art from Fort Marion* (Norman: University of Oklahoma Press, 1971), plate 43.

20. Bull and Blish, *Pictographic History*, nos. 103, 129, 170, 177, 197, 198, 218, 297, and 301.

21. Bull and Blish, *Pictographic History*, no. 197.

22. Sitting Rabbit (Mandan), untitled painting on canvas of the Missouri River from Standing Rock Reservation to the Yellowstone confluence, MS [1906–1907], in seven sections, in all 44 × 707 cm, no. 679, State Historical Society of North Dakota, Bismarck.

23. Thiessen, Wood, and Jones, "Sitting Rabbit 1907 Map."

24. Gero-Schunu-Wy-Ha (Oto), a map of the middle Missouri and upper Arkansas rivers with names and legends added by Captain William Armstrong, or a member of his military party of 1825. One legend states, "This Map was Sketched by an Otto Indian Called in that language Gero-Schunu-wy-ha, i.e. the man that is very Sorry—He was a member of the war party traced hereon—Aug. 12, 1825." MS, 53 × 42 cm., RG 75, Map 931, Cartographic Branch, National Archives.

25. W. Raymond Wood, *Notes on the Historical Cartography of the Upper Knife-Heart Region* (Columbia: American Archaeology Division, College of Arts and Science, University of Missouri, 1978), pp. 62–63.

26. This map, made by an Indian named Miguel, perhaps from within what is now southern Texas, has proved difficult to interpret, but it certainly includes San Gabriel in the Rio Grande valley, New Mexico, and possibly Mexico City; it almost certainly represents rivers, trails, and settlements on the Texas coastal plain and southern Great Plains. The map is endorsed "Pintura que por man^{do} de Don Fran^{co} Valverde . . . ," 1602. MS, 31 × 43 cm., Estante 1, Cajon 1, Legajo 3/22, Ramo 4, Archivo General de Indias, Seville.

27. George P. Hammond and Agapito Rey, "Don Juan de Oñate, Colonizer of New Mexico, 1595–1628," part 2, *Coronado Cuarto Centennial Publication, 1540–1940*, vol. 6 (Albuquerque: University of New Mexico Press, 1953), pp. 872–73.

28. Waldo R. Wedel, "Archaeological Remains in Central Kansas and Their Possible Bearing on the Location of Quivira," *Smithsonian Miscellaneous Collections* 101, no. 7 (1942): 18–20.

29. Lawrence van den Bosh, untitled pen-and-ink manuscript map of the lower Mississippi valley that was accompanied by a letter dated North Sassifrix, 19 October 1694. MS, 32 × 38 cm., no. 59, Edward E. Ayer Collection, Newberry Library, Chicago.

30. Cacique (Chickasaw?), "A Map Describing the Situation of the several Nations of Indians between South Carolina and the Mississippi; was Copyed from a Draught Drawn upon a Deer Skin by an Indian Cacique and Presented to Francis Nicholson Esqr. Governor of Carolina." MS (ca. 1720), 114 × 142 cm., CO 700, North American Colonies General, No. 6 (2), Public Record Office, London.

31. Christian Brun, "Dobbs and the Passage," *The Beaver*, Autumn 1958, pp. 27–29.

32. "A New Map of Part of North America from the Latitude of 40 to 68 Degrees. Including . . . the Western Rivers and Lakes falling into Nelson River in Hudson's Bay as described by Joseph La France a French Canadese Indian who Traveled thro those Countries and Lakes for 3 Years from 1739 to 1742." in Arthur Dobbs, *An Account of the Countries Adjoining to Hudson's Bay* (London: J. Robinson, 1744).

33. "Continuation of the Report of the Sieur de la Vérendrye touching upon the discovery of the Western Sea (Annexed to the Letter of M. de Beauharnois, of October 10, 1730)," French and English versions in Lawrence J. Burpee,

ed., "Journals and letters of Pierre Gaultier de Varennes de la Vérendrye and His Sons," *Publications of the Champlain Society* no. 16 (Toronto, 1927), pp. 43–63.

34. Untitled manuscript map of the rivers and lakes between Lake Superior and Lake Winnipeg of which the parts adjacent to Lake Winnipeg are indicated as "Carte Tracée Par Les Cris," undated but supposedly 1728–29. Original in the Archives Nationales, Paris; photographic copy in the National Map Collection, Ottawa, H2/902–19 (1728–29).

35. Philippe Buache, "Reduction de la Carte tracée par le Sauvage Ochagach et autres, . . ." an inset of "Carte Physique de Terreins les plus eleves de la Partie Occidentale du Canada . . . ," in *Considerations geographiques et physiques sur les nouvelles decouvertes au nord de la grande mer . . .* (Paris: Imprimerie de Ballard, 1753–54), part 3.

36. Guillaume Delisle, manuscript map of North America with pasted-on copy of an Indian map, undated but supposedly 1730, Bibliothèque Nationale, Départment des Estampes, Paris, Serie Vd, vol. 22.

37. Jacques N. Bellin, "Carte de l'Amerique Septentrionale . . . ," in Pierre F. X. de Charlevoix, *Histoire et déscription de la Nouvelle France . . .* , vol. 1 (Paris: Nyon, 1744).

38. John Franklin, *Narrative of a Journey to the Shores of the Polar Sea in the Years 1819, 20, 21 and 22*, vol. 1 (London: John Murray, 1823), p. 63.

39. Richard I. Ruggles of Queens University, Kingston, Ontario, has recently undertaken a major program of research on the history of the cartography of the Hudson's Bay Company between 1670 and 1870. Soon to be published as a Hudson's Bay Record Society publication, this includes an authoritative review of Indian maps and mapping on and around the Canadian prairies. See also Chapter 5 below. More information on this subject should become available within the next few years with the publication by the University of Nebraska Press of a new edition of the journals of the Lewis and Clark Expedition. Although nineteenth-century accounts by American explorers contain many references to overlanders soliciting route information from Indians, they are often unspecific as to the geographical extent and content of the information and the mode in which it was transmitted: John D. Unruh, Jr., *The Plains Across* (Urbana: University of Illinois Press, 1979), pp. 156–57. Unruh's doctoral dissertation treats the same subject in somewhat greater detail: "The Plains Across: The Overland Emigrant and the Trans-Mississippi West, 1840–1860" (Ph.D. diss., University of Kansas, 1975), pp. 224–25.

40. Five transcripts of maps made by Blackfeet Indians in 1801–1802, Peter Fidler's manuscript journal, E3/2, fol. 103–107, Hudson's Bay Company Archives, Winnipeg; two transcripts by William Clark of maps made by Indians in the winter or early spring of 1805, both showing the Missouri River from the Mandan villages to the Yellowstone River, Coe Collection, Beinecke Library, Yale University.

41. Manuscript map in ink on paper with the watermark "I. Taylor 1801" and endorsed on the verso, "Indian Chart Rocky Mountains," 38 × 50 cm., Manuscript Collection, Royal Commonwealth Society, London.

42. "Copy of a Map Presented To the Congress by Peter Pond a Native of Milford in the State of Connecticut . . . New York 1st March 1785 . . . ," MS, 71 × 53 cm., Add. MS 15, 332-C, Department of Manuscripts, British Museum, London.

43. "Map of the Missouri . . . ," in Georges H. V. Collot, *A Journey in North America, . . .* (Paris: A Berstrand, 1826), atlas volume.

44. Truteau, "Remarks on Manners of Indians."

5. "A CHART IN HIS WAY"

INDIAN CARTOGRAPHY AND THE LEWIS AND CLARK EXPEDITION

James P. Ronda

THE sixteenth of January 1805 was not the kind of day Lewis and Clark would have chosen for calm deliberation and the thoughtful exchange of cartographic information. On that cold Dakota day, Fort Mandan was the scene of angry words and hostile gestures as Mandans and Hidatsas traded jeers and insults. While Lewis and Clark watched helplessly, Hidatsa warriors from the village of Menetarra charged Mandans with spreading malicious rumors designed to breed fear and keep Hidatsas away from the expedition. As the tough talk flew higher, the expedition's hopes for diplomacy sank. But in the midst of the bitterness and harangue a remarkable event took place—something both important for the immediate needs of the expedition and symbolic of one of the most valuable relations between native people and the explorers. Among the Hidatsas at Fort Mandan was a young war chief intent on mounting a horse-stealing raid against the Shoshonis. Most of what passed between the eager warrior and the edgy explorers centered on an attempt to dissuade him from the proposed raid. Almost as an afterthought. William Clark noted that "this War Chief gave us a Chart in his Way of the Missourie."[1]

That map and the telling phrase "in his Way" typify the substantial cartographic contribution made by native people to the Lewis and Clark expedition. Throughout its nearly two and one-half years in the field, the expedition actively sought out Indian maps and mapmakers. That search brought Lewis and Clark more than thirty of what Malcolm Lewis has so aptly termed "cartographic devices."[2] But more important than the quest for Indian maps was the effort by the Corps of Discovery, and especially William Clark, to understand both the structure and substance of those documents. Lewis and Clark did not pursue Indian mapmakers just to obtain travel information from native sources. They knew Indian maps represented a vital part of a broader encounter, an attempt to communicate important ideas and experiences across the cultural divide. This essay seeks to evaluate expedition Indian maps within the framework of that encounter. The questions posed here are aimed at illuminating the maps, their makers, and the ways Lewis and Clark struggled to use those cartographic devices.

When the Hidatsa warrior offered Lewis and Clark a chart of the upper Missouri, he did it "in his Way." That way may have been a relief map constructed with heaps of dirt and marks on the ground or a river channel drawn with charcoal on a piece of hide. But whatever means were employed, we are reminded that native cartographic information came to and was preserved by Lewis and Clark in a variety of ways: described in words, drawn on hides or on the ground, or constructed topographically in sand—and preserved or redrawn by Lewis and Clark as distinctly Indian productions, or incorporated wholly within Lewis and Clark maps.

First, there were maps created by Indians either

verbally or graphically and then drawn or traced by Lewis and Clark as distinctively Indian maps. This describes a murky historical and cartographic process that can be clarified with two examples.

Early in January 1805, the Mandan chief Sheheke, or Big White, made one of his frequent visits to expedition quarters. After dinner, Big White offered what Clark described as "a Scetch of the Country as far as the High Mountains, and on the south side of the River Rejone [Yellowstone]" (fig. 5.1). Big White may well have drawn an outline of the Yellowstone and its tributaries and then the map was copied by Clark. But "sketch" does not necessarily mean a graphic representation. Big White might have given Clark simply a verbal description of the Yellowstone country. In fact, Clark records just such a description, noting the Indian's words about the tributaries of the Yellowstone, the general character of the terrain, and the presence of "great numbers of beavers." And of course, it is equally possible that Big White produced both a graphic map and a verbal description of the river.[3] But whatever the process, the map that emerged was plainly an Indian production and recognized as such by the explorers.

A second example of an Indian map produced either verbally or graphically and then drawn or traced by Lewis and Clark was received at the end of April 1806, when the expedition was in present-day eastern Washington with the Walula, or Walla Walla, Indians. The Walula chief Yelleppit had been especially friendly to the explorers on their westward trek, and now on the return journey he offered food, horses, and vital route information, part of which came in a map prepared by Yelleppit for the captains. Bearing Lewis's notation, "Sketch given us by Yellept the principal Chief of the Wallah wallah Nation," the map portrays the region around the Columbia-Snake confluence.[4] The maps that Big White and Yelleppit produced, in either verbal or graphic form, were preserved by the explorers as distinctively Indian products. Along with the maps of the Willamette River obtained from Multnomah Indians and the charts drawn by Nez Perce headman Hohots Ilppilp, these maps represent the first category of native cartography made available to Lewis and Clark.

When William Clark and Nicholas Biddle talked about native maps and map-making techniques in 1810, Clark suggested a second category. Indian maps, he explained, were "sometimes in sand, hills designated by raising sand, rivers by hollow." Characterizing these maps by their ephemeral nature, Clark observed that "Indian maps made on skins or mats may be given to you, by individuals, but are not kept permanently among them."[5] On at least nine occasions Lewis and Clark obtained such ephemeral maps from native sources. They ranged from simple charcoal-on-hide outlines of river channels to the very elaborate relief map of the Willamette country constructed by an elderly Multnomah Indian. Such maps presented a unique challenge to expedition patience as well as intellect; they brought the expedition face to face with traditional cartographic practices and conventions. Three examples of ephemeral maps tell us important things about the ways native people made maps and about how the expedition used them.

The most short-lived of the ephemeral maps produced for Lewis and Clark were those drawn on the ground. Such maps were not hastily made scratches in the dirt. Rather, they were often elaborate relief creations portraying mountain ranges and river systems. William Clark learned about the complexity of those maps firsthand on 20 August 1805. Camped with a northern Shoshoni band along the Lemhi River in what is now Idaho, Clark prevailed on the band headman, Cameahwait, to instruct him "with rispect to the geography of his country." Cameahwait's lesson was enhanced by a superb map, constructed on the ground, depicting the courses of the Lemhi and Salmon rivers. With heaps of sand the Shoshoni skillfully laid out the "vast mountains of rock eternally covered with snow."[6] What Cameahwait's tutorial in geography revealed was not especially good news for the expedition. Yet the map itself was a masterful and largely successful attempt to communicate complex geographical realities across cultural barriers. But no matter how accurate the map, its physical structure destined it to a short life.

Clark also knew that there were ephemeral maps having some chance for a longer life. Maps of that kind were drawn on hides or whitened skins. The expedition had perhaps its first look at such a map in late September 1805. While camped temporarily

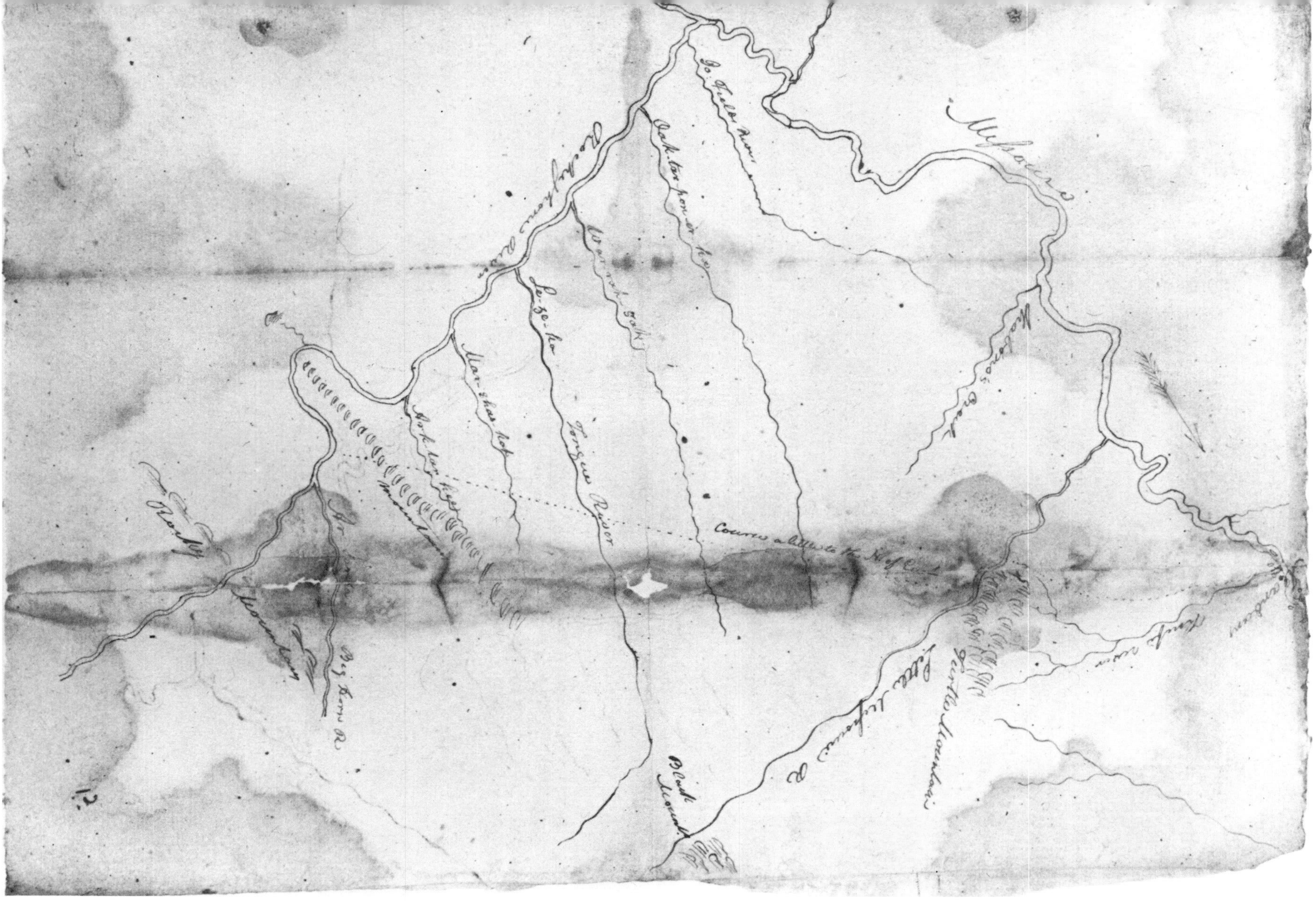

Fig. 5.1. Big White's map of the Yellowstone River and tributaries, 1805. See list of Indian maps, no. 2. (Western Americana Collection, the Beinecke Rare Book and Manuscript Library, Yale University)

at Weippe Prairie outside present-day Weippe, Idaho, Clark convinced the Nez Perce chief Twisted Hair to draft a map of the Clearwater River down to the Snake-Columbia confluence and as far west as Celilo Falls. Working with what Clark described as "great Cherfulness," Twisted Hair marked a white elk skin with the appropriate river courses. The Indian's map evidently impressed Clark and persuaded him to seek additional native cartography for the region. The explorer pursued several Nez Perce elders, all of whom gave "maps of the Country and river with the Situation of Indians [and] Towns." When Clark carefully compared the maps, he found little variation among them—something he took as a sign of their accuracy.[7]

Perhaps the best documented hide map made available to the explorers came to them in mid–October 1805. As the Corps of Discovery tarried a day or two at the Snake-Columbia junction, resting and preparing to challenge the Great River of the West, Lewis and Clark spent considerable time with Yakima and Wanapam Indians. It was from these people that the captains learned much about the physical and human geography of the middle Columbia. On 18 October, the Wanapam chief Cutssahnem, one of the Nez Perce guides (either Tetoharsky or Twisted Hair), and an unnamed Yakima drew an elaborate map of the Snake-Columbia confluence, the middle reach of the Columbia, and the Tapeteet, or Yakima, River. Cutssahnem and the others drew the map on a piece of hide with charcoal. Sufficiently impressed with the map for both its cartographic and ethnographic significance, Clark made a special point to copy the chart and save the original. That original hide map survived until 1895, when it was consumed in a fire at the University of Virginia.[8]

Indian maps plainly labeled as such by Lewis and Clark as well as those ephemeral productions mentioned in the journals make up the two largest categories of native cartographic devices. But there is a third category—more elusive, but nonetheless real. These are maps and verbal descriptions that have wholly disappeared within existing Lewis and Clark maps—Indian components incorporated within expedition maps. These elements are sometimes recognizable, either by structure or quality of information, as distinctively Indian in origin. In the process of gathering data and drafting his 1805 map of western North America, Clark noted that he was employing "the information of Traders, Indians, and my own observation & Ideas." That range of sources was verified by the North West Company trader François-Antoine Larocque, who noted that the explorers were busy with maps and charts founded on information "they had from the Indians."[9]

Clark made his telling comment acknowledging his sources on the same day that Big White offered his sketch of the Yellowstone but before the Hidatsa Missouri chart was available. Big White's contribution is readily identifiable in the way Clark drew the Yellowstone and its tributaries. But Clark's own words can lead us to other, now lost, Indian contributions. That information may well have come from Nor' Wester Hugh Heney. In December 1804 Heney gave Clark sketches he had obtained "from the Indians to the West of this place."[10] Today, just who those western Indians were, what the sketches portrayed, and their very nature are all unclear. What is plain is that Clark praised Heney as "a Verry intelligent man," eagerly sought him out, and promptly incorporated whatever Heney presented into expedition maps.

Lewis and Clark were not the only North American explorers to employ Indian maps as an essential part of the exploratory process. Explorers from Champlain to Coronado, from John Smith to the Vérendryes, all made use of native cartography. But it may be fair to say that no other expedition so actively looked for and attempted to use Indian maps. The Lewis and Clark search for native charts went far beyond what one might expect from an expedition already charged with so many complex missions.

There appear to be four distinct reasons for what amounted to a very productive quest for Indian maps. First, Lewis and Clark readily recognized that Indians as first-comers to the land had an unparalleled grasp of the terrain. The explorers knew that in the western wilderness native people held the key to understanding the face of the land. Indian maps could facilitate expedition travel. Maps like the upper Missouri chart from the Hidatsas, and the relief maps produced by Cameahwait and expedition guide Old Toby, allowed Lewis and Clark to make intelligent route decisions—decisions that would have been much more difficult without the maps. Employing Indian cartography to expand their own maps and observations was a second reason for Lewis and Clark to seek native mapmakers. Because the expedition did not venture far from Fort Clatsop during the winter of 1805–1806, a map of the Oregon coast from the Columbia down to Tillamook Bay made by a Clatsop Indian furnished Clark with important geographic and ethnographic information (fig. 5.2).[11] Having such maps gave the Corps of Discovery an extra reach to its mapping arm.

But gathering and evaluating Indians' maps was based on more than the expedition's need for native route and travel information. If exploration is a programmed enterprise—discovery by design—then much of what Lewis and Clark accomplished as the result of specific instructions from Thomas Jefferson, a third reason for their search for Indian maps. The president never directly ordered his explorers to collect Indian maps, but two sections in the instructions do suggest that sort of activity, at least by implication: "Altho' your route will be along the channel of the Missouri, yet you will endeavor to inform yourself, by enquiry, of the Character & extent of the Country watered by it's branches, & especially on it's Southern side."[12] Two phrases stand out in that order. The explorers were specifically commanded to engage in geographic inquiry, an undertaking that was certain to include maps. At the same time, Jefferson was especially interested in tributaries south of either the Missouri or the Columbia. Clark's search for the Multnomah, or Willamette, and the several Indian maps he obtained of the Willamette country, were a direct response to Jefferson's instructions.[13]

Finally, also reflecting a concern of Jefferson's, the expedition was to gather Indian maps for their ethnographic significance. A line in Jefferson's instructions to Lewis about Indian "language, traditions, [and] monuments" does not specify cartographic devices, but Nicholas Biddle, Lewis and Clark's first editor, understood that maps were included in that sequence. Writing about the Wanapam–Nez Perce–Yakima hide map that the explorers carefully traced on paper, the editor observed that "it exhibited a valuable specimen of Indian delineation."[14]

Lewis and Clark had at their disposal a substantial body of native geographical information. Much of that data was solicited by and offered to the expedition in the form of maps. That the explorers had such materials can be plainly demonstrated; what remains more important and more challenging is understanding the ways in which Lewis and Clark struggled to find meaning in those maps. Cameahwait's relief map, Cutssahnem's hide chart, and the verbal descriptions of many unnamed informants challenged the skill and imagination of the expedition's leaders.

In their recent book *The Nature of Maps,* Arthur Robinson and Barbara Petchenik devote considerable space to the notion of maps as communication systems. If maps are "a way of graphically expressing mental concepts and images," then all maps, mapmakers, and map users are conditioned by cultural values, concerns, and life experiences.[15] Indian maps were as much a cultural product as any ritual or object. When Euro-Americans tried to use such maps they were confronted with something both familiar and yet strangely unsettling. Unlike the stick maps from the Pacific Marshall Islands that record complex ocean currents, Indian maps were readily identifiable to non-Indian eyes as maps. At the same time, Indian maps represented conceptions of distance, space, and time that were often fundamentally different from those commonly held by the bearded strangers. Equally bewildering were the symbols and conventions used to express those cultural considerations. In this context the map becomes not only a communication system but also an arena for yet another part of the American encounter.

When Lewis and Clark sought to interpret and then use Indian maps, they faced a whole battery of problems. On one level the physical structure and expression of many Indian maps may have daunted someone like William Clark. As a young officer in Gen. Anthony Wayne's Legion of the United States, he was schooled in the conventions of European cartography, in which flat maps have North at the top, locations are plotted by a system of latitude and longitude grid lines, and distances are measured in miles or leagues. But when Clark examined a native chart, he saw a device at once recognizable as a map and yet unfamiliar in structure and expression. As Malcolm Lewis's research has shown, Indians

Fig. 5.2. Clatsop Indian map of the Oregon coast south to Tillamook Bay, ca. 1806. See list of Indian maps, no. 15. (Western Americana Collection, the Beinecke Rare Book and Manuscript Library, Yale University)

often oriented their maps along sunrise and sunset lines or toward the direction of travel. Distances were measured in terms of travel time, the term "sleeps" or "days" often appearing in expedition records when Clark copied an Indian map. Language itself posed a problem since the translation of simple words like "above," "below," "to," and "from" could be critical for interpreting a map or verbal description. As Peter Fidler, long-time Hudson's Bay Company surveyor, observed in 1802, "The Indian map conveys much information where European documents fail; and on some occasions are of much use, especially as they shew [where] such & such rivers & other remarkable places are, tho' they are utterly unacquainted with any proportion in drawing them."[16] It was those culturally determined "proportions" that would challenge Clark and others who sought to understand Indian maps.

As expedition cartographers soon learned, the symbols Indians used to express map information were often quite different from those found in Euro-American maps. While Western cartography in general was moving toward the use of mimetic pictures to indicate mountains, rivers, or settlements, Indian mapmakers continued to rely on arbitrary symbols to communicate meaning. Lines drawn in the sand or on a hide might mean creeks, game trails, or often-used war trails. Those arbitrary symbols stood for a general perception of reality; they were not meant to express every twist and turn of the trail. Because Indians mapped what had relevance in their own lives, the concerns of an exploring party bent on neither hunting nor raiding must have sometimes seemed bewildering. As land travelers, Hidatsas could give a good account of their raiding paths to Three Forks, and Clark dutifully noted that route in his map of 1805. But the Hidatsas must have found it difficult to respond to questions about navigation on the Missouri or the amount of time necessary to portage heavy baggage around the Great Falls of the Missouri. As always, perspective and experience are everything.

William Clark confronted more than strange symbols and unfamiliar conventions in Indian maps. Beneath those lines, marks, and heaps of sand was a different way of seeing the material world. Seeing has long been understood as something more than a physical, optical act. Seeing is organizing and giving meaning to disparate and disconnected bits of shadow and substance. Thus the meaning and graphic depiction of a terrain feature will differ markedly from culture to culture. When that Hidatsa warrior gave Clark "a Chart in His way," the map was grounded in native concepts of time, distance, and space. Because all that now survives of most Lewis and Clark Indian maps are the redrawings done by the explorers, we can only guess at how those ideas were expressed, or at the process of converting them to suit expedition needs.

Despite arbitrary symbols not part of their learning and lore, and despite divergent ways of understanding the physical world, Lewis and Clark did succeed in gaining important information from Indian maps. Big White's portrayal of the Yellowstone found its way into Clark's map of 1805; the relief maps made by Cameahwait and Old Toby became part of several maps showing routes across the Continental Divide; maps of the Willamette country drafted by Multnomah Indians plainly influenced Clark's 1810 master map of western North America. As the expedition's leading cartographer, William Clark obviously made a major effort to comprehend Indian maps, which surely expanded his imagination and tested his talents.

What can be described with less certainty are the ways native cartographers understood their part in cartographic cultural contact. If Clark had to alter his angle of vision to cope with that Hidatsa chart, what was involved when the young warrior tried to comply with the explorer's request? Did his methods change to meet the needs of the bearded stranger? Was the map he constructed for Clark different from those he had done before, or was this his first attempt to graphically depict land he knew from previous raids? Or what of Hohastillpilp, Lewis and Clark's fractured rendering of the name of the Nez Perce diplomat and warrior better known as Hohots Ilppilp, or the Bloody Chief? This mapmaker gave the expedition a fine large map in two parts of routes over the Great Divide and on to the plains.[17] This map reflected the experiences of a man widely traveled in the plateau world. Did it also mirror a long tradition of Nez Perce mapping? In what ways did the Bloody Chief struggle to fit his perceptions to the needs of the expedition? There do not seem to be ready answers to these

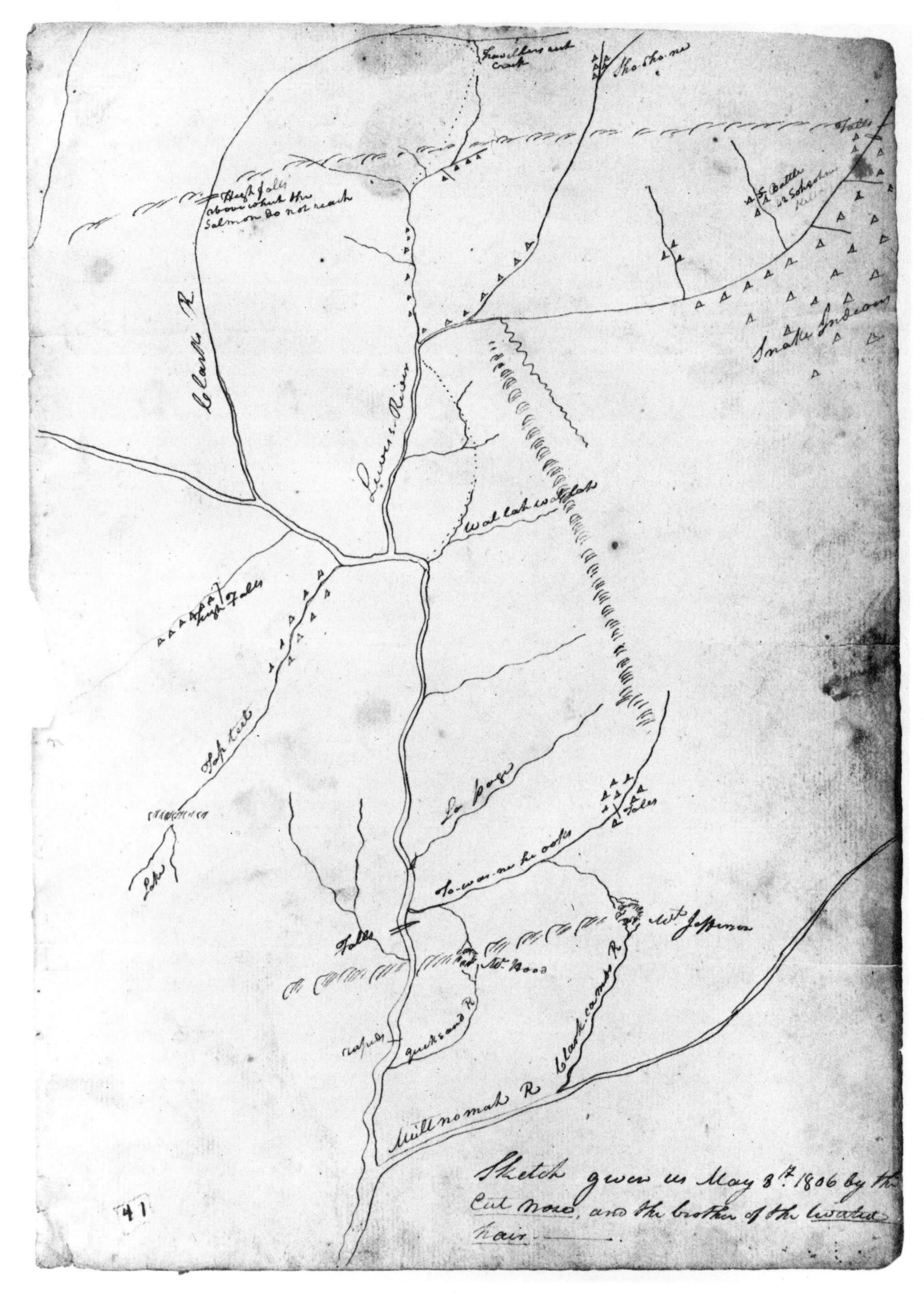

Fig. 5.3. Nez Perce map of the middle Columbia River, 8 May 1806. See list of Indian maps, no. 25. (Western Americana Collection, the Beinecke Rare Book and Manuscript Library, Yale University)

questions. What we do know about the significance of Indian maps for Lewis and Clark is implicit in William Clark's important observation about the whole body of Indian information given to the expedition: "Our information is altogether from Indians collected at different times and entitled to some credit."[18]

The story of Lewis and Clark Indian cartography has a curious and revealing epilogue. On 11 May 1806, Lewis and Clark undertook a full day of diplomacy with Nez Perce chiefs and elders. The negotiations began when the explorers drew a map of the Clearwater country to help explain American policy. But this was no ordinary map. It was made with charcoal on a mat, as Lewis put it, "in their way."[19] The mapping ways of Hidatsas and Nez Perces had become at least partially an expedition way. Maps once formidable in structure and design could now be made and understood by the explorers themselves. Effort and understanding had made map encounters into common ground.

Indian Maps in the Records of the Lewis and Clark Expedition

This list represents maps that are plainly Indian productions as well as those having identifiable native components. Maps mentioned in expedition journals are included even if they did not survive.

Abbreviations:

Moulton	Gary E. Moulton, ed., *Atlas of the Lewis and Clark Expedition*
Thwaites	Reuben Gold Thwaites, ed., *Original Journals of the Lewis and Clark Expedition*

1. *17 December 1804*
"Some sketches from him [Hugh Heney], which he obtained from the Indians to the West of this place." Thwaites, 1: 239.
2. *7 January 1805*
The Sheheke map of the Yellowstone River and its tributaries. Printed: Thwaites, vol. 8, map 12; Moulton, maps 31a and b. (See fig. 5.1.)
3. *16 January 1805*
"This War Chief gave us a Chart in his Way of the Missourie." Thwaites, 1: 249.
4. *20 August 1805*
Relief map by Lemhi Shoshoni chief Cameahwait. Thwaites, 2: 380.
5. *23 August 1805*
Two sand relief maps of the Salmon River country made by expedition guide Old Toby. Thwaites, 3: 27; "Biddle's Notes," Jackson, *Letters*, 2: 545.
6. *21 September 1805*
Map of the Clearwater River drawn by an unnamed Nez Perce chief for William Clark while at Weippe Prairie. Thwaites, 3: 81.
7. *22 September 1805*
Twisted Hair's map of the Clearwater and Snake River country. Thwaites, 3: 85. Note: Information from the 21 and 22 September Nez Perce maps is contained in a sketch printed in Thwaites, 3: 102.
8. *22 September 1805*
"I [William Clark] precured maps of the Country and river with the Situation of Indians [and] Towns from Several men of note Seperately which varied verry little." Thwaites, 3: 85.
9. *10 October 1805*
Sketch map drawn by William Clark from Nez Perce sources showing the Clearwater and south fork of the Snake River. Printed: Thwaites, 3: 102.
10. *14 October 1805*
Sketch map drawn by William Clark from Nez Perce sources showing Indian camps along the Clearwater and Snake rivers. Printed: Thwaites, 3: 114.
11. *18 October 1805*
"A Sketch of the Columbia and its waters and The Situation of the Fishing establishments of the Natives above the Enterance of Lewis's River, given by the Chopunnish, Sokulk, and Chimnapum Indians." Printed: Thwaites, vol. 3, facing p. 118. See also Clark's sketches, Thwaites, vol. 3, facing p. 130, 184.
12. *September–October 1805*
An Indian map from Columbia River sources showing band locations for native people in the Bitterroot–Clark Fork–Pend Oreille country. Printed: Thwaites, vol. 3, facing p. 116.
13. *10 October 1805*
A sketch of Plateau Indian locations along the Snake and Columbia rivers. Printed: Thwaites, 3: 184.
14. *Mid-October 1805*
A sketch by William Clark of band and village sites on the Columbia River, from Indian sources. Printed: Thwaites, vol. 3, facing p. 168.
15. *Early 1806*
Map of the Pacific coast from the Clatsop villages to Tillamook Bay, by a Clatsop Indian. Printed: Thwaites, vol. 8, map 39; Moulton, map 94. (See fig. 5.2.)
16. *2 April 1806*
A map of the Willamette country by two Multnomah Indians. Thwaites, 4: 235–36.
17. *3 April 1806*
A map of the Willamette country and Indian sites drawn by an elderly Multnomah Indian. Thwaites, 4: 241. Note: Maps 16 and 17 may be the basis for map 18.

18. *2–3 April 1806*
"A Sketch of the Moltnomer River given by several different Tribes of Indians near its enterance into the Columbia." Printed: Thwaites, vol. 4, facing p. 242. Note: In the period 17–20 April, when the expedition was at The Dalles, William Clark gathered a number of important Indian maps.
19. *17 April 1806*
"A Sketch of the Columbia as also Clarks River." Thwaites (4: 292) notes that this is his map 40. April 17, however, is not the date found on the map.
20. *18 April 1806*
Indian map of the Plateau-Dalles region. Printed: Moulton, map 95. This is a preliminary stage of Thwaites, vol. 8, map 40, and Moulton, map 96.
21. *18 April 1806*
"This Sketch was given by a Skaddot chief, a Chopunnish and a Skillute Several other Indians at the Great Narrows of Columbia." Printed: Thwaites, vol. 8, map 40; Moulton, map 96. Note: This is evidently an earlier and simpler version of the following map.
22. *20 April 1806*
An Indian sketch of the Columbia and its branches. Printed: Thwaites, vol. 4, facing p. 308.
23. *ca. 27 April 1806*
"Sketch given us by Yellept." Printed: Allen, *Passage through the Garden,* p. 340.
24. *8 May 1806*
A preliminary draft of the following map. Printed: Moulton, map 97.
25. *8 May 1806*
"The relation of the Twisted Hair and Neeshneparkkeook [Cut Nose] gave us a sketch of the principal water courses West of the Rocky Mountains a copy of which I [William Clark] preserved." Thwaites, 5: 5. Printed: Thwaites, vol. 8, map 41; Moulton, map 98. (See fig. 5.3.)
26. *29 May 1806*
Map of trails over the Continental Divide by Hohots Ilppilp. Printed: Thwaites, vol. 8, map, 42; Moulton, map 99.
27. *29 May 1806*
An Indian sketch of the Snake River country. Moulton notes that this is by Hohots Ilppilp. Printed: Thwaites, vol. 8, map 44; Moulton, map 100. Note: The two Hohots Ilppilp maps have recently been found to be two parts of one large map.
28. *Late May 1806*
Possibly a sketch preliminary to the following map. Printed: Moulton, map 102.
29. *29–31 May 1806*
"This Sketch was given by Sundry Indians of the Chopunnish Nation." The map shows trails and villages from the Clark Fork to the Three Forks of the Misssouri. There are two stages of this map. The earlier version is printed in Thwaites, vol. 8, map 43 and in Moulton, map 101. The second stage is printed in Thwaites, vol. 5, frontispiece.
30. *Late 1806*
A map by William Clark of Indian trails from Three Forks to the Yellowstone River. This map has native information in it although Clark's returning party met no Indians on this leg of the journey. Printed: Thwaites, vol. 8, map 48; Moulton, map 106.

Notes

The author gratefully acknowledges the valuable information provided for this paper by Gary E. Moulton.

1. Reuben Gold Thwaites, ed., *Original Journals of the Lewis and Clark Expedition, 1804–1806,* 8 vols. (New York: Dodd, Mead, 1904–1905), 1: 249.

2. G. Malcolm Lewis, "Indian Maps," in *Old Trails and New Directions: Papers of the Third North American Fur Trade Conference,* ed. Carole M. Judd and Arthur J. Ray (Toronto: University of Toronto Press, 1980), p. 11.

3. Thwaites, *Original Journals,* 1: 245. The maps are printed in Thwaites, *Original Journals,* vol. 8, map 12, and in Gary E. Moulton, ed., *The Atlas of the Lewis and Clark Expedition* (Lincoln: University of Nebraska Press, 1983), maps 31a and b.

4. The Yelleppit map is part of the collections of the State Historical Society of Missouri at Columbia. It is conveniently reproduced in John L. Allen, *Passage through the Garden: Lewis and Clark and the Image of the American Northwest* (Urbana: University of Illinois Press, 1975), p. 340.

5. Biddle's notes [c. April 1810], Donald Jackson, ed., *Letters of the Lewis and Clark Expedition with Related Documents, 1783–1854,* 2d ed., 2 vols. (Urbana: University of Illinois Press, 1978), 2: 532.

6. Thwaites, *Original Journals,* 2: 380.

7. Thwaites, *Original Journals,* 3: 85.

8. Thwaites, *Original Journals,* 3: 130. The map is printed in Thwaites, *Original Journals,* vol. 3, facing p. 118. See also Elliott Coues, ed., *History of the Expedition under the Command of Lewis and Clark,* 4 vols. in 3 (1893; reprint, New York: Dover Publications, 1964), 2: 643. In addition to this

formal copy of the map, Clark made a second copy in his field notebook. See Thwaites, *Original Journals*, vol. 3, facing p. 130.

9. Thwaites, *Original Journals*, 1: 246; François-Antoine Larocque, "The Missouri Journal," in L. R. Masson, ed., *Les Bourgeois de la Compagnie du Nord-Ouest*, 2 vols. (1889–90; reprint, New York: Antiquarian Press, 1960), 1: 310.

10. Thwaites, *Original Journals*, 1: 238. A fragment of a map based on Heney's information can be found in Ernest S. Osgood, ed., *The Field Notes of Captain William Clark, 1803–1805* (New Haven: Yale University Press, 1964), p. 324. This sketch reveals the routes used by Indian and white traders in the Dakota Rendezvous trade system.

11. Thwaites, *Original Journals*, vol. 8, map 39; Moulton, *Atlas*, map 94.

12. Jefferson's Instructions to Lewis [20 June 1803], Jackson, *Letters*, 1: 63.

13. Thwaites, *Original Journals*, 4: 235–36, 241, facing p. 242, 254–55.

14. Jefferson's Instructions to Lewis [20 June 1803], Jackson, *Letters*, 1: 62; Coues, *History*, 2: 643.

15. Arthur H. Robinson and Barbara B. Petchenik, *The Nature of Maps: Essays toward Understanding Maps and Mapping* (Chicago: University of Chicago Press, 1976), p. 1 and chapter 2.

16. Lewis, "Indian Maps," pp. 18–19; Peter Fidler to the London Committee, Oxford House, 10 July 1802, as quoted in Eric Ross, *Beyond the River and the Bay: Some Observtions on the State of the Canadian Northwest in 1811* (Toronto: University of Toronto Press, 1973), p. 6.

17. Thwaites, *Original Journals*, vol. 8, maps 42 and 44; Moulton, *Atlas*, maps 99 and 100.

18. Clark to William Croghan, 2 April 1805, Jackson, *Letters*, 1: 230.

19. Thwaites, *Original Journals*, 5: 19.

6. THE SCIENTIFIC INSTRUMENTS OF THE LEWIS AND CLARK EXPEDITION

Silvio A. Bedini

THE Lewis and Clark expedition, called "the most consequential and romantic peace-time achievement in American history," had its genesis in the mind of Thomas Jefferson fully two decades before the exploring party departed from Pittsburgh on August 31, 1803.[1] The need to determine the character and true expanse of the western regions of the continent lingered in his mind, and during the intervening years he encouraged three unsuccessful attempts to achieve that goal. It was not until after he had assumed the presidency in 1801 that he was provided with an opportunity to bring his dream to realization. The venture was not only to accomplish all that Jefferson had hoped, but also to prove to be the first and one of the most important exercises in the application of scientific practices and instrumentation attempted by the young republic.[2]

The third president was eminently suited to plan such a project, for he was unquestionably the best informed individual in the United States on national geography. Not only had he spent many years collecting and studying all that had been written and published about the subject, but the public offices he had held had also provided ample opportunity for him to meet Indians and others who had traveled in the West, and he recorded all that he could learn from them. Furthermore, he was knowledgeable about scientific practices and instruments in general and was personally experienced in surveying and mapping and making astronomical observations, all of which would be required to record the regions to be explored. As president also of the American Philosophical Society, he could call upon the most eminent men of science for advice on all the subjects with which the proposed expedition would be concerned.[3]

Having selected his personal secretary, Meriwether Lewis, to lead the exploring party, Jefferson gave him the task of compiling lists of the needs and estimates of costs of such an undertaking even before submitting the proposal to Congress. Lewis was to prove to be an excellent choice, for he was self-taught in the natural sciences and a lover of the outdoors, and his previous army career had made him experienced in the handling of men. He also had the advantage of being familiar with the lands beyond the Allegheny Mountains.[4]

Jefferson made available his own instruments and scientific library to Lewis and personally instructed Lewis in the use of the instruments required for surveying and determining latitude. By his own admission, Jefferson was not a practicing astronomer and unfamiliar with the use of the octant, but he nonetheless had a firm understanding of the principles involved. Lewis practiced particularly with the octant, an instrument designed primarily for use at sea but applicable also on land to observe altitudes of the sun or a star for determining latitude (fig. 6.1).[5]

Octants were made in a triangular shape from a dark, close-grained tropical wood, preferably ebony. A moveable arm, or index, was pivoted from the apex of the triangle and pointed to a scale of degrees on the arc opposite the apex. A mirror attached to the apex moved with the index arm. A sight, with two sets of colored glass shades for observing the sun, was attached to the right limb of the triangle. A horizon glass, half-mirrored and half-clear, was situated on the left limb.

With the octant, the observer looked through the sight to observe simultaneously the horizon, visible through the unsilvered portion of the horizon glass, and the object, the sun or a star, reflected by the mirror at the apex of the index arm to the silvered portion of the horizon glass and then to his eye. To achieve this alignment the angle between the two mirrors had to be one-half the altitude of the object being sighted. The angle through which the index mirror moved from the position of parallelism was determined by the movement of the index arm; double this angle was read on the arc, which was calibrated to ninety degrees.[6]

The sextant closely resembled the octant in appearance and function and was based on the same principles of optics (cf. figs. 6.1 and 6.6). However, it was made entirely of cast brass instead of wood and was provided with an arc of 120 degrees, which enabled it to measure lunar distances.[7]

Although Lewis apparently became reasonably competent in the use of the octant, it was obvious that more professional training was required. Jefferson thereupon enlisted the cooperation of fellow members of the American Philosophical Society. As consultants on scientific instrumentation he selected Robert Patterson, professor of mathematics at the University of Pennsylvania, and Maj. Andrew Ellicott, secretary of the Pennsylvania Land Commission at Lancaster. Patterson was an authority on scientific principles and instruments, and Ellicott, as the nation's foremost surveyor, had considerable field experience in their use.[8]

For advice and instruction in the scientific data to be collected by the exploring party, the president sought out the botanist Benjamin Smith Barton; Caspar Wistar, anatomist with a wide knowledge of zoology; and Dr. Benjamin Rush of the Pennsylvania Hospital, who advised on medical practices and supplies.[9]

In anticipation of Lewis's visit, Patterson immediately began to prepare astronomical formulas for field use, including one for computing the longitude from observations of lunar distances, and another for calculating the time, altitudes, etc., expressed in the same manner by algebraic signs, which, he assured Jefferson, would make it "easy enough for boys or sailors to use."[10]

In response to the president's request for his cooperation with Lewis's instruction in the use of surveying and astronomical instruments, Ellicott stressed the necessity for Lewis to "acquire a facility, and dexterity, in making the observations, which can only be attained by practice." He also cautioned that the final calculations of latitude and longitude could not be made in the field, but would have to be computed after the expedition's return because of the considerable quantity of astronomical tables and other reference material that was needed.[11] He noted further that the instrumentation required by the exploring party was identified in his published account of the survey of the southern boundary with the Spanish territory, a copy of which he had given to the president. In this work he had specified that all that was required for determining both latitude and longitude was "a good sextant, a well made watch with seconds, and the artificial horizon, the whole of which may be packed up in a box 12 inches in length, 8 in width, and 4 in depth."[12]

Ellicott went on to provide specific instructions for determining the meridian altitude of the sun by means of a method of taking equal altitudes that he had used successfully for years, although it was generally not practiced by other American surveyors of his time.[13]

Ellicott also enclosed detailed instructions for making an artificial horizon of his own preference, using water instead of the other liquids commonly used, and with the trough made separate from the cover to avoid possible disturbance of the liquid by the wind.

In order to take altitudes with the octant, a clear and distinct horizon was essential, but such was not always available because of fog, mist, or inclement weather. The artificial horizon served as a substitute and enabled the observer using the instrument to dispense with the visible horizon when taking sights of the sun or a star. It featured a trough of mercury, the surface of which served as a

Fig. 6.1. Octant, second half of eighteenth century, made by Thomas Ripley, London, of ebony and brass. (Smithsonian Institution)

reflector, with two sheets of glass attached gablewise over the trough to prevent movement of the mercury by the wind. With the artificial horizon, the instrument measured the angle between the sun and its image on the mercury's surface, this angle being equal to twice the sun's apparent altitude.[14]

On 14 March 1803 Lewis began his journey of instruction, departing from Washington en route to Harper's Ferry, Lancaster, and Philadelphia to meet and work with the various advisers. William Irvine and Israel Whelen, the purveyor of public stores, had been notified by the War Office that Lewis was to be provided with such items as he requested from military supplies. Whelen was to purchase whatever Lewis needed that could not be obtained from that source, inasmuch as all preparations for the expedition were to be vested in the War Office.[15]

At Harper's Ferry, Lewis arranged to obtain weapons and supervised the construction of an iron-framed folding boat to be used on the upper Missouri River.[16] These duties kept him there for almost a month before he could move on to Lancaster. At Lancaster he spent part of each day with Ellicott, who drew from his own field experience, particularly of the difficulties and hazards to be anticipated in uninhabited and unknown country. During his many years surveying the wilderness, Ellicott had frequently experienced every type of privation and unexpected danger and had learned how to deal with them. His wise counsel was to prove of immeasurable value to Lewis and his companions in the next several years. Lewis remained at Lancaster for two weeks and practiced using Ellicott's field instruments, including the octant and sextant, profiting greatly from Ellicott's instruction and the experience with the instruments. In retrospect, however, it became apparent that the training period allowed was much too brief to achieve the competence required for the level of scientific responsibility imposed by Jefferson.[17]

In Philadelphia, Lewis lost no time consulting with the other advisers, in accordance with Jefferson's arrangements, and collecting and purchasing supplies for the expedition. Most of them were readily available from military stores or could be purchased, but acquisition of the instruments presented difficulties because some of them, such as thermometers, sextants, and chronometers, were not yet being made in the United States. They had to be imported from England, and high-quality ones were not always available.[18]

Jefferson had definite opinions not only about the scientific data to be collected, but also concerning the instrumentation to be used for the purpose. He had concluded that the best means of determining the longitude in the field was by the measurement of lunar distances, to be accomplished with a theodolite or possibly a portable equatorial instrument. The establishment of latitude was relatively simple, requiring only an octant, but the determination of the longitude was considerably more difficult.

At that time there were two common methods used for determining longitude. One required observation of the times that one of Jupiter's major satellites entered or exited from the shadow of the planet. Inasmuch as four major satellites were readily visible, observations could be made with relative frequency, except when the planet was not favorably situated for observation. The difficulty in using this method resulted from the uncertainty of the times of the satellites's appearances, and because for long periods the planet was not easily observed.

The second method consisted of measurement of the lunar distances—determining the local time of the moon's transit and comparing it with its time of transit at a prime meridian. Basically, this method used the moon's movement around the earth as a clock, with the moon functioning as the hand or index and the sun, planets, and stars serving as the markers or indicators. The method using the Jovian satellites was easier to calculate, but their appearances were infrequent and uncertain, whereas the moon was visible quite often. Considering the advantages and disadvantages of both methods, Jefferson selected lunar observation for the expedition.[19]

In addition to a sextant or other astronomical instrument to observe lunar distances, a precise timekeeper was required to establish the times of observation. Jefferson opposed taking the usual timepiece, an astronomical regulator in the form of a tall case clock or a pocket chronometer, into the field, noting that being "fearful that the loss or derangement of his watch on which these lunar observations were to depend, might lose us this great object of his journey, I endeavored to devise some method of ascertaining the longitude by the moon's motion without a timepiece."[20]

Jefferson proposed employing a theodolite or a universal equatorial instrument (fig. 6.2 and 6.3). He owned examples of both instruments and was inordinately fond of them. Like the octant and sextant, they were used to measure angles and were most accurate for measuring horizontal angles such as those between a celestial body and a fixed point on the earth. The universal equatorial instrument was made portable, with an elaborate mounting which could be set by clockwork to follow the course of an observed celestial body across the sky to provide a continuous record, and it was the most sophisticated astronomical instrument of the time. He later noted that his proposal for observing lunar distances "was founded too in the use of the Equatorial the only instrument with which I have any familiarity. I never used the Quadrant at all; and had thought of importing three or four Equatorials for the use of those parties. They get over all difficulty in finding a meridian." It was true, he admitted, that Jesse Ramsden, the maker of the theodolite, specified that it was necessary to use a timekeeper with it, but it was Jefferson's opinion that "this cannot be necessary, for the margin of the equatorial circle of this instrument being divided into time by hours, minutes, and seconds, supplies the main function of the time-keeper, and for measuring merely the interval of the observations, is such as not to be neglected. A portable pendulum for counting, by an assistant, would fully answer the purpose."[21]

Equatorial instruments were extremely expensive, however, while theodolites were more readily available. The latter consists of a telescopic sight and two circles, mounted at right angles to each other, each divided into degrees and readable in minutes. The pedestal stand supporting the apparatus could be leveled and adjusted.

In Jefferson's opinion, both the equatorial and the theodolite had distinct advantages over the sextant and the Borda reflecting circle because their telescopes were more flexible and their apparatus for correcting refraction and parallax rendered the notations of altitude unnecessary, and even dispensed with the need for the timekeepr or portable pendulum.[22]

So convinced was Jefferson that these instruments of his selection could satisfactorily eliminate the need for a timekeeper for observing lunar distances, that he communicated his proposal not only to Ellicott and Patterson, but to others as well. Among them were William Dunbar, John Garnett, William Lambert, Isaac Briggs, Joshua Moore, and possibly others to whom he had written concerning the expedition.[23] In response, Ellicott and Patterson suggested alternative methods for determining longitude, each requiring timekeepers, but diplomatically did not comment on the use of the equatorial or theodolite.[24]

Jefferson now began to have some doubts of his own, for in a letter to Lewis he commented, "I would wish that nothing that passed between us here should prevent your following his [Ellicott's] advice, which is certainly the best. Should a time-piece be requisite, it is possible Mr. Arnold could furnish you one. Neither Ellicott not Garnet [sic] have given me their opinion on the substituting the meridian at land instead of observations of time, for ascertaining longitude by lunar motions. I presume, therefore, it will not answer."[25]

Lewis reported that both Ellicott and Patterson disagreed with the use of the equatorial or theodolite in the field. They believed that such instruments were too delicate for the rough use they would be given, difficult to transport, and easily put out of order. In short, Lewis went on, referring to the theodolite, "in its application to my observations for obtaining the longitude, it would be liable to many objections, and to much more inaccuracy than the sextant." Instead, both Patterson and Ellicott recommended that Lewis be equipped with two sextants; one or two artificial horizons; a good "Arnald's watch" (chronometer); a plain surveying compass, with ball-and-socket joint; a two-pole chain; and a set of drafting instruments. "As a perfect knowledge of the time will be of the first importance in all my Astronomical Observations," Lewis went on to explain to the president, "it is necessary that the time-keeper intended for this expedition should be put in the best possible order, if therefore Sir, one has been procured for me and you are not perfectly assured of her being in good order, it would be best perhaps to send her to me by safe hand." He explained that the Philadelphia clock-

Fig. 6.2. Theodolite made by Jesse Ramsden of London and owned by Thomas Jefferson. (Thomas Jefferson Memorial Foundation)

Fig. 6.3. Universal equatorial instrument, made by Jesse Ramsden of London, similar to one owned by Thomas Jefferson. (National Museum of American History, Smithsonian Institution)

maker Henry Voigt had offered to clean it and Ellicott would regulate it.[26]

With all the advice that Jefferson had sought from members of the American Philosophical Society and others, there was inevitably some overlap. John Vaughan, librarian of the society, had also been consulted by someone to whom Jefferson had written, and he in turn sought advice from Ellicott. The latter informed him that for use in the field a brass sextant was infinitely superior to a wooden octant, and that one of the best quality might be purchased for between eighty and one hundred dollars. Ellicott himself planned to make an artificial horizon for the expedition, a project that required a slice of talc he wished to obtain from a block in the Society's museum.[27] Ellicott's use of talc for the reflecting surface of the instrument was an innovation evidently not used by others. Soon after his arrival in Philadelphia, Lewis visited Vaughan, bringing an artificial horizon made by Ellicott and a letter from him. Lewis also called upon Patterson with a letter from Ellicott asking him to provide Lewis with his own formulas for longitude.[28]

Lewis had done his homework well, for the list he had compiled with Jefferson's cooperation while he was still in Washington included most of the instruments later recommended by Ellicott and Patterson, as well as others they did not consider necessary. Each instrument listed would fulfill a particular aspect of the expedition's mission. On Lewis's earlier list, but not included on the surviving record of instruments acquired for the expedition, were a microscope for study of plant life and minerals; a theodolite, originally intended for astronomical observation but useful also for surveying; hydrometers for determining the amount of water vapor in the atmosphere; a brass rule; magnetic needles; and a measuring tape. Some of these items were probably deleted by Ellicott and Patterson as impractical, and others may have been eliminated simply because they were not available in Philadelphia at that time. Or some of them could have been purchased from other makers or dealers, from whom no records of purchase have survived.[29]

Additions made to the list included a spirit level for surveying, one plated and three brass pocket compasses, a magnet for "touching" the compass needles when they lost their magnetism, a sextant, spare talc for the artificial horizon, a plain survey-

Fig. 6.4. Cased measuring tape, bound with a memorandum book for field notes, English, 1846. (Science Museum, South Kensington, London)

ing compass for surveying through dense woodland and underbrush, a circular protractor with index arm for map-making, a six-inch pocket telescope, a log line and reel, and a log-ship.

Particularly interesting on Lewis's original list is an

> Instrument for measuring made of tape with feet & inches marked on it, confined within a circular lethern [sic] box of sufficient thickness to admit the width of the tape which has one of its ends confined to an axis of metal passing through the center of the box, around which and within the box it is readily wound by means of a small crank on the other side of the box which forms a part of the axis, its tape when necessary is drawn out with the same facility & ease with which it is wound up."[30]

This instrument described is the common surveyor's measuring tape. Such tapes were not commercially produced until almost the mid-nineteenth century. Among the earliest examples of a commercial tape measure known is one manufactured in England in about 1846 (fig. 6.4). The unusually detailed nature of the description suggests that a commercial tape measure did not then exist. Jefferson may have proposed that such an item would be useful and should be ordered to be made by one of the Philadelphia instrument makers. Expedition records mention a "tape line" used to measure the size of fishes, indicating that some form of the tape measure had been obtained.

In 1800, Philadelphia was a major American shipping center, with many makers and dealers of mathematical instruments. Some specialized in the production of navigational and surveying instruments, while others imported and sold optical and meteorological instruments from England and France.[32] With the assistance of Patterson and Whelen, Lewis was able to obtain most of the instruments he required, the majority purchased from Thomas Whitney's shop on Water Street (fig. 6.5). In addition to making and selling instruments, Whitney also modified them as required by his clients. For example, he attached a small high-powered reading lens, which he called a "microscope," to the index arm of the sextant that Lewis purchased.[33] The vernier scale was so minute that it required magnification to be read, and at that time sextants were equipped with separate hand-held magnifiers that fitted into the field case. It was not until a decade or more later that a lens attached directly to the index arm of the sextant became a standard feature.

Whitney provided a total of fifteen instruments to the expedition, at a cost of $162.20. From him Lewis purchased a late-model octant with tangent screw, probably English, for $22. It had a fourteen-inch radius and a vernier scale, and it was capable of making back observations. For ninety dollars Lewis also obtained a sextant of a ten-inch radius and equipped with a vernier scale and three eyepieces: a hollow tube and two telescopes, one of which reversed the image. A cased set of plotting, or drafting, instruments was needed for map-making; the spirit level and two-pole chain were standard surveying equipment. The log line, reel, and log-ship were used with a timekeeper to determine the speed

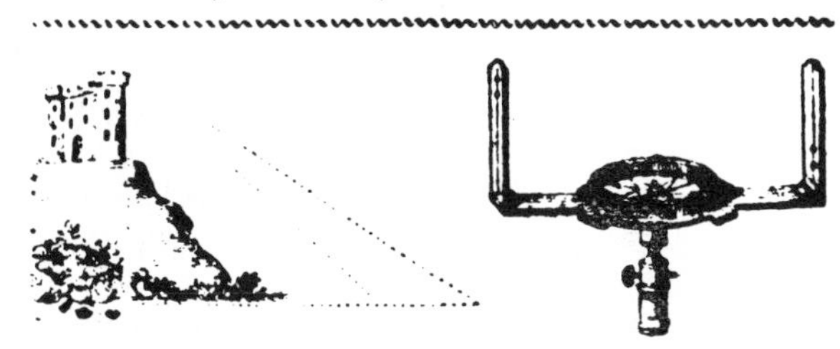

THOMAS WHITNEY,

Mathematical Instrument Maker,

North Sixth street, near the Mill Pond, one mile from Philadelphia,

Presents his sincere thanks to his friends and the public, and respectfully soliciting the continuation of their favours, wishes to inform them, that he has devoted his attention principally to the making of

SURVEYING COMPASSES,

For more than thirteen years past, and has made about *five hundred* of them, the good qualities of which are well known to many Surveyors, in at least sixteen of the States and Territories of the Union.

THOMAS WHITNEY, continues to make SURVEYING COMPASSES, of all the constructions in general use, and also for Vertical and horizontal angles, leveling, &c. newly projected, and other Instruments made to drawings or descriptions; improved Protractors, Gunner's Calibers, and Quadrants, Standard Measures, Surveying Chains, Magnets, Pocket Compasses, with various other Instruments and apparatus.

Also has for Sale,

Cases of Drawing Instruments, Pasometers, or Reflecting Semicircles, Polygonagraphs, Globes, &c.

N. B. Orders will be thankfully received, and promptly attended to, if left at No. 105, North Second street, at Messrs. M'Allisters, No. 48, Chesnut street. where the Instruments are for sale, or at his House and Manufactory, North Sixth street.

Instruments carefully clean'd and repaired.

Fig. 6.5. Advertisement of Thomas Whitney, Philadelphia, maker of mathematical instruments, who furnished a large number of the instruments taken on the expedition. (From *Whitely's Philadelphia Annual Advertiser,* 1820)

of a ship or boat or the distance it had run in a given period of time. A brass boat or marine compass was a basic necessity for navigating. Lewis purchased one silver-plated pocket compass, probably for his own use, and three others of brass. Since a compass was an item that was frequently damaged or lost, he made certain that he would have replacements. From Whitney he also obtained a spare parallel glass and a slab of talc as replacements for the artificial horizon prepared for the expedition by Ellicott.[34]

One of Lewis's most important purchases was a gold-cased chronometer from the Philadelphia watch and clockmaker, Thomas Parker.[35] The chronometer, or "Arnald's watch" as Jefferson had called it, cost $250, with an additional seventy-five cents for the winding key. As yet no chronometers were being manufactured in the United States, and the expedition's schedule did not allow time to order one from England, as Jefferson had suggested. The only alternative was to purchase such an English-made timepiece that was already available in Philadelphia. As Lewis described it, the chronometer he purchased had a balance wheel and escapement of "the most improved construction."[36]

For cleaning and adjustment, Lewis brought the chronometer to the shop of Henry Voigt, the foremost clockmaker of the time. Voigt also constructed a protective mahogany case, in which he suspended the chronometer by means of "an universal joint," by which Lewis may have meant gimbals. Voigt also cleaned and adjusted Lewis's own silver pair-case watch with second hand, which Lewis planned to take on the expedition. Voigt's total bill for the work was $7.37.[37] Lewis sent the chronometer by Dr. Barton to Ellicott to be regulated and rated. Ellicott checked the chronometer's rate for two weeks until he felt assured that it was properly adjusted.

Meanwhile Patterson had begun work on a "Statistical Table," on which Lewis was to record each astronomical observation as he made it. Patterson originally had planned to furnish Lewis with just a sketch from which he could develop such a table, but not having completed the sketch in time, Patterson compiled the table and sent it on to Lewis after his departure. As he assured Jefferson, it was "an expedient that would save a great deal of time, and be productive of many advantages."[38]

Jefferson's final instructions to Lewis, submitted on 30 June, specified, in addition to a multitude of other requirements, the instrumentation to be used "for ascertaining by celestial observations the geography of the country thro' which you will pass." Lewis was instructed to make observations, beginning at the mouth of the Missouri, of latitude and longitude at all notable points of the river, and especially at distinguishable points, such as mouths of rivers, rapids, islands, and natural landmarks that would be recognizable again later. He was asked to determine the river courses between such points by means of the compass, the log-line, and observations of time that were to be corrected by the celestial observations themselves. Variations of the compass at various places also were to be noted. Points of interest on the portage between the heads of the Missouri and water offering the best communication with the Pacific were also to be fixed by means of observations.

Jefferson was particularly insistent that all "observations are to be taken with great pains & accuracy" and that they were to be recorded distinctly and in such a manner that they would be comprehensible to others as well as the observers, who would be able to establish the latitude and longitude for the locations at which they had been taken by using the requisite tables. These records were to be submitted to the War Office so that final calculations could be made later by several qualified individuals working at the same time. To ensure against possible loss of records, Jefferson directed the explorers to make several copies of their notes during their leisure and give them into the safekeeping of their most trustworthy men. "A further guard," he wrote, "would be that one of these copies be on the paper of birch, as less liable to injury than common paper. . . ."[39] Jefferson also required the explorers to record climatological data similar to the records he personally had maintained daily since he was a law student at Williamsburg.[40]

Realizing the need for a second leader if an accident or illness should befall Lewis, Jefferson left the selection to Lewis, who chose William Clark, one of his former army friends and a younger brother of the explorer George Rogers Clark. Clark's considerable military experience and familiarity with the region where the Ohio and Mississippi rivers met later proved him to have been an excellent choice. Clark had no role in planning the expedition, for there was little time between his selection and Lewis's departure on 5 July for Pittsburgh. After further unexpected delays in procuring equipment, the exploring party finally departed from Pittsburgh at the end of August 1803. Clark would join the party later.[41]

The scientific entries in the surviving journals for the early part of the expedition proper apparently were kept only by Clark. He carefully compiled tables of courses, times, and distances during which the party followed the river, noting such landmarks as islands, sandbars, towns, river openings, unusual items observed along the riverbanks, weather, and observations made with the octant.

Lewis began making his scientific entries in the spring of 1805, after the party left the Mandan villages, and continued consistently for the remainder of the expedition. From the time that the exploring party wintered with the Mandans, the explorers adopted the procedure of having Clark make a copy of all of Lewis's scientific entries in his own journal. The decision to undertake this duplication may have resulted from the accident on 14 May when "some of the papers and nearly all of the books got wet, but not altogether spoiled" when their white pirogue capsized. Lewis's entries were copied by Clark, frequently without alteration but often dated earlier for no apparent reason.[42]

The earliest temperature records were kept by Lewis as the party progressed down the Ohio River in autumn 1803, and Clark continued the practice at the Wood River camp in early 1804. The last of their thermometers was accidentally struck against a tree and broken in September 1805, however, and no temperatures could be recorded thereafter. The thermometers were probably of a type similar to those described by Jefferson to Isaac Briggs. "The kind preferred," he wrote, "is that on a lackered plate slid into a mahogany case with a glass sliding cover, these being best weather."[43]

Although Lewis is usually considered the scientific specialist of the expedition, Clark made many more significant scientific contributions than are generally realized, particularly relating to natural history. Lewis may have been given more credit because his journal entries reflected superior qualities of writing and expression. Because Lewis had been trained to use the octant and Clark had not, Lewis probably instructed him in its use in the field, for Clark practiced with it constantly at the Wood River camp and may even have become more proficient in its use than Lewis. His previous experience as a surveyor served him well, and he kept a notebook in which he had written problems of celestial navigation with their solutions. Clark was responsible for recording navigational data as well, and he also prepared sketches and observational notes. His maps proved to be remarkable for their fine quality and accuracy.[44]

In an entry for 22 July 1804 of the journal recording astronomical observations, Lewis provided a detailed description of the instruments he used for this purpose: a sextant (fig. 6.6), an octant, three artificial horizons, a chronometer, and a surveying compass. Lewis preferred the reversing telescope and used it in all his observations. He found the octant particularly useful for making back observations when the sun's altitude at noon proved too great to be observed with the sextant.

Of the three artificial horizons with which the exploring party was equipped, the one prepared by Ellicott utilizing water as the reflecting surface turned out to be useful "when the object observed was sufficiently bright to reflect a distinct immage." The one provided by Patterson, consisting of a glass pane cemented to the side of a wooden ball and adjusted by a spirit level and platform, was more adaptable for taking altitudes of the moon and stars and of the sun under dull conditions. The third horizon, utilizing the index mirror of a sextant attached to a flat board and adjusted also by a spirit level and platform, was the most adaptable with bright objects such as the stars.

The chronometer, protected in the field case made for it by Voigt, was, according to Lewis's account, wound up each day at noon and its "rate of going" confirmed by observations made by Lewis. He found it to be fifteen and five-tenths seconds too slow within a twenty-four hour period on mean solar time, which was very close to the rate established for it by Ellicott. For taking the magnetic azimuth of the sun and the pole star, Lewis used the surveying compass (fig. 6.7), and he found it useful also for taking a traverse of the river. From these compass bearings, combined with the distances from point to point that he estimated, he was able to chart the Missouri River. From the onset of the expedition to the winter of 1805, Lewis

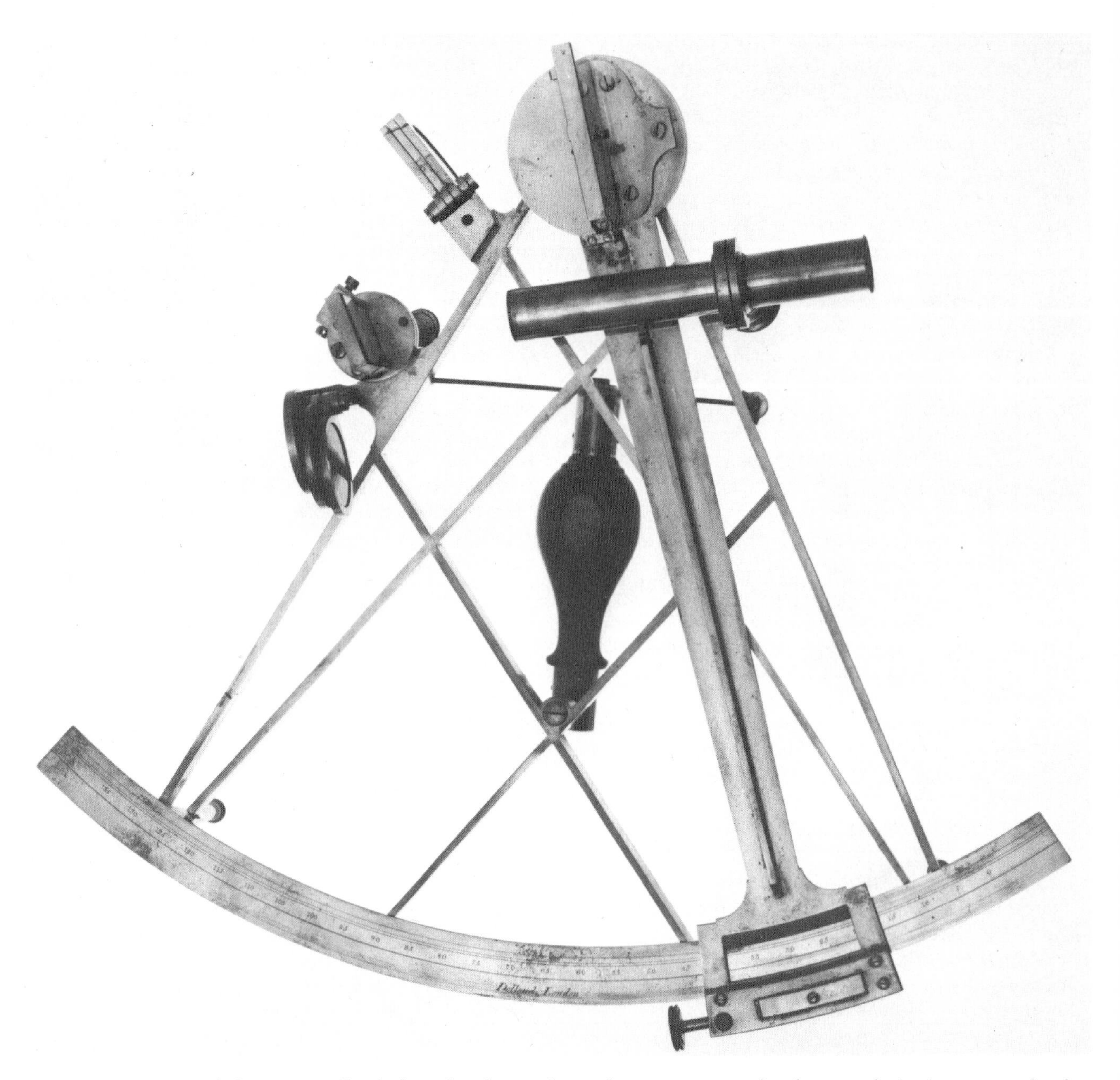

Fig. 6.6. Sextant, made by Peter Dollond of London, late eighteenth century. Note the absence of a built-in magnifier for the vernier scale.

made as many as seven or eight observations for each attempt to determine longitude, the number dependent on the degree of visibility of the object or objects observed.[45]

A careful study of the astronomical data in the expedition's journals suggests that any deficiencies in the observations made by Lewis and Clark probably resulted as much from their lack of skill and experience with instruments as from the difficult conditions under which the observations had to be

Fig. 6.7. Plain surveying compass, brass, eighteenth century, made by Benjamin Rittenhouse of Philadelphia. (Division of Physical Sciences, National Museum of American History, Smithsonian Institution)

made. It is doubtful whether even an expert astronomer or surveyor such as Major Ellicott would have been able to make a set of lunar observations without the help of several assistants. Singlehandedly, an observer would have to take several altitudes of the moon and of the planet or star in rapid succession, then proceed to observe several lunar distances and, finally, take several altitudes of the moon and planet or star once more.

A less experienced observer, however, such as Lewis or Clark, probably required as many as three assistants. As the principal observer measured the angle between the moon's limb and a star or planet, one assistant measured the moon's altitude, another the altitude of the star or planet being observed, and the third assistant recorded the times of observations with the chronometer. The three required angles had to be measured simultaneously and the observations repeated four or five times; then the mean of each group of observations had to be calcualted. Finally, the sums of the lunar distances, each of the two sets of altitudes, and the times were to be divided by the number of sets to eliminate or reduce minor errors of observation.[46]

The explorers' relative inexperience with sophisticated scientific instrumentation was compounded not only by the difficult field conditions under which they had to make their observations but also by the vagaries of their chronometer. Although the timepiece had been carefully regulated first at Philadelphia and later at Lancaster, and its "rate of going" established, Lewis and Clark neglected it occasionally and permitted it to run down. Consequently, the original "rate of going" established in Pennsylvania no longer applied, and local time had to be obtained by observing equal altitudes of the sun to establish the moment of noon. Lewis faithfully recorded longitude by celestial observation as well as by dead reckoning, but the observations rarely proved to be correct, and the most useful data were the records made by dead reckoning, especially those showing their westering, combined with the observations for latitude. Lacking the time, convenience, and expertise to make the necessary calculations of longitude in the field, the task was left for later, as Ellicott had urged.[47]

Lewis had been instructed to submit his records of observations to the War Office after the exploring party's return so that the necessary calculations for longitude and latitude could be made from his data. Ferdinand Rudolph Hassler, instructor in mathematics at West Point, was selected for the task, and he attempted to correct the longitudes by making additional calculations. After considerable trial and error, he reported that he could make nothing of the observations. Errors made by the explorers may have been partly to blame, but it is also likely that Hassler did not have clear knowledge of the procedures that had been used in the field. For example, he could not have known that although Lewis regulated the chronometer on mean solar time, he entered the observations on local time. Writing to Patterson in 1810 after having received a chart of Lewis's calculations from Vaughan, Hassler stated that he had compared these with results he had obtained before, but continued to find discrepancies. Although he had been promised all the journals to study, he complained that he had received "only one, in a fair copy, which I see has many faults in writing."[48]

The problem was undoubtedly compounded by the fact that Lewis and Clark had compiled the journals after their return from rough notes made in the field, notes which were later discarded or lost, except for a few pages, so that comparisons could not be made. Although for the most part the explorers were careful to date their observations in the journals according to the original drafts, occasionally they inserted data wherever space permitted, and not necessarily related to the dates on which they had occurred.

More than a decade after the expedition's return, Jefferson commented on Lewis's observations for longitude and latitude in a letter to José Correa da Serra, newly appointed Spanish minister to the United States, with whom he shared scientific interests. He wrote, "altho', having with him the Nautical almanacs [for the three years during which they were in the field], he could and did calculate some of his latitudes, yet the longitudes were taken merely from estimates by the log-line, time, and course. So that it is only as to latitudes that his map may be considered as tolerably correct; not as to its longitudes."[49]

The many kinds of observations of the country that the explorers were required to record, their camp duties, and the rigors of survival under difficult conditions left Lewis and Clark insufficient time for difficult scientific observations. Despite the shortcomings of their astronomical records, the explorers returned with a remarkable corpus of information on flora, fauna, ethnology, and the geography of the regions they had traversed. Unfortunately, neither Jefferson nor Congress had made any special provisions to preserve the appurtenances of the expedition, the collections, or the records kept by the explorers on the nation's first scientific endeavor. Jefferson's lack of foresight is surprising in view of the considerable concern for the preservation of records he had always demonstrated since his student days at Williamsburg.[50]

When the exploring party returned to St. Louis, preserved materials were stored temporarily, but some botanical specimens were shipped to Jefferson, and some were permanently lost. No arrangement was made for the classification and preservation of the scientific collections, notes, and reports sent or brought back by Lewis and Clark; consequently, the materials were dispersed to various repositories, with inadequate record keeping and inevitable loss.

Jefferson allowed Lewis and Clark to keep their original journals so that they could benefit financially from their publications. Many of the journals were subsequently dispersed and lost. It was not until 1904 that presumably all the written records of the expedition were finally located and brought together for publication, a full century after Lewis and Clark had arrived at their preliminary camp.[51]

Fig. 6.8. Detailed view of the dial of William Clark's pocket directional compass made by Thomas Whitney of Philadelphia, of mahogany and brass, with paper dial and leather field case. (Division of Political History, National Museum of American History, Smithsonian Institution)

Particularly regrettable is the dispersal of the scientific instruments used on the nation's first organized scientific exploration. All that survive are a handful of unimportant personally owned items, few of which can be satisfactorily documented as having been used on the expedition.[52] The appurtenances of the trip—knives, tomahawks, fishing gear, weapons, scientific instruments, keelboat and canoes, handmade clothing, and many other items used and collected by the exploring party—would have been of enormous interest and value to the public at that time and in future generations. In the absence of a request for Congress to appropriate funds to preserve the items, they were sold at public auction at St. Louis in the autumn of 1806 for $408.62.[52]

In retrospect, in addition to the important specific information the explorers brought back about the new lands in the West, the Lewis and Clark expedition had two important results. It had provided Jefferson with an opportunity to enlist government support of science for the first time, and it served as a precedent for future exploring expeditions to support the American position in the West. As Jefferson wrote to William Dunbar in 1805, while the expedition was still in progress, "The work we are now doing, is, I trust, done for posterity, in such a way that they need not repeat it. . . . We shall delineate with correctness the great arteries of this great country: those who come after us will extend the ramifications as they become acquainted with them, and fill up the canvas we begin."[54]

Notes

1. Seymour Adelman, "Equipping the Lewis and Clark Expedition," *American Philosophical Society Bulletin for 1945* (1946): 39.

2. Paul Russell Cutright, *A History of the Lewis and Clark Expedition* (Norman: University of Oklahoma Press, n.d.), p. 3; Reuben Gold Thwaites, ed., *Original Journals of the Lewis and Clark Expedition, 1804–06*, 8 vols. (New York: Dodd, Mead & Co., 1904–1905; reprint, New York: Arno Press, 1969), 7: 193–205.

3. Donald Jackson, *Thomas Jefferson and the Stony Mountains* (Urbana: University of Illinois Press, 1981), pp. 86–97; Silvio A. Bedini, "Jefferson: Man of Science," *Frontiers* (Annual of the American Academy of Natural Sciences of Philadelphia) 3 (1981–82): 10–23.

4. John E. Bakeless, "Lewis and Clark's Background for Exploration, "*Journal of the Washington Academy of Sciences* 44, no. 11 (November 1954); 334–38; Dumas Malone, *Jefferson the President: First Term, 1801–1805* (Boston: Little Brown and Company, 1970), pp. 43–44, 275–76; Jackson, *Thomas Jefferson*, pp. 117–21.

5. Jefferson to Robert Patterson, 2 March 1803, in *Letters of the Lewis and Clark Expedition, With Related Documents, 1783–1854*, 2d ed., ed. Donald Jackson, 2 vols. (Urbana: University of Illinois Press, 1962), 1: 21; Herman R. Friis, "Cartographic and Geographic Activities of the Lewis and Clark Expedition," *Journal of the Washington Academy of Sciences* 44, no. 11 (November 1954): 343–46.

6. The octant, or Hadley Reflecting Quadrant, as it was originally known, was simultaneously invented about 1730 in Philadelphia by a plumber and self-taught man of science named Thomas Godfrey and in England by John Hadley, a mathematician and mechanician employed by the admiralty. The instrument incorporated two principles of optics: first, that the angle of coincidence equaled the angle of reflection in a plane containing the normal to the reflecting surface at the point of reflection, and second, that if a ray of light suffered two successive reflections in the same plane by two mirrors, the angle between the first and last direction of the ray was twice the angle between the mirrors. Because the angle of the mirrors was one-half the altitude of the object observed, double the angle would be read on the arc when the mirror on the index arm moved from the parallel through the angle. In this manner the arc would read to 90 degrees, although it was in itself one-eighth of a circle, or 45 degrees.

Within a short period after its invention, the octant was improved by the addition of a vernier scale, a small sliding scale for making more accurate fractional readings, and a small telescope that replaced the earlier pinnule sight, or that was offered as an alternative. After 1750 the wooden arm, or index, was replaced with brass, and by 1775 the instrument was substantially reduced in size so that it was easier to use. See H. O. Paget-Hill and E. W. Tomlinson, *Instruments of Navigation* (London: H.M.S.O., 1958), pp. 13–14; Silvio A. Bedini, *Thinkers and Tinkers: Early American Men of Science* (New York: Charles Scribner's Sons, 1975; reprint, Rancho Cordova, Calif.: Landmark Enterprises, Inc., 1983), pp. 118–23; "The Description of a New Instrument for Taking Angles. By John Hadley, Esq., Vice-Pr. R. S. Communicated to the Society on May 13, 1731," *Philosophical Transactions of the Royal society for August and September 1731* 37, no. 420 (1733–34): 147–57, pl. 13.

7. The invention of the sextant in about 1757 is attributed to Captain John Campbell, a British naval officer, and resulted from his experiments to find a more accurate means of measuring lunar distances. See E. G. R. Taylor, *The Mathematical Practitioners of Hanoverian England, 1714–1840* (Cambridge: The University Press, 1965), pp. 32, 45, 199; Charles H. Cotter, *A History of Nautical Astronomy* (New York: American Elsevier Publishing Company, 1968), pp. 87–91.

8. Cotter, *Nautical Astronomy,* pp. 91–96; E. G. R. Taylor and M. W. Richey, *The Geometrical Seaman* (London: Hollis and Carter, 1962), pp. 79–81.

9. *Dictionary of American Biography,* s.v. "Robert Patterson"; Silvio A. Bedini, "Andrew Ellicott, Surveyor of the Wilderness," *Surveying and Mapping* 36, no. 2 (June 1976): 113–35; Catherine Van Cortlandt Mathews, *Andrew Ellicott: His Life and Letters* (New York: Grafton Press, 1908), pp. 50–79; Bedini, *Thinkers and Tinkers,* pp. 160–61.

10. Paul Russell Cutright, "Contributions of Philadelphia to Lewis and Clark History," *We Proceeded On,* supplement no. 6 (July 1982): 1–18, previously published as "Meriwether Lewis Prepares for a Trip West, *Bulletin of the Missouri Historical Society* 23, no. 1 (October 1966): 3–20; Jackson, *Letters,* 1: 16–19.

11. Robert Patterson to Jefferson, 15 March 1803, Jackson, *Letters,* 1: 28–31.

12. Andrew Ellicott to Jefferson, 6 March 1803, Jackson, *Letters,* 1: 23–25. Jefferson's letter to Ellicott of 28 February 1803 has not survived.

13. [Andrew Ellicott], *The Journal of Andrew Ellicott, Late Commissioner on Behalf of the United States . . . for Determining the Boundary Between the United States and the Possessions of his Catholic Majesty in America* (Philadelphia: 1803; reprint, Chicago: Quadrangle Books, 1962), app., p. 42.

14. Silvio A. Bedini, *The Life of Benjamin Banneker* (New York: Charles Scribner's Sons, 1972; reprint, Rancho Cordova, Calif.: Landmark Enterprises, 1984), pp. 113–14; Bedini, "Andrew Ellicott," pp. 121, 124.

15. Thwaites, *Original Journals,* 6: 231–37; Adelman, "Equipping the Expedition," pp. 40–41; Jackson, *Letters,* 1: 75–77.

16. Jackson, *Letters,* 1: 37–40, 75–76; Thwaites, *Original Journals,* 7: 217; Donald W. Rose, "Captain Lewis's Iron Boat: 'The Experiment,'" *We Proceeded On* 7, no. 2 (May 1981): 4–7.

17. Lewis to Jefferson, 9 April 1803, in Jackson, *Letters,* 1: 48–49.

18. Jackson, *Letters,* 1: 48–55, 69–97.

19. Cotter, *Nautical Astronomy,* pp. 180–267; Bedini, *Thinkers and Tinkers,* pp. 346–52. Longitude is measured in an east-west direction from an arbitrary point. To calculate the difference in time and of the meridian of longitude, the observer must know the time at the arbitrary point at the moment of the sun's meridian at his position. See A. Pannekoek, *A History of Astronomy* (New York: Interscience Publishers, 1961), pp. 276–81; Cotter, *Nautical Astronomy,* pp. 189–92.

20. Jefferson to Patterson, 16 November 1805, in Jackson, *Letters,* p. 270.

21. Jefferson to Robert Patterson, 29 December 1805, The Papers of Thomas Jefferson, Manuscripts Division, Library of Congress. Jefferson owned a theodolite and a universal equatorial instrument, both made by the noted Jesse Ramsden of London, and the latter described in pamphlets by Ramsden entitled *Description of a New Universal Equatorial Instrument* (1771) and *Description of the Universal Equatorial, and of the New Refraction Apparatus, Much Improved by Mr. Ramsden* (1791).

22. The Borda reflecting circle was considered to be one of the most accurate instruments for measuring lunar distances as well as for taking altitudes. Invented in 1752 as a replacement for the octant by Tobias Mayer, a German astronomer, the reflecting circle was constructed on the same basic principle as the octant but was circular in shape, with two mirrors that could be moved alternately so that altitudes could be taken successively. To establish a mean, the altitudes totaled by the instrument were divided by the number of observations made. The instrument was named for Chevalier de Borda, a French inventor who greatly improved the instrument and published a description of it in 1787. The instrument was produced by various English and French makers. Cotter, *Nautical Astronomy,* pp. 83–87.

23. All were Jefferson's correspondents on scientific matters over a period of years. William Dunbar (1749–1810) was a Scottish planter and man of science who settled near Natchez in 1792 and was surveyor of the district. He served as representative of the Spanish government in the establishment of a boundary between the United States and Spanish possessions. John Garnett (ca. 1751–1820) of New Brunswick, New Jersey, was a publisher of astronomical tables and nautical almanacs and also imported and sold navigational instruments. William Lambert (fl. 1790–1820) of Washington, D.C., was a clerk in the War Department and amateur astronomer with his own observatory. He corresponded with Jefferson on astronomical subjects and later avidly supported the establishment of a national observatory in Washington. Isaac Briggs (1763–1825), a Quaker from Sandy Spring, Maryland, was one of Ellicott's assistant surveyors in the survey of the Federal Territory (District of Columbia). In 1803 he undertook the survey of the Mississippi Territory for the federal government. See particularly Jefferson to Dunbar, 25 May 1805; Dunbar to Jefferson, 9 July 1805; Jefferson to William Lambert, 27 December 1804; and Jefferson to Dunbar, 12 January 1806, in Jackson, *Letters,* 1: 55–56, 244–46, 250–51, 290.

24. Ellicott to Jefferson, 6 March 1803, and Patterson to Jefferson, 15 March 1803, in Jackson, *Letters,* 1: 28–31.

25. Jefferson to Lewis, 30 April 1803, in Jackson, *Letters,* 1: 44–45.

26. Lewis to Jefferson, 14 May 1803, in Jackson, *Letters,* 1: 48–49. An "Arnald's watch," or chronometer, was designed to keep time with considerable precision, having a compensated balance to overcome irregularity because of temperature changes. It was invented in the third quarter of the eighteenth century by John Harrison, an English carpenter turned clockmaker. It was modified and improved by several other English clockmakers: Larcum Kendal, Thomas Earnshaw, and John Arnold. Arnold's improvements included a bimetallic compensation balance, an improved pivoted detent escapement, and a helical balance spring. See Rupert T. Gould, *John Harrison and His Timekeepers* (London: National Maritime Museum, Greenwich, 1958), and R. Good, "John Harrison's Last Timekeeper of 1770," *Pioneers of Precision Timekeeping,* Monograph No. 3 (London: Antiquarian Horological Society, n.d.), pp. 7–8, 19–29.

27. Ellicott to John Vaughan, 16 April 1803, Andrew Ellicott Papers, Manuscript Division, Library of Congress, Washington, D.C.; Ellicott to Jefferson, 18 April 1803, in

Jackson, *Letters,* 1: 36–37. John Vaughan (1756–1841), a Philadelphia merchant, was treasurer and librarian of the American Philosophical Society.

28. Ellicott to John Vaughan, 7 May 1803, and Ellicott to Robert Patterson, 7 May 1803, in Jackson, *Letters,* 1: 45–46.

29. See Jackson, *Letters,* 1: 69–97, for records relating to the acquisition of supplies by Lewis.

30. Jackson, *Letters,* 1: 69–97.

31. In the collections of the Science Museum, South Kensington, London.

32. Silvio A. Bedini, *Early American Scientific Instruments and Their Makers* (Washington, D.C.: GPO, 1964), pp. 30–33, 5864; Bedini, *Thinkers and Tinkers,* pp. 354–56.

33. Invoice of Thomas Whitney, 13 May 1803, in Jackson, *Letters,* 1: 82. Thomas Whitney (? –1823) was an English-born maker of mathematical instruments who established himself in Philadelphia about 1797. He specialized in surveying instruments of all types; an 1819 advertisement claimed he had made more than five hundred surveying compasses. See Bedini, *Early American Scientific Instruments,* p. 30.

34. Jackson, *Letters,* 1: 82; Cutright, "Contributions," pp. 14–16; Thwaites, *Original Journals,* pp. 231–46.

35. Invoice of Thomas Parker, 19 May 1803, in Jackson, *Letters,* 1: 88. Thomas Parker (1761–1833), trained as a clockmaker by David Rittenhouse and John Wood, established his own shop in 1783 at 13 South Third street, Philadelphia, and produced tallcase and shelf clocks.

36. Lewis to Jefferson, 14 May 1803, and Lewis to Ellicott, 27 May 1803, in Jackson, *Letters,* 1: 48–49, 51.

37. Invoice of Henry Voigt, 19 June 1803, in Jackson, *Letters,* 1: 91. Henry Voigt (1738–1814) was a German-born clockmaker and mechanician who operated a wire mill in Reading, Pennsylvania. From about 1780 he worked as a clockmaker and mathematical instrument maker. He was well known to Jefferson, for whom he repaired clocks and watches over a period of years. He moved to Philadelphia about 1791. Sebastian Voight, who was also a clockmaker in Philadelphia in the same period, may have been a brother. Henry Voigt's son, Thomas Voight, also worked as a clockmaker in Philadelphia from 1811 to about 1835. See Bedini, *Thinkers and Tinkers,* pp. 326–27.

38. Patterson to Jefferson, 15 March 1803 and 18 June 1803, in Jackson, *Letters,* 1: 28–31, 51.

39. Jefferson's instructions to Lewis, 20 June 1803, in Jackson, *Letters,* 1: 61–66; Paul Cutright, "Jefferson's Instructions to Lewis and Clark," *Bulletin of the Missouri Historical Society* 22, no. 3 (April 1966): 302–20.

40. Fred. J. Randolph and Fred. L. Francis. "Thomas Jefferson as Meteorologist," *Monthly Weather Review,* December 1895, pp. 456–58; Alexander McAdie, "A Colonial Weather Service," *Popular Science Monthly,* 7 July 1894, pp. 39–45.

41. Lewis to William Clark, 19 June 1803, in Jackson, *Letters,* 1: 57–58; John Louis Loos, "William Clark's Part in the Preparation of the Lewis and Clark Expedition," *Bulletin of the Missouri Historical Society* 10, no. 6 (July 1954): 490–511; Jerome O. Steffen, *William Clark, Jeffersonian Man on the Frontier* (Norman: University of Oklahoma Press, 1977), pp. 44–46.

42. Jackson, *Letters,* 1: 173–76; Friis, "Cartographic Activities," pp. 349–50; Reuben Gold Thwaites, "The Story of Lewis and Clark's Journals," *Annual Report of the American Historical Association for the Year 1903* 1 (1904): 107–29; Paul Russell Cutright, *A History of the Lewis and Clark Journals* (Norman: University of Oklahoma Press, 1976), pp. 8–15.

43. Jefferson to Isaac Briggs, 5 June 1804, the Papers of Thomas Jefferson, Manuscripts Division, Library of Congress, Washington, D.C. Jefferson purchased some of his thermometers from a Philadelphia stationer named Sparhawk. An example of the type of thermometer he described to Briggs and which he once owned is in the collections of the Historical Society of Pennsylvania.

44. Cutright, *History of the Journals,* pp. 3–15; Friis, "Cartographic Activities," pp. 349–51; Thwaites, *Original Journals,* 2: 131–32.

45. Thwaites, *Original Journals,* 6: 230–65.

46. Cotter, *Nautical Astronomy,* pp. 189–92.

47. Jefferson to José Correa da Serra, 26 April 1816, in Jackson, *Letters,* 1: 611–12.

48. Clark to Ferdinand Rudolph Hassler, 26 January 1810, and Hassler to Patterson, 12 August 1810, in Jackson, *Letters,* 1: 491–92, 556–59.

49. Jefferson to José Correa da Serra, 26 April 1816, in Jackson, *Letters,* 1: 611–12.

50. Jefferson to the Hon. St. George Tucker, 9 May 1798, and Jefferson to George Wythe, 16 January 1798, the Papers of Thomas Jefferson, Manuscripts Division, Library of Congress; Helen Bullock, "The Papers of Thomas Jefferson," *American Archivist* 4 (January–October 1941): 243–44.

51. E. G. Chuinard, "Thomas Jefferson and the Corps of Discovery: Could He Have Done More?" *American West* 12 (November–December 1975): 4–13; Dumas Malone, *Jefferson, the President: Second Term, 1805–1809* (Boston: Little Brown and Company, 1974), pp. 208–12.

52. Jan Snow, "Lewis and Clark in the Museum Collections of the Missouri Historical Society," *Gateway Heritage* 2, no. 2 (Fall 1981): 36–41. The watch does not bear a maker's signature or other identification, but is believed to be the silver paircase watch owned by Lewis and carried on the expedition. A pocket compass made by Thomas Whitney with a leather case, claimed by family tradition as having been owned by Clark and carried on the expedition, with some documentation, is in the collections of the National Museum of American History of the Smithsonian Institution, Accession No. 122,864 (see fig. 6.8).

53. Roy E. Appleman, *Lewis and Clark: Historic Places Associated with Their Transcontinental Exploration (1804–06)* (Washington, D.C.: U.S. Department of the Interior, National Park Service, 1975), pp. 235, 375, n. 150.

54. Jefferson to William Dunbar, 25 May 1805, in Jackson, *Letters,* 1: 245.

7. PRACTICAL MILITARY GEOGRAPHERS AND MAPPERS OF THE TRANS-MISSOURI WEST, 1820–1860

John B. Garver, Jr.

THE contributions of the U.S. Army to the mapping and resource assessment of the American West have most often been identified and chronicled with the Corps of Topographical Engineers.[1] The names Maj. Stephen H. Long, Lt. John C. Frémont, Lt. William H. Emory, and Lt. Gouverneur K. Warren are representative of a long list of prominent army topographical engineers who made their indelible mark describing, surveying, evaluating, and mapping the North American plains in the pre–Civil War nineteenth century.[2] In the context of military mapping and surveying methods at that time, members of the elite Corps of Topographical Engineers could be referred to as "professional" military geographers and cartographers whose journals, scientific reports, and carefully constructed maps provided detailed coverage of narrow passageways through and selected areas in the trans-Missouri West.

There were other contemporary frontier military officers and War Department surveyors, however, who also explored, described, assessed, and mapped the ordinary plains landscape. Their contributions as "practical" military geographers and mappers were perhaps as important or even more important to the future settlement of eastern emigrant Indians and Anglo-Americans in the North American plains country than were the sweeping and well-documented trans-area surveys of the Topographical Engineers. While little known today, the names of such practical War Department surveyors and military geographers and mappers—surveyor Isaac McCoy, Lt. Enoch Steen, Capt. Nathan Boone, Capt. Philip St. George Cooke, Maj. Clifton Wharton, and Capt. Seth Eastman, to name only a few—deserve recognition for increasing geographical knowledge of the trans-Missouri lands through their journals, reports, and maps.

War Department Exploratory Surveys: 1828–1834

Geographical Reports of Isaac McCoy

In conjunction with plans presented in 1825 by President James Monroe for the removal of eastern Indian tribes to the trans-Missouri West, a number of exploratory surveys of the "proposed Indian country" were conducted by the War Department.[3] The reports of these surveys and accompanying maps, particularly those prepared by Isaac McCoy, greatly expanded geographical knowledge of the region by providing both general and detailed views of the land upon which to appraise its potential for settlement. Appointed by the War Department in 1828 to accompany an exploring party of eastern Indians to examine lands west of Missouri and as a member of Capt. George H. Kennerly's expedition, McCoy spent the months of August through December 1828 touring the area of eastern Kansas, south of the Kansas River, "to become acquainted with the fitness of the country for habitation."[4]

In his letter of instructions, Col. Thomas L. McKenney, commissioner of Indian affairs, War Department, requested that Gen. William Clark, superintendent of Indian affairs, St. Louis, appoint a topographist to accompany the 1828 Kennerly expedition. "By topographist, as the word is meant to apply to this undertaking," wrote McKenney, "is meant one who can map, or sketch, and bring home geographical and other information of the country through which they may pass."[5] As a result, army topographer Lt. Washington Hood was appointed and, with his assistant, John Bell, accompanied the Kennerly-McCoy expedition.[6] Both McCoy and Hood maintained journals and later prepared maps of the trans-Missouri country based on their notes and observations and on information available from earlier explorations and maps of the area.

McCoy's first definitive report on the eastern Kansas area was submitted to the secretary of war and published by Congress in 1829.[7] He noted that the country was characteristically high rolling prairie, differing from most prairie lands in Ohio, Indiana, and Illinois. "In those countries, prairie lands are usually too flat, with too little stone; often accompanied with quagmires and ponds, and consequently unfavorable to health. Here it is quite the reverse; scarcely a quagmire is to be found." The entire country was "clothed with grass." His main concern, however, was that "Timber is too scarce. This is the greatest defect observable."[8] In view of the natural meadows available and the obvious shortage of timber for fencing and fuel, he suggested that future inhabitants occupy themselves mostly with raising cattle, sheep, horses, and mules, reinforcing the conclusion of Lt. Zebulon Pike some twenty years earlier.[9]

In the fall of 1830, McCoy, joined by a military escort from Cantonment Leavenworth, surveyed the Delaware lands and the north line of the Delaware outlet to a distance of 210 miles west of the Missouri state line. McCoy was also instructed by Secretary of War John W. Eaton to "furnish a description of the country through which you pass; its soil, productions—animal, vegetable and mineral—and general face of it."[10] In his detailed report, McCoy repeated his earlier appraisal of the country, remarking that for 200 miles west of the State of Missouri and Territory of Arkansas, the country was favorable for settlement: "Water, wood, soil, and stone are such as to warrant this conclusion."[11]

Continuing his exploratory surveys in the summer of 1831 with a military escort of two officers and twenty-five soldiers from Fort Gibson, McCoy examined the Osage Indian country as far as the Arkansas River. He described it as "good country, of woodland and prairie, with some large creeks." Wood was restricted to the lowlands, but it was adequate for a "considerable population." He concluded "once and for all, that, excepting bottom lands, the whole of this Indian territory, say 600 miles north and south, and 200 miles east and west, is high and undulating, and certainly, in point of healthiness, is not surpassed by any district in the western country." In his February 1832 report to Secretary of War Lewis Cass, McCoy emphasized the geographical nature of his four years of survey work, remarking that having conducted "more than half a dozen excursions, I have acquired a pretty thorough knowledge of the character of the country." The fourteen-page report, exceptional in its careful description of the geography of the area explored, was ordered printed by Congress.[12] McCoy later wrote that he had seven hundred additional copies printed at his own expense "for gratuitous distribution." The report to the secretary of war, he noted, "was also accompanied by a large map."[13]

Isaac McCoy's Maps of the Proposed Indian Territory

In addition to his descriptive reports on the physical geography of the Indian country, McCoy prepared and forwarded detailed maps of the western lands to government officials. One such map, titled "Map of the Proposed Indian Territory," encompassing the whole of the country "from Red river on the south, to the distance of about six hundred miles north, and west as far as the Rocky Mountains, exhibiting at one view the probable extent of habitable country, and sketches of the several tracts of land which have been assigned to different tribes," was submitted to Secretary of War Lewis Cass on 6 March 1832 (fig. 7.1).[14]

Compared with its most immediate predecessor, Maj. Stephen Long's 1823 "Map of the Country Drained by the Mississippi, Western Section," McCoy's map was a marked improvement in scale, accuracy, and detail and was effectively used both in the field and in Washington by men who planned and executed national strategies for Indian removal

Fig. 7.1. Isaac McCoy's "Map of the Proposed Indian Territory," 1832. (National Archives)

to the western country.[15] Carl Wheat concludes that "McCoys' maps had a great influence on the maps of the day and were of basic character. So far as is known McCoy never printed any maps, but his data were shown on many maps."[16]

In his various reports on the nature of the country, McCoy had characteristically described the landscape as "high rolling prairie," becoming more level in the western portions; thus, he probably saw no value in map portrayal of the relatively flat surface and concentrated his effort on more correctly delineating the numerous river systems and proposed reservations which would provide an organizational framework for actual Indian resettlement. McCoy's "Map of the Proposed Indian Territory" was the first to show with reasonable accuracy the true courses of the Arkansas, Kansas, and Platte rivers, and their main tributaries, within the eastern portion of the trans-Missouri country. The inaccurate portrayal of river patterns in outlying areas (where he admittedly lacked personal knowledge) probably resulted from reference to older, generalized maps.

Some of the tributaries shown were perhaps conjectural. Yet much of McCoy's cartographic work, especially in the vicinity of the Kansas River, was detailed ground survey. In addition, he had conducted exploratory surveys in the country extending from the northern boundary of the Delaware lands and outlet and as far south as the Canadian River. Boundaries of the several established and proposed Indian reservations were identified; distances between most major landscape features could be measured with a new degree of accuracy, although the main channel of the Osage (Marais des Cygnes) River was drawn somewhat south of its true course through the eastern Kansas area. Names of major tributaries of the Kansas River were shown, as were the relative locations of Osage, Kansa, Shawnee, Delaware, Pawnee, Omaha, and Puncah Indian villages. It was perhaps the first map available to government officials that located Cantonments Leavenworth, Gibson, and Towson and the sites of Harmony Mission, Union Mission, Clermont's (Osage) Town, and Council Bluffs.

Two hundred miles west of the Missouri border, the "Western Limit of Habitable Country" was clearly marked. In locating his "habitable line" slightly west of the 98th meridian, McCoy extended the potential of arable land some one hundred miles west of Major Long's notation at the 96th meridian, although the "habitable line" generally corresponded with the eastern limit of the Great American Desert and the area of "Deep Sandy Alluvion" shown on Long's map. Interestingly, the proposed locations for eastern emigrant Indian reservations were well to the east of the "habitable line," directly inferring the existence of favorable country for resettlement purposes.[17]

McCoy also prepared maps of a larger scale based on boundary line surveys of designated Indian reservations which were highly accurate in their presentation of landscape detail. One fine example was his large-scale map of the eastern Kansas area completed in September 1833 (fig. 7.2). The map delineated precisely the surveyed boundaries of eastern emigrant Indian reservations in the vicinity of the Kansas and Marais des Cygnes rivers, and in tracing a more northern course for the Marais des Cygnes, the map corrected a serious error of McCoy's 1832 "Map of the Proposed Indian Territory." Within the reservations, villages of the Piankeshaw, Wea, Peoria, Shawnee, Delaware, and Kansa Indians were given pinpoint lcoations. The Shawnee and Kansa Indian agencies were marked. The surveyed reservation of Cantonment Leavenworth was shown by boundary lines. Major rivers were identified by name, some reflecting an Indian tradition (Wackarusa, Kansa, Okeetsa), others the early French influence (Marais des Cygnes, Sauterelle), and still others a transformation to Anglicized names (Soldier, Stranger). Several were identified by two names (Okeetsa or Stranger, Sauterelle or Neesh-cosh-cosh-che-ba), indicating that while some features reflected an older Indian culture, a new landscape terminology was being superimposed by the early European-American intruders.

War Department Commission Survey, 1832–1834

A special three-man commission was appointed by Secretary of War Lewis Cass in 1832 to

> visit and examine the country set apart for the emigrating Indians, west of the Mississippi. . . . the general object is to locate them all in as favorable positions as possible, in districts sufficiently fertile, salubrious and extensive, and with boundaries, either natural or ar-

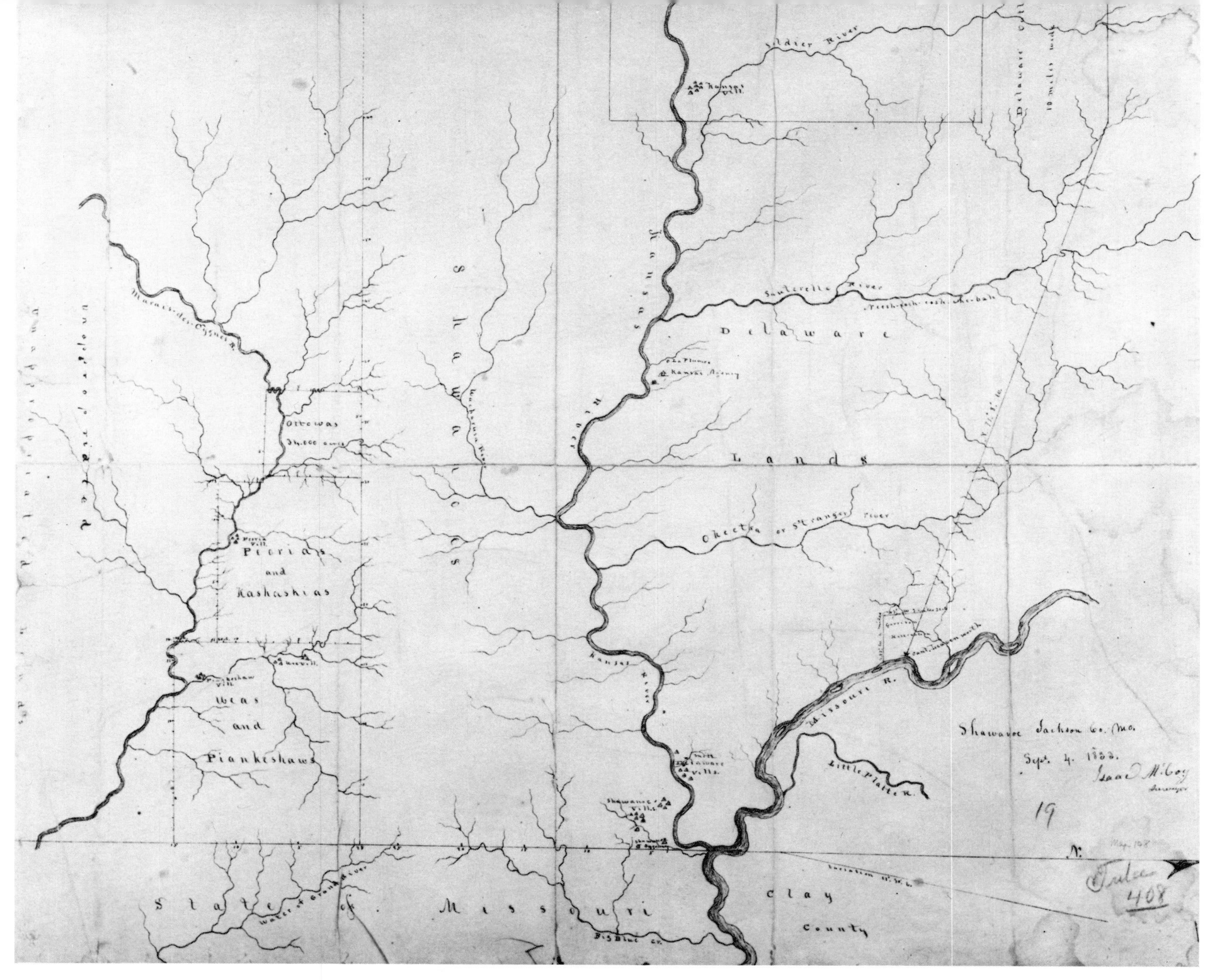

Fig. 7.2. Isaac McCoy's map of the eastern emigrant Indian reservations in the Kansas area, 1833. (National Archives)

> tificial, so clearly defined as to preclude the possibility of dispute. There is country enough for all, and more than all; and the President [Andrew Jackson] is anxious that full justice should be done to each [tribe].[18]

Informed of the commission's appointment and desirous to participate, Isaac McCoy wrote to "The Commissioners West" suggesting that such surveys would greatly improve "our knowledge of the geography and resources of the country," as they would be "doubtless permanent." Enclosed with the letter was a copy of his map of the "Proposed Indian Territory," the original having been forwarded to the secretary of war seven months earlier.[19]

Armed with McCoy's detailed reports and map of the western country, and aided by the frontier military, the commissioners busied themselves in 1832 and 1833 collecting first-hand information on the lands where the eastern tribes would be settled permanently. Their extensive report of fifty-three pages, "Regulating the Indian Department," was submitted to the secretary of war in February 1834 and provided much information regarding the climate, soil, physical resources, and agricultural potential of the land. The report supported, by and large, the earlier favorable appraisals of Isaac McCoy.[20] The image of a "Great American Desert" as the future permanent homeland for eastern emigrant Indian tribes was implicitly rejected by the commissioners: their report made no mention of "deserts" or "barren wastelands" in the area planned for resettlement. Instead, the land was represented as mostly prairie, with numerous watercourses skirted with timber and quite suitable for agriculture and stock raising.

Lt. Washington Hood's Map of the Western Territory

Perhaps of equal importance to the commissioners' report was the "Map of the Western Territory etc.," published with the report by Congress in 1834 (fig. 7.3). Prepared by Lt. Washington Hood of the army's Topographical Bureau and measuring seventeen inches by eighteen inches, the map encompassed the country from the Red River on the south to the Black Hills on the northwest and extended from the borders of Missouri and Arkansas on the east to the Rocky Mountains on the west.[21]

The Black Hills were severely elongated from north to south but were better positioned, well east of the Yellowstone and Powder rivers, with respect to tributaries of the Missouri River drainage system than on Major Long's 1823 map. Major roads within the Indian country, including the Sante Fe Road from the Missouri River to Santa Fe, and the locations of frontier military forts were shown, as well as Isaac McCoy's "habitable line," distinctly marked two hundred miles west of the Missouri frontier. Hood's map was the first published which showed the proposed Indian reservations in the trans-Missouri West. Boundary lines for the indigenous and emigrant Indian reservations were correctly delimited, giving in one view the vast extent and "permanence" of the frontier zone between white and Indian country. The colorfully marked reservation boundaries, together with the overprint of names of the designated Indian occupants: Kickapoos, Delawares, Shawnees, Peorias, Weas, Osages, Pottawatomies, etc., lent an air of legitimacy and reality to the Indian removal plan.

Expansion of Geographical Knowledge by the Military

Practical Military Geographers

"A knowledge of the topography of every part of the national frontier and its vicinity, is essential to every officer and soldier," wrote Maj. Gen. Edmund Gaines, commander of the Western Department, to his subordinate commanders in 1834, ordering all officers in command of marches in the western country to enter in a journal or field book

> concisely, the position and military character of the country over which they travel, the nature and quality of the surface, the courses and distances travelled . . . noting the actual or estimated width, depth, general courses and velocity of the principal rivers and creeks, the character of the soil, the mineral and fossil appearances, the timber and other productions, particularly such as belong to the military resources of the country; . . . with sketches such as are usually and most conveniently made with a pen or pencil in active reconnoissance [sic].[22]

Accordingly, subsequent journals and reports of dragoon and military expeditions in the western country, some ranging to the Rocky Mountains and beyond, included geographical descriptions of the country traversed and often maps as well. Thus, the government's collection of geographical information

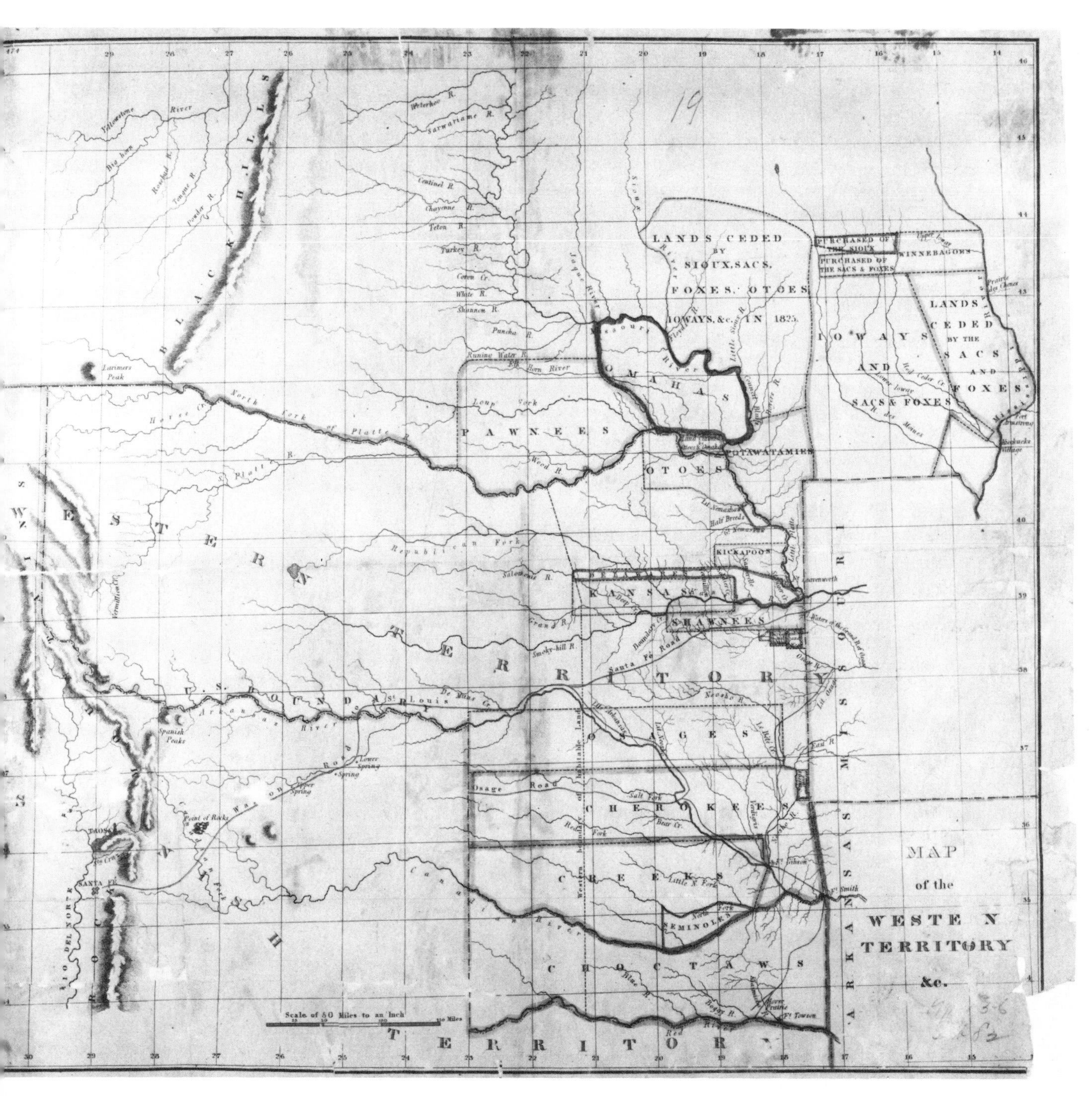

Fig. 7.3. Lt. Washington Hood's "Map of the Western Territory," 1834. (National Archives)

on the trans-Missouri West did not end with the War Department exploratory surveys conducted from 1828 to 1834. While these important surveys were directed toward evaluation of the agricultural potential of the eastern portion of the trans-Missouri lands for Indian resettlement, the frontier army, in conducting shows of force across the vaguely known country and in establishing new posts in isolated locations, also contributed to knowledge of the geography of the vast "Western Territory." Particularly significant were the reports and maps of dragoon expeditions printed in government documents and occasionally published commercially in books, magazines, and newspapers.[23] Excerpts from the reports and journals of the 1835 expedition of Col. Henry Dodge to the Rocky Mountains, the 1843 western reconnaissance of Capt. Nathan Boone, and the 1844 march of Maj. Clifton Wharton to the Pawnee villages on the Platte provide examples of this expanded effort by the military to more accurately observe, describe, and map the surface features and resources of the trans-Missouri country.

Dodge Expedition, 1835

In the summer of 1835, Colonel Dodge, with three companies of dragoons, conducted an extended U.S. military expedition through the Kansas area—a march from Fort Leavenworth to the villages of the Otoes and Pawnees on the Platte, west along the Platte River to the Rocky Mountains, then south to the Arkansas River, east to the Santa Fe Trace, and return along the trace to Fort Leavenworth.[24] Lt. Enoch Steen, in charge of ordnance for the expedition, was given the additional duty of preparing a map of the country traversed. Steen's well-executed map was published with the report of the expedition (fig. 7.4). Wheat suggests that the map was based on Hood's 1834 "Map of the Western Territory" (fig. 7.3) and that "doubtless the advice of Isaac McCoy was sought and obtained."[25] Steen's, however, was perhaps the first published map to label major tributaries of the Arkansas River which the Santa Fe Road crossed and to identify the general locations of the Pawnee and Otoe villages on the Platte River, Bent's newly established trading house on the upper Arkansas, and Council Grove and Pawnee Rock along the Sante Fe route.[26] The map showed general locations of the Kiowa, Comanche, and Cheyenne Plains Indian tribes but incorrectly placed the "Grosventres Indians of the Prairie" and the Blackfeet Indians far south and east of their recognized tribal territories. While Steen erred in several instances, he did portray a number of cultural features on the landscape as well as identify important physical features.

The campaign, acclaimed a brilliant success by higher headquarters, did much to dispel unfavorable public opinions as to the effectiveness of a mounted force in the plains environment which had surfaced following Dodge's inglorious 1834 dragoon expedition to the Pawnee Pict village.[27] Impressed with the overall success of Dodge's extended expedition, Secretary of War Lewis Cass wrote in his 1835 report that "The regiment of dragoons has been usefully employed in penetrating into the Indian country, . . . and in adding to our geographical knowledge of those remote regions."[28]

Captain Boone's Journal, 1843

Instructed to make a reconnaissance of the western prairies in the early summer of 1843, Capt. Nathan Boone, commanding a small detachment of dragoons, departed Fort Gibson on May 14 and explored the area between the Arkansas and Cimarron rivers, west to the Great Bend of the Arkansas, returning to Fort Gibson on July 31. Boone's journal included much geographical and geological information on the little-known territory.

Crossing to the south side of the Arkansas about seventy-five miles northwest of Fort Gibson, Boone reported the following:

> Vegetation is somewhat different from the north side of the Arkansas, saw red oak, and Bur oak. . . . The country crossed over was rolling and in some places hilly, with timber a few hundred yards on the creeks and their tributaries. . . . After getting on the limestone land a great change was observed in the vegetation, the grass was finer, the trees of different character, Linden, Hackberry, Black Ash, hickory, Sycamore, Cotton wood, elm, grapevines etc. were observed. On the prairies the vegetation was the same, but more luxuriant.[29]

The journal then described passing across a level salt plain with a thin surface film of white "crystalized salt." Frequently mentioned were "high gypsum hills of very fine soil" or a stratum of red sandstone capped with gypsum. Turning south from the Great Bend of the Arkansas, the expedition crossed

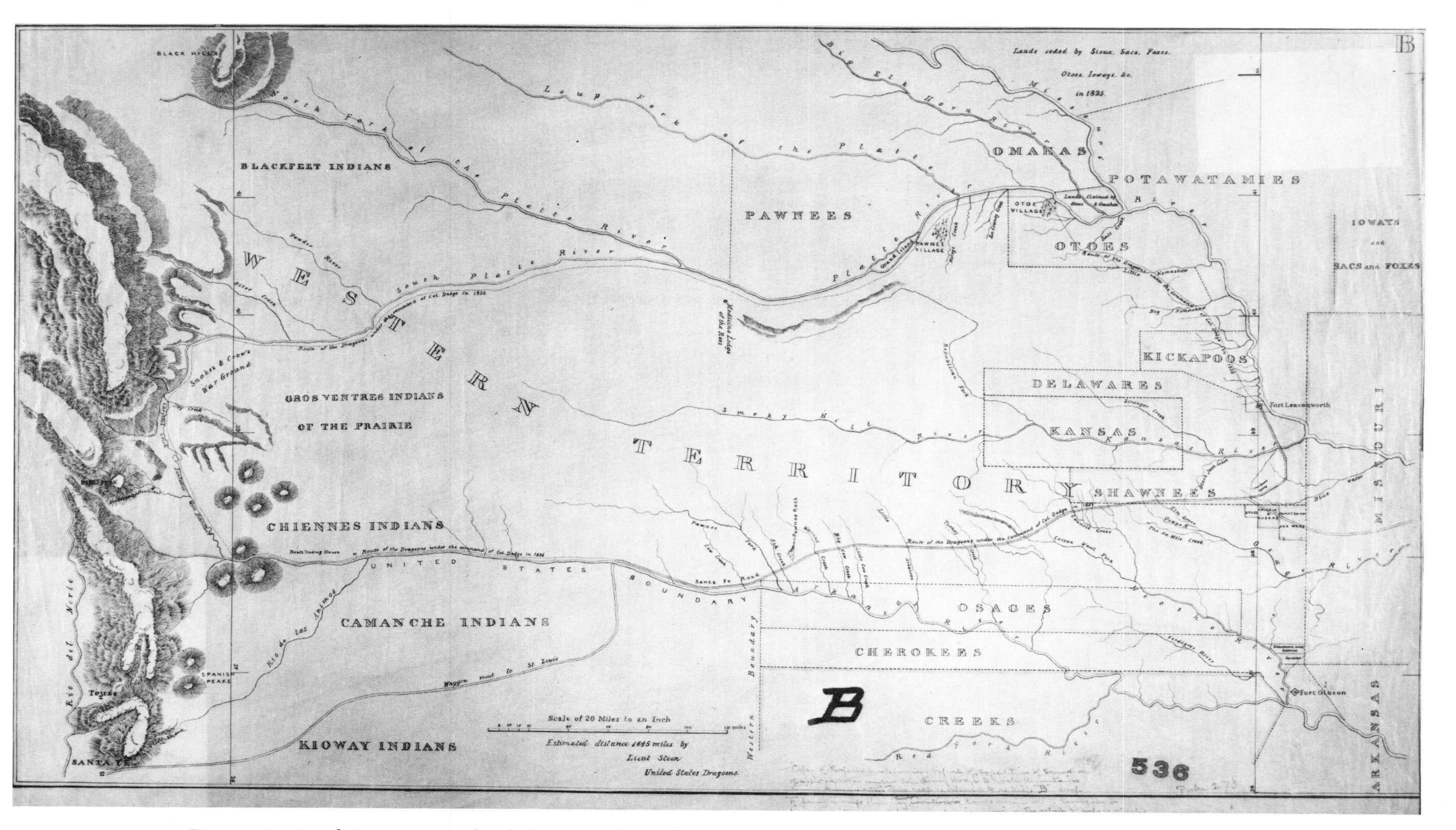

Fig. 7.4. Lt. Enoch Steen's map of Col. Henry Dodge's expedition through the Western Territory, 1835. (National Archives)

"a light sand soil, the rest clayey," "country very level"; "saw a grand sight of perhaps 10,000 Buffalo feeding on the plain below as far as the eye could reach." Near the Cimarron River, the journal noted another salt plain:

> The rock Salt appears to lay near the surface of the water here and springs boiling up through it cover the surface with a concentrated solution which at once begins to deposit chrystals . . . in many places an inch in thickness, and is easily obtained, perfectly clean and white. . . . The quantity of Salt appears to be unlimited.[30]

Crossing the Cimarron on 6 July, the expedition continued the march south through sandy country, "The vegetation changing somewhat with the soil." Boone's journal and map of the march route were forwarded to the War Department by Gen. Zachary Taylor, who noted in his letter of transmittal that the report contained "much valuable and curious information, particularly in relation to the Salt region. . . ."[31] Considering the almost total lack of knowledge existing on the area traversed, the journal was, in fact, a remarkable description of a complex geographic and geologic landscape.

Captain Cooke's Santa Fe Caravan Dragoon Escort, 1843

In the summer of 1843, Capt. Philip St. George Cooke, with four companies of U.S. Dragoons from Fort Leavenworth, accompanied a mixed caravan of American and Mexican traders to the border crossing of the Arkansas with the mission of providing protection against a "company" of armed Texans. Intercepting the Texans on the south side of the Arkansas River and east of the 100th meridian, Cooke escorted Mexican traders returning to Santa Fe from Missouri as far as the Crossing of the Arkansas "without incident."[32]

Upon his return to Fort Leavenworth, Captain Cooke prepared a "Map of Santa Fe Trace from Independence to the Crossing of the Arkansas" (fig. 7.5). Although never published, the map was undoubtedly referred to by senior officers in Col. Stephen W. Kearny's Army of the West in its 1846 march from Fort Leavenworth to the Crossing of the Arkansas en route to Sante Fe in the Mexican War. Measuring a distance of 375 miles from Fort Leavenworth to the Crossing, it was the most reliable map of that portion of the Santa Fe Road since surveyor Joseph C. Brown's 1827 manuscript map of the route from Fort Osage to Santa Fe.[33] Stream crossings along the route were clearly marked, as were those "spots" where trees for fuel were available. In addition, Cooke's map was the first to show the recently surveyed and constructed military road from Fort Leavenworth south to Fort Scott, established in 1842 on the Marmaton River.[34] Cooke's map compares favorably to Topographical Engineer Lt. William H. Emory's detailed "Map of a Military Reconnaissance" from Fort Leavenworth to Santa Fe, published by Congress in 1847.[35]

Wharton Expedition, 1844

Intermittent tribal conflict resulting in raids by Sioux war parties on Pawnee villages and hunting encampments in the early 1840s was viewed by western military leaders as a potential threat to the continued safe movement of traders and emigrants following the Platte River route to "Oregon country" and California. Accordingly, in May 1844, Colonel Kearny requested and received permission to "send the Dragoons [stationed at Forts Leavenworth and Scott] on a two months campaign, with instructions to visit the Otoes and Pawnees on the Platte, & to endeavor to reconcile the difficulties now existing between the latter Indians & the Sioux."[36] Specific orders were issued to Major Clifton Wharton, who was directed to conduct the expedition with five companies of dragoons. Waiting for the Pawnees to return to their villages from the summer hunt, the command did not depart Fort Leavenworth until August. Desirous of establishing a more direct route to the Pawnee villages, Wharton left the Council Bluffs road after two days' march and struck northwest across uncharted country (fig. 7.6). He wrote in his journal of the many difficulties in marking and traveling over the "experimental" route, noting that it presented "greater difficulties than were anticipated owing to our ignorance of the country through which it passes." He was convinced, however, that it was an improvement over the old Council Bluffs route, for

> there can be no doubt that the march to the Platte, striking it where we did, nearly opposite the mouth of the *Loup-fork*, with a better knowledge of the country can be made in ten or eleven days . . . one thing is certain, This route is a practicable one, as well as a short one, when the old, and lower route may be rendered impracticable by heavy rains.[37]

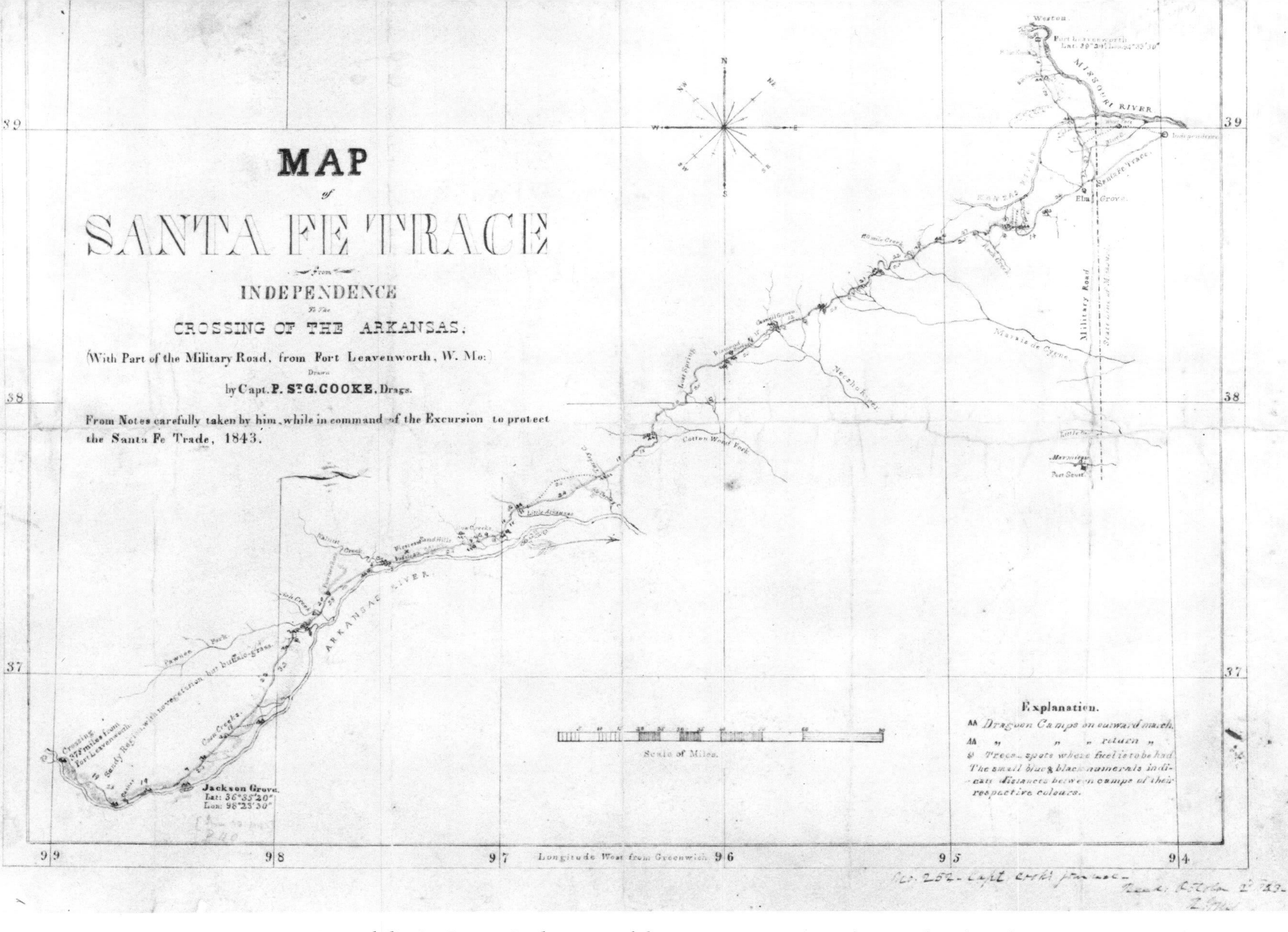

Fig. 7.5. Capt. Philip St. George Cooke's map of the Santa Fe Trace, 1843. (National Archives)

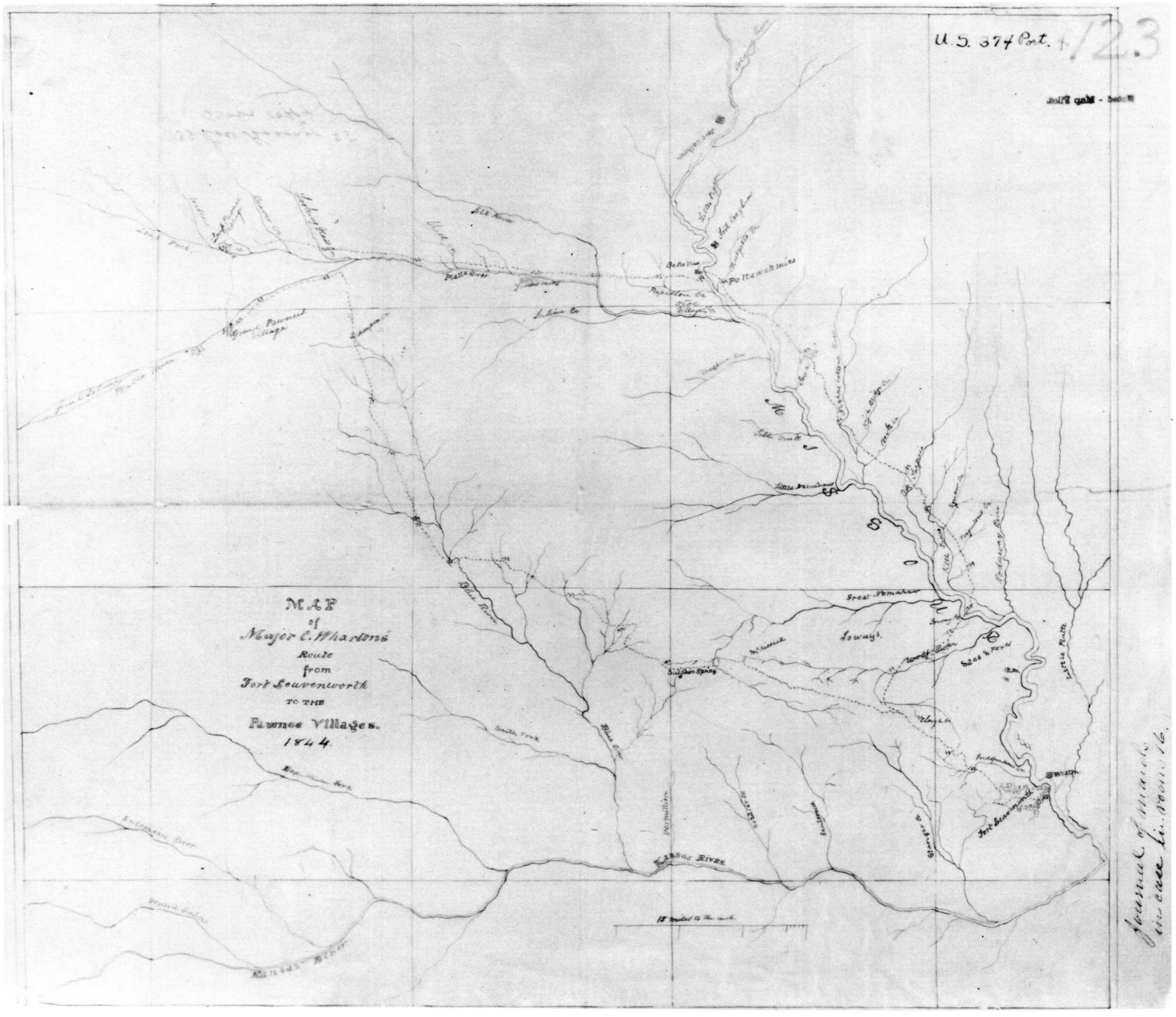

MAP
of
Major C. Wharton's
Route
from
Fort Leavenworth
TO THE
Pawnee Villages.
1844.

U.S. 374 Port. + /23

Journal of march
in case Li. Room 16.

Platte River

Elk Horn

Bellevue

Pottawatamies

Papillon Cr.

Little Nemahaw

Great Nemahaw

Ioways

Sulphur Spring

Wolf River

Sac & Foxes

Blue River

South Fork

Republican Fork

Vermillion

Kansas River

Fort Leavenworth

Weston

Little Platte

Grand Pawnee Village

18 miles to the inch

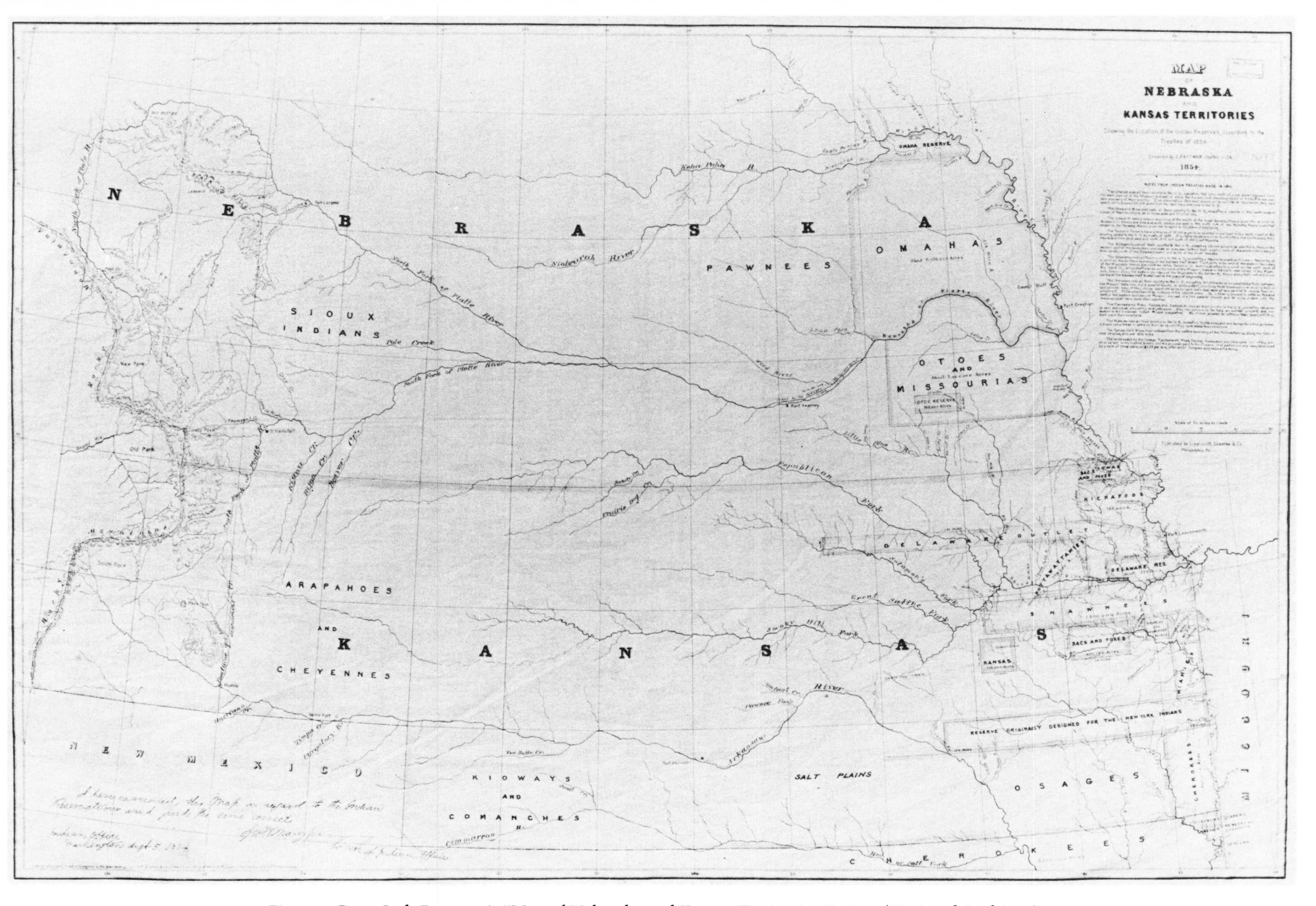

Fig. 7.7. Capt. Seth Eastman's "Map of Nebraska and Kansas Territories," 1854. (National Archives)

A manuscript map of Major Wharton's "Route from Fort Leavenworth to the Pawnee Villages" was prepared and forwarded with a journal of the expedition to the War Department. The map showed the major drainage in the area of reconnaissance and named many of the rivers and streams along the route. The villages of the Grand Pawnee, Loup Pawnee, and Otoe Indians were marked, as was the Council Bluffs Agency at Bellevue. It was perhaps the first map to locate the temporary military posts of Fort Croghan and Fort Calhoun on the banks of the Missouri River.

Captain Eastman's Map of 1854 Indian Treaties

For a far different purpose than the reconnaissance maps of Steen, Boone, Cooke, and Wharton in the 1830s and 1840s, a "Map of Nebraska and Kansas Territories showing the locations of the Indian Reserves, according to the Treaties of 1854" was prepared by Capt. Seth Eastman, U.S. Army (fig. 7.7). The map was authenticated by Indian Commissioner George C. Manypenny in August 1854 and commercially published by Lippincott, Grambe & Co. of Philadelphia. The maximum extent of the old reservations and approximate boundaries of the now diminished reserves of the Omahas, Otoes and Missouris, Iowas, Delawares, and Shawnees were clearly marked on the map, as were the general locations of the Plains Indians shown far removed from the prospective area for Anglo-American settlement. The map included notes from the 1854 Indian treaties describing the location and size of the new reserves and mentioned those lands which were not available for settlement until after the Indians had made their selections. If not the first, it was one of the earliest maps of the newly created territories which visually displayed the vast areas in their eastern portions formerly reserved for Indian homelands that were now "open" to white settlement. "The lands thus acquired," announced Manypenny in November 1854, "are of excellent quality, eligibly situated, and are now being rapidly settled and will soon be brought under cultivation by that portion of our population who intend to make these Territories their future homes."[38]

Many War Department surveyors and frontier military officers were involved in exploring, describing, and mapping strategic routes and areas in the vast and basically uncharted trans-Missouri country. Surveyor McCoy, Lieutenant Steen, Major Wharton, and Captains Boone, Cooke, and Eastman are only representative of this larger group. Through their descriptive journals, reports, and maps, these "practical" military geographers and mappers made significant contributions to increased knowledge of the North American Plains landscape in the first half of the nineteenth century.

Notes

1. Recent and generalized works which cover the U.S. Army Corps of Topographical Engineers in the West are William H. Goetzmann, *Army Exploration in the American West, 1803–1863* (New Haven: Yale University Press, 1959); W. Turrentine Jackson, *Wagon Roads West: A Study of Federal Road Surveys and Construction in the Trans-Mississippi West, 1846–1869* (Berkeley: University of California Press, 1952); Frank N. Shubert, *Vanguard of Expansion: Army Engineers in the Trans-Mississippi West, 1819–1879* (Washington, D.C.: GPO, 1980); Carl I. Wheat, *Mapping the Transmississippi West, 1540–1861*, 5 vols. (San Francisco: Institute of Historical Cartography, 1957–63).

2. Herman R. Friis, "The Role of the United States Topographical Engineers in Compiling a Cartographic Image of the Plains Region," in Brian W. Blouet and Merlin P. Lawson, eds., *Images of the Plains: The Role of Human Nature in Settlement* (Lincoln: University of Nebraska Press, 1975), pp. 27–33.

3. U.S. Congress, House, 19th Cong., 1st sess., 1825–26, H. Doc. 1, Serial 131, p. 91.

4. U.S. Congress, House, McCoy to Porter, 29 January 1829, 20th Cong., 2d sess., 1828–29, H. Rept. 87, Serial 190, p. 8.

5. McKenny to Clark, 10 June 1828, Microcopy No. 711, Main Series, Adjutant General Letters Received, 1812–1889, R.G. 94, roll 6, National Archives, Washington, D.C. (hereafter cited as AGLR 6:5).

6. Clark to McKenney, 4 July 1828, AGLR 6:5. Clark noted that "Mr. Hood, (latterly from West Point), a young man of some cleverness, is selected [as topographer]." Hood's map of the "Western Territory" was published in 1834.

7. House, McCoy to Porter, 29 January 1829, 20th Cong. 2d sess., H. Rept. 87, pp. 5–24. Commenting on this explora-

tory tour some years later, McCoy remarked that although his appointed duties as a member of the exploratory party did not require him personally to prepare an extensive report on the "suitableness of the country for settlement," nevertheless he did so, because "I was exceedingly desirous that the character of the country should be known to the people of the United States. . . ." The report, "covering eighteen large pages," was printed by the House Committee on Indian Affairs. To allow wider circulation, McCoy had an additional one thousand copies printed at his own expense, which he presented to influential government officials and private citizens "in order to direct public attention to the subject." Isaac McCoy, *History of Baptist Indian Missions* (Washington, D.C.: William M. Morrison, 1840; reprint, with Introduction by Robert F. Berkhofer, Jr., New York: Johnson Reprint Corporation, 1970), pp. 371, 376.

8. Ibid., pp. 9–10.

9. Donald Jackson, ed., *The Journals of Zebulon Montgomery Pike*, 2 vols. (Norman: University of Oklahoma Press, 1966), 2: 26. In his 29 January 1829 letter of transmittal to Secretary of War P. B. Porter, McCoy noted that in addition to his report, "a description of the country," he had enclosed a map of the country explored "extending west to the Rocky Mountains and north beyond what may probably be the limits of the Indian territory. . . . It also exhibits the claims of the several tribes now in that country, and the amount of unappropriated lands." House, McCoy to Parker, 29 January 1829, 20th Cong., 2d sess., H. Rept. 87, pp. 5–6.

10. U.S. Congress, Senate, Eaton to McCoy, 3 June 1830, 23d Cong., 1st sess., 1833–34, S. Doc. 512, Serial 245, 2: 6.

11. McCoy to Eaton, April 1831, ibid., pp. 435–36.

12. U.S. Congress, House, McCoy to Cass, Country for Indians West of the Mississippi, 1 February 1832, 22d Cong. 1st sess., 1831–32, H. Doc. 172, Serial 219, pp. 2–7.

13. McCoy, *Indian Missions*, p. 439.

14. McCoy to Cass, 6 March 1832, AGLR 8:8. McCoy had prepared a map of the proposed territory four years earlier, as he noted in his journal of Tuesday, 5 August 1828: "This day I completed my map of the country proposed for Indian territory. It is 2 feet 7 inches by about 3 feet." Lela Barnes, "Journal of Isaac McCoy for the Exploring Expedition of 1828," *Kansas Historical Quarterly* 5 (1936): 236.

15. In his 1832 report, McCoy wrote that "with a map of this country spread before him," invariably the emigrant Indian responds favorably to the plan for resettlement. House, 22d Cong., 1st sess., H. Doc. 172, p. 11. A copy of the map with accompanying field notes was requested by Gen. William Clark, superintendent of Indian affairs in St. Louis, to examine in connection with the decision on the "claim of the Panis to land assigned to the Delawares." Additional copies were prepared by the War Department Topographical Bureau for use by government agencies and Indian commissioners involved with resettlement strategies. Senate, 23d Cong., 1st sess., S. Doc. 512, 2: 870, 925.

16. Wheat, *Mapping the Transmississippi West*, 2: 142.

17. Apparently McCoy considered this map an important and continuing contribution to increased knowledge of western geography, for in 1838 he wrote that he had forwarded to the War Department "a corrected copy of the large map of the Indian territory. It was believed that this map had contributed somewhat to the fixing of the bounds of the Indian territory in the public mind." McCoy, *Indian Missions*, p. 541. McCoy's 1838 "Map of the Indian Territory" (National Archives, Record Group No. 75, no. 216), however, does not mark the locations of the various Indian villages, agencies, or missions (as did his 1832 map). Slightly updated, it denotes reservations' boundaries as of 1838, yet depicts much the same information as Hood's 1834 "Map of the Western Territory, etc.," including the north-south line denoting the "Limit of Habitable Country."

18. Senate, 23d Cong, 1st sess., S. Doc. 512, 2: 870, 872.

19. McCoy to The Commissioners West, 15 October 1832, ibid., 3: 486–87.

20. The report was printed in U.S. Congress, House, "Regulating the Indian Department," 23d Cong., 1st sess., 1833–34, H. Rept. 474, Serial 263, pp. 78–131.

21. Hood's map, prepared in Washington, D.C., was a compilation effort and undoubtedly relied heavily on the field reports, plats, and maps submitted by Isaac McCoy to the War Department. McCoy's 1832 "Map of the Proposed Indian Territory" was very likely the basic reference map used by Hood in drafting his "Western Territory" map. Wheat suggests that Isaac McCoy personally assisted Hood in making the latter. Wheat, *Mapping the Transmississippi West*, 2: 149. Hood, however, was not totally unfamiliar with western geography, as he had accompanied the Kennerly-McCoy 1828 expedition as topographer.

22. U.S. Congress, House, 25th Cong., 2d sess., 1837–38, H. Doc. 311, Serial 329, pp. 2–3. A copy of the journal or field book with sketches was "in every case" to be forwarded to department headquarters, with a second copy to be retained at the post where the officer in command of the march was stationed. Examples of maps prepared from field sketches are Dodge's 1835 expedition (fig. 7.4), Cooke's 1843 excursion (fig. 7.5), and Wharton's 1844 march (fig. 7.6).

23. Known contemporary published accounts, excluding government documents, were James Hildreth, *Dragoon Campaigns to the Rocky Mountains, Being a History of the Enlistment, Organization and First Campaigns of the Regiment of United States Dragoons* (New York, 1836); Stephen W. Kearny, "Official Letters on the Army of the West," *Niles Weekly Register* 72: 170–71; J. Henry Carleton, "Dragoon Campaign to the Pawnee Villages in 1844," *Spirit of the Times* (New York), 9 November 1844 to 12 April 1845; *Bangor* (Maine) *Daily Courier, Weston* (Missouri) *Western Democrat*, and "A Dragoon Campaign to the Rocky Mountains in 1845," *Spirit of the Times* (New York), 27 December 1845 to 30 May 1846; Philip St. G. Cooke, *Scenes and Adventures in the Army, or, Romance of Military Life* (Philadelphia, 1857).

24. The original plan for the 1834 march of the dragoons called for an excursion into the upper Platte and upper Missouri country but was abandoned because of the late start of the Dodge expedition to the Pawnee Pict village and the serious health problems which developed in the regiment. Atkinson to Jones, 8 August 1834, AGLR 9: 10.

25. Wheat, *Mapping the Transmississippi West*, 2: 149–50.

26. Bent's Fort was constructed in 1833–34 under the supervision of William Bent of the Bent and St. Vrain firm for the purpose of trading with the various Plains Indian tribes in the general area. Louise Barry, *Beginning of the*

West: Annals of the Kansas Gateway to the American West, 1540–1854 (Topeka: Kansas State Historical Society, 1972), pp. 256–57.

27. Louis Pelzer, *Marches of the Dragoons in the Mississippi Valley* (Iowa City: State Historical Society of Iowa, 1917), pp. 34–47.

28. U.S. Congress, Senate, 24th Cong., 1st sess., 1835–36, S. Doc. 2, Serial 279, p. 39. The map accompanying Dodge's report of the expedition was prepared from field sketches by the Army Topographical Bureau.

29. Pelzer, "Captain Boone's Journal of an Expedition Over the Western Prairies," *Marches of the Dragoons,* pp. 192–93.

30. Ibid., pp. 202–203, 214–19.

31. Ibid., pp. 224–25, 182.

32. Leo E. Oliva, *Soldiers on the Santa Fe Trail* (Norman: University of Oklahoma Press, 1967), pp. 44–52; Francis P. Prucha, *The Sword of the Republic: The United States Army on the Frontier, 1783–1846* (New York: Macmillan, 1969), pp. 372–79; Josiah Gregg, *Commerce of the Prairies,* 2 vols. (New York: Henry G. Langley, 1844; reprint, Philadelphia: J. B. Lippincott, 1962), p. 331; Barry, *Beginning of the West,* p. 568.

33. Kansas State Historical Society, *Eighteenth Biennial Report, July 1, 1910–June 30, 1912* (Topeka: Kansas State Historical Society), pp. 107–16; see also Kate L. Gregg, ed., *The Road to Santa Fe: The Journal and Diaries of George Champlin Sibley and Others. . . .* (Albuquerque: University of New Mexico Press, 1952).

34. Documentary research by Eloise Robbins indicates that the Fort Leavenworth to Fort Scott (1842) section of the military road crossed into Missouri in two places, but otherwise traced "in an irregular fashion" along the Kansas side of the boundary. Eloise F. Robbins, "The Original Military Post Road Between Fort Leavenworth and Fort Scott," *Kansas History: A Journal of the Central Plains,* Summer, 1978, pp. 90–100. Erna Risch suggests that except for a small northern section which was left unimproved, the road "was completed under quartermaster supervision by 1841." Erna Risch, *Quartermaster Support of the Army: A History of the Corps, 1775–1939* (Washington, D.C.: Quartermaster Historian's Office, Office of the Quartermaster General, 1962), p. 217.

35. U.S. Congress, House, "Notes of a Military Reconnaissance, from Fort Leavenworth, in Missouri, to San Diego, in California," Col. W. H. Emory, 30th Cong., 1st sess., 1847–48, H. Doc. 41, Serial 517, pp. 1–416.

36. Kearny to Asst. Adj. General, Western Division, 9 May 1844, AGLR 17: 20; Barry, *Beginning of the West,* p. 489. Wharton was instructed to keep a journal of the expedition and note "The military features of the country over which you pass—its resources—water courses—general topography. . . ." Turner to Wharton, 12 June 1844, AGLR 17: 20.

37. W. E. Connelley, ed., "The Expedition of Major Clifton Wharton in 1844," *Kansas Historical Collections* 16 (1923–25): 283.

38. U.S. Congress, House, 33d Cong., 2d sess., 1854–55, H. Doc. 1, Serial 777, p. 214.

8. MAPPING KANSAS AND NEBRASKA

THE ROLE OF THE GENERAL LAND OFFICE

Ronald E. Grim

THE rectangular alignment of fields, farmsteads, and roads is one of the most striking characteristics of the settlement pattern of the Great Plains. As most students of this region's cultural landscape are aware, the dominant factor in the formation of this regular, geometric pattern was the federal government's rectangular survey system. The basic features of this survey system (base lines, principal meridians, thirty-six-square-mile townships, sections, and quarter sections) have been outlined in introductory geography and cartography textbooks, while historical and cultural geographers have examined the system's effect on the landscape.[1] In addition, much has been written about the land alienation process and the development of the General Land Office, the federal agency that administered the newly developed cadastral system.[2] Few writers, however, have specifically addressed the mapping activities that were an integral part of the land survey and disposal system.[3]

This article examines the mapping activities of the General Land Office in Kansas and Nebraska. By focusing on these two states, it is possible to identify the personnel and procedures involved in the cartographic process, the geographical and temporal progress of the mapping program, and the resulting manuscript and published cartographic records. The General Land Office's role in Kansas and Nebraska does not represent a unique situation in the agency's mapping activities, but it does provide an example of the General Land Office's role in mapping the Great Plains.

Previous cartographic portrayals of this region were limited to the basic elements of the landscape documenting the experiences of early explorers and military expeditions.[4] On the other hand, the General Land Office's mapping activities represent a new phase in the cartographic representation of the Great Plains, because those activities resulted in the first comprehensive topographic mapping of the area.

The Cartographic Process in Kansas and Nebraska

By the time the General Land Office surveys reached the Great Plains, a fairly well defined system of surveying and mapping had been established, complete with its own bureaucracy and standardized procedures. Surveying of the public domain began in Ohio following the passage of the Land Ordinance of 1785 and proceeded westward, paralleling the westward expansion of the settlement frontier. In Kansas and Nebraska, General Land Office activities commenced in 1854 after the Kansas-Nebraska Act established the two territories. Both territories were combined under one surveyor general until 1867. Thereafter, they were separate offices until the Kansas surveys were completed in 1876 and the Nebraska surveys in 1895.

Surveying activities in Kansas and Nebraska, as well as in other public land states, were directed by a surveyor general, who reported directly to the commissioner of the General Land Office in Washington, D.C. During the period when Kansas and

Nebraska were combined under one office, there were five surveyors general, each of whom served an average term of three years. The surveyors for Kansas and Nebraska were John Calhoun (August 1854–July 1858), Ward B. Burnett (July 1858–April 1861), Mark W. Delahay (April 1861–October 1863), Daniel W. Wilder (October 1863–March 1865), and Hiram S. Sleeper (March 1865–July 1867). Sleeper continued in the Kansas office until April 1869, when he was replaced by Carmi W. Babcock, who served until the office was closed in 1876. All were political appointees, and only Calhoun and Sleeper had any previous surveying experience.[5]

The surveyor general had a small staff consisting of a chief clerk, a principal draftsman, assistant draftsmen, copyists, clerks, accountants, and messengers. The chief clerk occupied the highest paid position, serving as the surveyor general's assistant and supervising the other employees in the surveyor general's absence. The other employees assisted in the review and transcription of field notes, the compilation and copying of plats, and the preparation and copying of contracts and bonds. The tenure of the permanent staff was relatively short, although there was some upward mobility within the office. For example, during the first ten years of the Kansas-Nebraska office, there were eleven chief clerks, none serving for more than two years but averaging terms of eleven months each. Of the eleven, four had previously held the position of principal draftsman.[6]

There was no permanent official surveying staff. The actual surveys were completed by deputy surveyors who worked under contract to the surveyor general. Each deputy surveyor was responsible for hiring his own surveying party, which consisted of the surveyors themselves as well as chainmen, flagmen, and axemen. Payment to the deputy surveyors was based on a mileage rate: twelve dollars for standard lines, seven dollars for township exterior lines, and five dollars for section lines. Although some deputy surveyors attempted to abuse the system, examiners were hired to check the quality of the surveys.[7]

Considering the short tenure of office employees and the contract nature of surveying, there is a remarkable consistency in the physical appearance of the cartographic records. Given the number of deputy surveyors involved, a degree of subjectivity and selectivity in the recording of the data was inevitable. However, the uniformity that existed in the surveying and mapping activities was provided by a manual of surveying instructions first published in 1855.[8] The instructions prescribed the method of surveying, the type of information to be recorded in the field notes, and the procedure for preparing plats.

The system of surveying prescribed was hierarchical, starting with the establishment of a base line and principal meridian as depicted in diagram A of the manual (fig. 8.1). The next level involved establishing a series of guide meridians (every eight ranges) and standard parallels or correction lines (every four tiers north of the base line and every five tiers south of the base line). The crossing of the guide meridians and parallels created a grid framework from which the exterior boundaries of a block of townships could be surveyed, after which the individual townships would be subdivided into sections. Surveying began in the southeast corner of each township between sections 35 and 36 and progressed in a standard fashion to the northwest corner, ending between sections 5 and 6. Any deficiencies from the standard section size of 640 acres were adjusted to exterior lots of varying sizes in the northern and westernmost half-miles of the township.

Other instructions established the mechanics of surveying, such as the specific instruments to be used, the proper procedure for chaining or measuring, and the type of mound or monument to be erected. An ordinary compass and a two-pole chain (thirty-three feet long) or a four-pole chain (sixty-six feet) were to be used for surveying township and subdivision lines. The four-pole chain could be used on level ground, but the two-pole chain was to be used on more uneven surfaces. Procedures for tallying (counting the chains) and leveling were also specified. A Burt's improved solar compass was to be used when variations in the compass were found.[9] Detailed instructions were also provided for the erection of stones and mounds for township, section, and quarter section corners.

Instructions for field notes were also very specific. The surveyors were warned:

> [The notes] must be a faithful, distinct and minute record of everything officially done and observed by the surveyor and his assistants, pursuant to instructions, in

relation to running, measuring, and marking lines, establishing boundary corners etc., and present, as far as possible, a full and complete topographical description of the country surveyed, as to every matter of useful information, or likely to gratify public curiosity.[10]

Specifically, the surveyors were requested to record the precise length of every line run, the kind and diameter of all "bearing trees," the kind of materials from which mounds were constructed, trees on line, intersections of the line with land and water objects (settlers' claims, improvements, prairies, rivers, creeks, bottoms), land surface (level, rolling, or broken), soil (first, second, or third rate), timber, coal deposits and other minerals, roads and trails, compass variation, and a general description of the township as a whole.[11]

The field notes were to be returned to the surveyor general's office, where a transcript would be prepared for transmittal to the General Land Office in Washington, D.C. In addition, three plats would be drawn according to a prescribed style. One copy was retained by the surveyor general, one copy was forwarded to the General Land Office in Washington, and the third copy was sent to a local land office.[12]

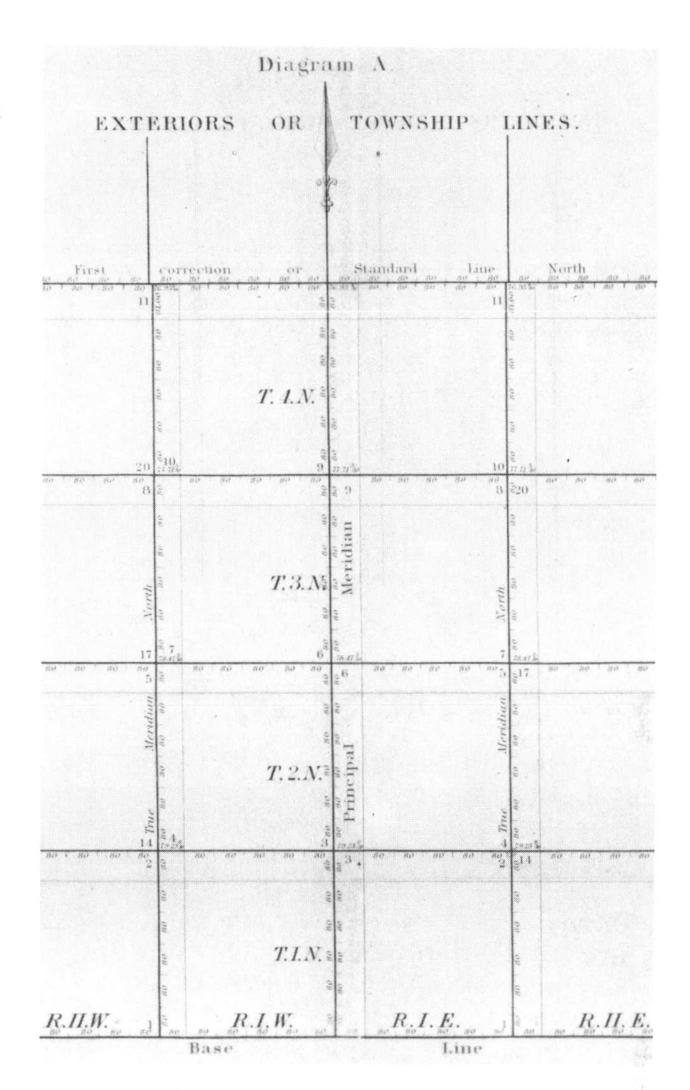

Fig. 8.1. Diagram, "Exteriors or Township Lines," illustrating the grid of base line, meridians, and correction lines, which provides the basic framework for township surveys. (U.S. General Land Office, *Instructions to the Surveyors General*, 1855)

Progress of Surveys in Kansas and Nebraska

The progress of surveys in Kansas and Nebraska from 1854 to 1895 can be traced in the annual reports of the surveyor general to the commissioner of the General Land Office.[13] Although the comments and types of statistics recorded from year to year were not consistent, it is possible to summarize the general progress of these surveys.

After opening a temporary office at Fort Leavenworth in October 1854, the surveyor general's first project was the surveying of the base line, which would also serve as the boundary between the two territories.[14] This initial contract was granted to John P. Johnson. He started at the point where 40° north latitude crosses the Missouri River and extended the base line westward 108 miles to the proposed intersection with the principal meridian. The examiner found that his surveys were so defective that they had to be resurveyed.[15] The second attempt was completed by the examiner, Charles A. Manners, but he surveyed only sixty miles of the base line to the proposed intersection with the first guide meridian. Additional surveys during the first year included the demarcation of the first guide meridian and standard correction lines from four north to four south. In addition, more than two hundred townships were in the process of being surveyed but were not completed when the 1855 annual report was submitted.

During 1856 the base line was extended westward to the intersection with the principal meridian, and the principal meridian and standard parallels from

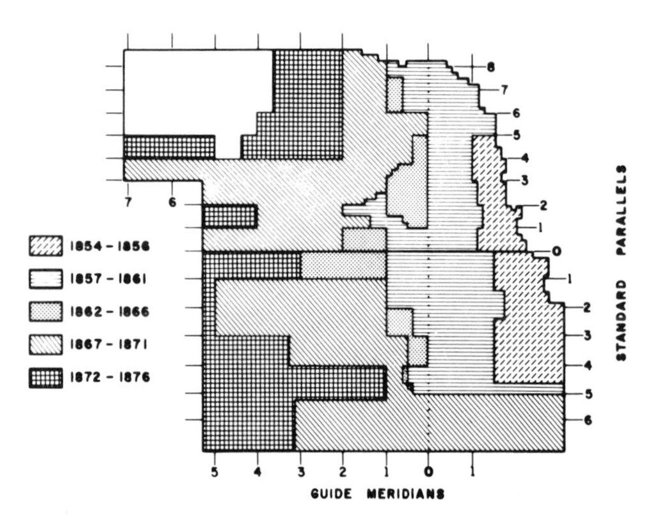

Fig. 8.2. Progress of township surveys, Kansas and Nebraska, 1854–76.

the northern boundary of Nebraska to the fifth standard parallel south were surveyed. Subdivided townships included most of the area in the northern two-thirds of Kansas, east of range 13, and the southern two-thirds of Nebraska, east of the first guide meridian. The surveys in this area were complicated by the presence of Indian reserves and trust lands. In Nebraska the surveys excluded the Omaha and Oto reserves, while in Kansas the surveys excluded the Kickapoo, Potawatomi, Delaware, Kansa, Sauk and Fox, and Ottawa reserves as well as the Osage, Cherokee, and New York Indian lands in the southern one-third of the state.

The geographic progress of township surveying and mapping is summarized in fig. 8.2, which shows the extent of completed townships at five-year intervals, beginning with the status as reported in the 1856 annual report.[16] By the beginning of the Civil War, all the townships east of the principal meridian and north of township 23 south had been surveyed. In addition, the surveys extended westward along the base line to the first guide meridian west and northwest to Fort Kearney in Nebraska. During the Civil War, surveying activity was greatly reduced, extending westward only along the base line to the third guide meridian west.

During the first five-year period after the Civil War, major expansion occurred. In Kansas the surveys in the southern one-third of the state corresponded to the opening of the Osage Indian lands, while the westward expansion between the first and third standard parallels took place in response to the location of the Butterfield Overland Trail and the Kansas Pacific Railroad. In Nebraska the surveys extended westward along the base line and between the second and fourth standard parallels, following the route of the Union Pacific Railroad. By 1876 the remainder of Kansas was surveyed, while the surveys in Nebraska extended to the north and west, leaving only the northwest quarter of the state unsurveyed. This area was finally completed in 1895.

The total acreage subdivided into townships in each state on a yearly basis is depicted in fig. 8.3.[17] It clearly shows that the surveys were not conducted at uniform rates. In general, the surveys in Nebraska progressed at a slower rate than did those in Kansas. In the early years, the difference was attributed to the desire to survey the Indian trust lands so that they could be sold quickly. Until 1859 the surveys progressed rapidly, with at least two million acres surveyed each year in each territory. From 1860 to 1866, progress was greatly retarded. In each territory the yearly average was less than one million acres. After the Civil War, there was a dramatic increase in the quantity of land surveyed, corresponding to the interest in the construction of the transcontinental railroads and anticipated settlement on railroad lands. In Kansas surveying activity peaked in 1871 with 7.1 million acres. In Nebraska it peaked in 1873 with 4.4 million acres.

Manuscript Survey Records

The primary cartographic records that resulted from this surveying activity were manuscript plats and field notes. Although separate plats were prepared as the base line, meridians, correction parallels, and exterior township lines were surveyed (figs. 8.4 and 8.5), the basic records were the three township plats: an original, which was retained by the surveyor general; a duplicate, which was sent to the commissioner of the General Land Office (fig. 8.6); and a triplicate, which was sent to the local land office (fig. 8.7).[18] These plats were compiled from the original field notes on preprinted forms at a scale

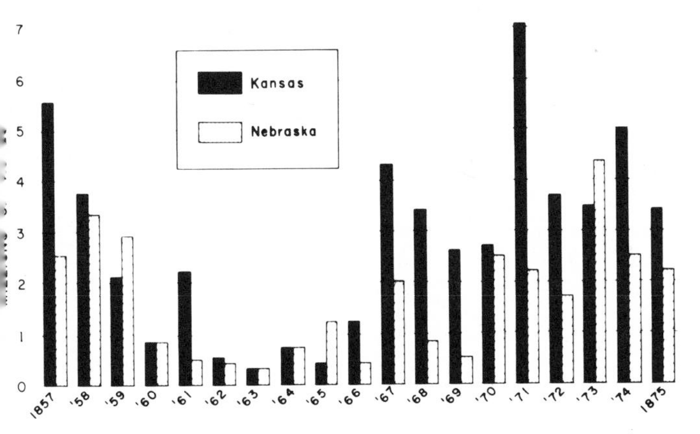

Fig. 8.3. Number of acres surveyed, Kansas and Nebraska, 1857–75.

of forty chains to an inch, or one inch equals one-half mile.

The creation and contents of the plats can be illustrated by the records for township 25 south, range 3 west, a township in Sedgwick County, on the Arkansas River, approximately twenty miles northwest of Wichita, Kansas. The authority statement in the lower margin of the township plat (fig. 8.6) indicates that several deputy surveyors were involved in surveying this township and that the process covered the period from October 1858 to March 1861. The southern boundary was surveyed by Frederick Hawn in October 1858, while the remaining exterior township lines were surveyed by James Withrow in September 1860. Isaac C. Stuck and William M. Hill surveyed the section lines from 20 to 24 November 1860. According to standard procedure, the field notes were submitted to the surveyor general, then located in Nebraska City, where the notes were examined and copied, and the plats compiled and copied. On 18 March 1861, the plats were certified by Ward Burnett, the surveyor general.

The surveys of the exterior boundaries, which were performed under two separate contracts, were recorded on two different plats. The plat for the fifth standard parallel (fig. 8.4), which includes the southern boundary of the township, illustrates that these plats showed a minimum of information, concentrating on the surveyed line and only those physical or cultural features that were crossed in the survey. The remaining exterior boundaries of the townships were surveyed with a block of forty township, bounded by the fourth and fifth standard parallels south, the principal meridian, and the first guide meridian west (fig. 8.5). Normally, plats of this type show only the surveyed lines and occasionally the basic drainage pattern or major trails and roads. In this manuscript plat, pencil annotations record the general condition of the land for each township. For township 25 south, range 3 west, the annotation indicates: "good farming lands, [second rate] soil, no overflow, level and rolling."[19]

The township plat itself represents the subdivision of the township into sections. Besides the survey lines, the plats normally show the variety of physical and cultural information observed during the survey of the section lines. The plat for township 25 south, range 3 west (fig. 8.6) looks rather plain, but the lack of descriptive detail is a result of the uniform land surface of the township. The only water bodies are the Arkansas River, with no tributary streams, and a pond on the line between sections 16 and 17. The only wooded area is a small grove along the Arkansas River in section 9. One road parallels the north side of the Arkansas River. No preexisting settlements are indicated and, by implication, most vegetation is prairie.[20]

This interpretation of the contents of the plat can be verified by the survey field notes. The original notes, which were recorded in the field as the surveys were conducted, were retained by the surveyor general, while a duplicate copy was sent to Washington, D.C. The field notes for the township subdivision provide a mile-by-mile description of the survey. For example, the notes for the line between sections 16 and 17 read:

> North between sections 16 and 17, variation 11°05′ East; 29.00 [chains] Enter pond (now dry), bears E & W; 38.00 [chains] Leave pond, bears E & W and about 15 chains long from E to W; 40.00 [chains] Set sandstone 15 in. long, 10 in. wide, 11 in. thick for quarter section corner; 80.00 [chains] Set sandstone 17 in. long, 9 in. wide, 11 in. thick for corner to sections 8, 9, 16 & 17; Gently rolling prairies, soil 2nd rate.[21]

The notes also include the survey of the meanders of the Arkansas River, a list of personnel involved in the survey, and a general description of

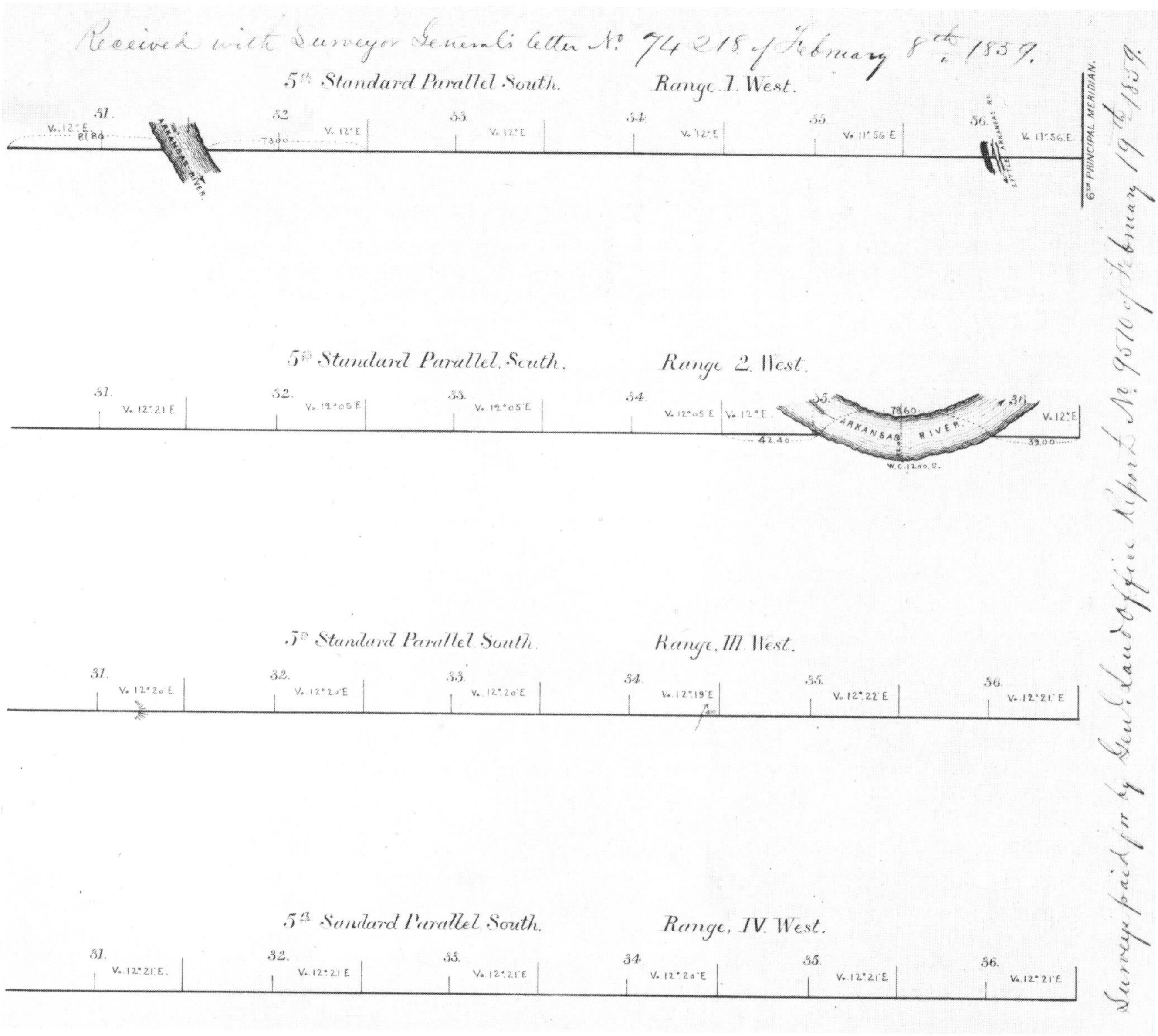

Received with Surveyor General's letter No. 74 218 of February 8th 1859.

5th Standard Parallel South. Range. I. West.

31. V. 12° E 21.84 ARKANSAS RIVER 32. 73.00 V. 12° E 33. V. 12° E 34. V. 12° E 35. V. 11° 56' E 36. LITTLE ARKANSAS R. V. 11° 56' E

6th PRINCIPAL MERIDIAN.

5th Standard Parallel. South. Range 2. West.

31. V. 12° 21' E 32. V. 12° 05' E 33. V. 12° 05' E 34. V. 12° 05' E V. 12° E. 42.40 35. 78.60 ARKANSAS RIVER W.C. 12.00 36. 39.00 V. 12° E

5th Standard Parallel South. Range. III. West.

31. V. 12° 20' E 32. V. 12° 20' E 33. V. 12° 20' E 34. V. 12° 19' E 35. V. 12° 22' E 36. V. 12° 21' E

5th Standard Parallel South. Range. IV. West.

31. V. 12° 21' E. 32. V. 12° 21' E 33. V. 12° 21' E 34. V. 12° 20' E 35. V. 12° 21' E 36. V. 12° 21' E

Survey paid for by Genl. Land Office Report No. 9510 of February 19th 1859.

Fig. 8.4. Detail of manuscript plat showing the fifth standard parallel south between the sixth principal meridian and the first guide meridian west as surveyed by Frederick Hawn under contract of 2 July 1858. (Cartographic and Architectural Branch, National Archives, Record Group 49, Kansas Exterior Boundaries, vol. 1, p. 129)

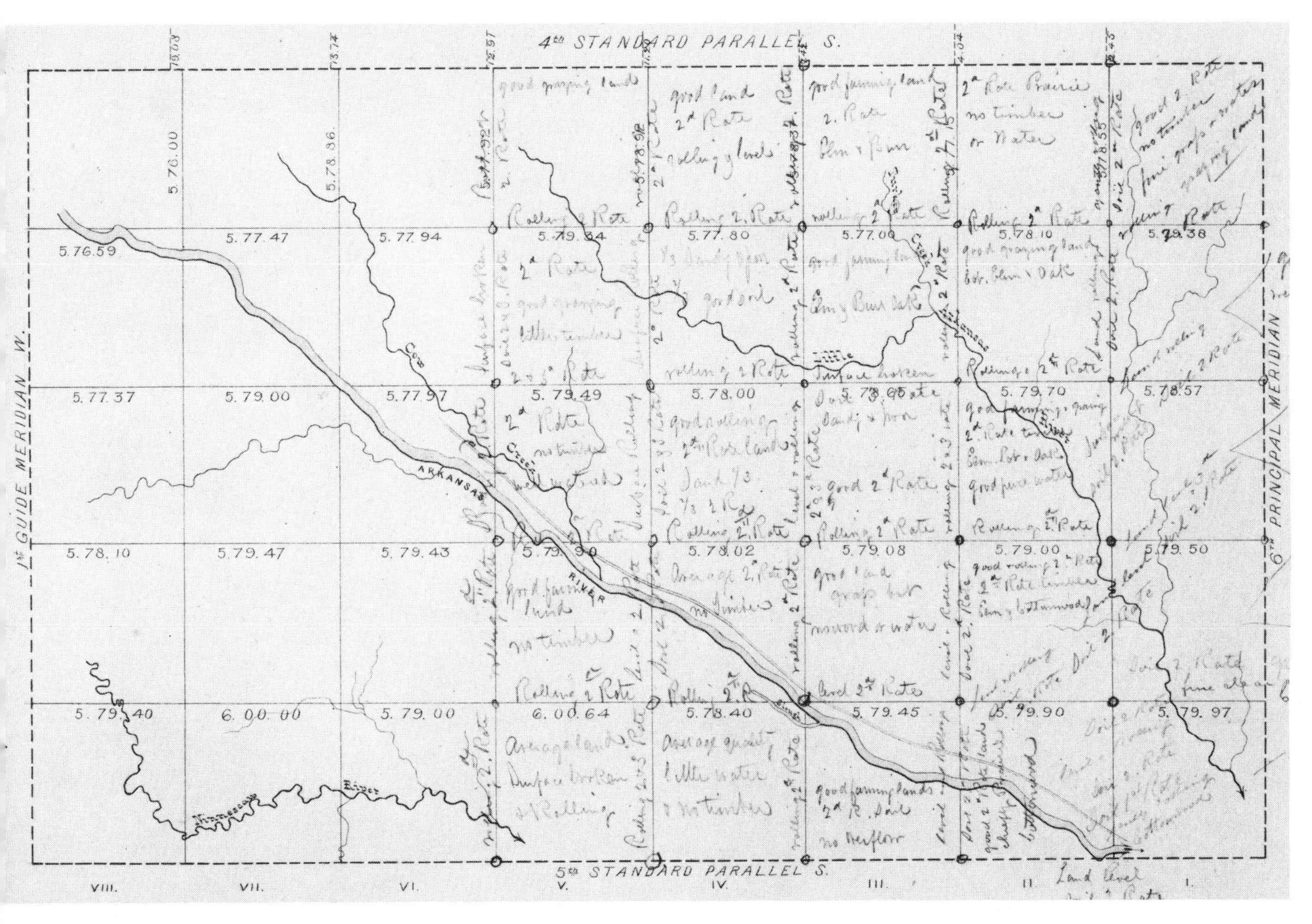

Fig. 8.5. Detail of manuscript plat showing township lines for townships 21–25 south, ranges 1–8 west, Kansas, surveyed by James Withrow under contract of 6 August 1860. (Cartographic and Architectural Branch, National Archives, Record Group 49, Kansas Exterior Boundaries, vol. 2, p. 3)

the township, which in this case confirms the original interpretation of the physical and cultural features depicted on the plat. The general description reads:

> Land in this township mostly 2nd rate. Usually slightly rolling prairie with considerable level bottom along the Arkansas River. The bottom is, however, very sandy and we should suppose would not produce abundantly as farming land. The Arkansas River runs entirely through it from West to East and is now completely dry and can be crossed at any point with teams. . . . The banks are almost destitute of timber and what there is, consists of cottonwood and willow, not of much use except firewood. Neither stone or mineral of any kind discovered in the township. No settlement in the township.[22]

The triplicate copy of the plat (fig. 8.7), is a valuable research tool, not for the information it gives about the survey process, since that information basically duplicates what is on the other two plats, but because it graphically records the General Land Office's initial disposal of the land.[23] The third copy was forwarded to the local land offices, and as the land was sold or disposed, the register recorded on the plat, as well as in a tract book, the disposition

TOWNSHIP Nº 25 SOUTH RANGE Nº III WEST OF THE 6TH PRINCIPAL MERIDIAN KAN. TER.

Sec. 1. 640.00
Sec. 2. 640.00
Sec. 3. 640.00
Sec. 4. 640.00
Sec. 5. 584.94
Sec. 6.
Sec. 7. 610.74
Sec. 8. 531.60
Sec. 9. 503.20
Sec. 10. 633.90
Sec. 11.
Sec. 12.
Sec. 13. 519.50
Sec. 14. 491.50
Sec. 15. 519.00
Sec. 16. 616.60
Sec. 17.
Sec. 18. 638.64
Sec. 19. 639.04
Sec. 20.
Sec. 21.
Sec. 22.
Sec. 23.
Sec. 24. 606.00
Sec. 25.
Sec. 26.
Sec. 27.
Sec. 28.
Sec. 29.
Sec. 30. 640.00
Sec. 31. 640.40
Sec. 32.
Sec. 33.
Sec. 34.
Sec. 35.
Sec. 36.

ARKANSAS RIVER

Road

Pond

5TH STANDARD PARALLEL SOUTH.

Scale 40 Chains to an Inch

Meanders of the Right & Left banks of the Arkansas River. V11°10'E								
Posts	Courses	Chs. Lks.	Posts	Courses	Chs. Lks.	Posts	Courses	Chs. Lks.
	(Left Bank)			Right Bank				
	(Section 13)			Section 6.				
1	N 55¼ W	16.40	11	S 35¾ E	29.00			
	N 54¾ W	14.80		S 21¼ E	8.00			
	N 34½ W	11.00		S 52¾ E	10.00			
	N 36¾ W	16.00	12	S 29° E	7.10			
	N 38° W	3.40						
	N 67° W	13.20		Section 7				
	N 47 W	8.00	12	S 71° E	12.00			
	S 84½ W	5.30		S 89½ E	9.00			
2	N 75° W	10.00		S 63½ E	13.00			
				N 65¾ E	7.00			
	Section 14.		13	S 70¾ E	10.50			
2	N 63½ W	12.40						
	N 81½ W	13.70		Section 8				
	N 83° W	8.70	13	S 78¾ E	17.00			
	S 73½ W	5.40		N 67¾ E	7.00			
	N 65¾ W	20.90		S 87½ E	16.00			
	N 70¾ W	4.70		S 73 E	13.00			
	N 86½ W	11.80		S 66¾ E	14.00			
3	N 73° W	5.90	14	S 64½ E	17.10			
	Section 10.			Section 9				
3	N 79° W	6.70	14	S 55½ E	14.00			
4	S 78¼ W	7.50		S 40¾ E	16.30			
				S 38° E	25.00			
	Section 15		15	S 40¾ E	17.70			
4	S 87° W	18.70						
	N 74¾ W	4.20		Section 16				
	S 68° W	6.60	15	S 46¼ E	8.00			
5	N 59½ W	4.60		S 74½ E	19.00			
			16	S 87½ E	6.60			
	Section 10							
5	N 57¼ W	3.30		Section 15				
	S 85° W	10.80	16	S 85½ E	10.60			
	S 89½ W	11.00		S 74¾ E	11.00			
	N 64½ W	5.50		S 88¾ E	16.50			
6	S 86½ W	4.20		S 83½ E	11.40			
				S 80¼ E	10.20			
	Section 9			S 89½ E	11.50			
6	N 58½ W	6.00		N 82½ E	3.20			
	N 66° W	19.00	17	S 81¼ E	6.40			
	N 44½ W	26.00						
	N 37° W	17.00		Section 14.				
	N 49½ W	9.00	17	S 71° E	8.10			
	N 35¾ W	13.00		N 88½ E	7.40			
7	N 58½ W	17.20		S 69¾ E	9.90			
				S 81¼ E	11.00			
	Section 8			S 71¼ E	16.60			
7&8	N 49¾ W	13.70		S 87½ E	20.30			
			18	S 64 E	9.10			
	Section 5.							
8	N 68¾ W	20.00		Section 13.				
	S 88° W	11.00	18	S 77½ E	12.10			
	N 75¾ W	11.00		S 69¾ E	17.10			
	N 85¾ W	19.00		S 40½ E	14.00			
9	S 80° W	10.50		S 22° E	8.20			
				S 40½ E	10.80			
	Section 6.		19	S 31° E	10.00			
9	N 81¼ W	16.00						
	N 66½ W	12.00		Section 24.				
	N 61½ W	18.00	19&20	S 50¼ E	37.20			
	N 73° W	8.00						
	N 42½ W	6.00						
	N 34¾ W	20.00						
	N 18¾ W	16.00						
	N 40¼ W	8.00						
10	N 87 W	3.60						

Total number of Acres 22.085.92

Estimated Area of River 347 acres.

Surveys Designated	By Whom Surveyed	Date of Contract	Amount of Surveys			When Surveyed
			M.	Chs.	Lks.	
Township lines	James Withrow	August 6th 1860		63		September 1860
Subdivisions	Stuck & Hill	September 21 1860	59	77	26	Novr 20–24 1860
5th Standard Parallel	Frederick Hawn	July 2nd 1858				October 1858
Meanders			14	19	60.	
Total			74	76	86	

The above Map of Township Nº 25 South of Range Nº III West of the 6th Principal Meridian Kansas Territory is strictly conformable to the field notes of the survey thereof on file in this Office, which have been examined and approved.

Surveyor Generals Office. Nebraska City, N.T. [illegible] 1861

Ward B. Burnett
Sur. Gen.

Fig. 8.6. Duplicate or headquarters copy of the plat for township 25 south, range 3 west, surveyed by Stuck and Hill, 20–24 November 1860. The original plat, which is currently in the custody of the Kansas secretary of state, is the same as the duplicate copy except that it does not include the meanders of the Arkansas River listed in the right margin of the plat. (Cartographic and Architectural Branch, National Archives, Record Group 49, Kansas Township Plats)

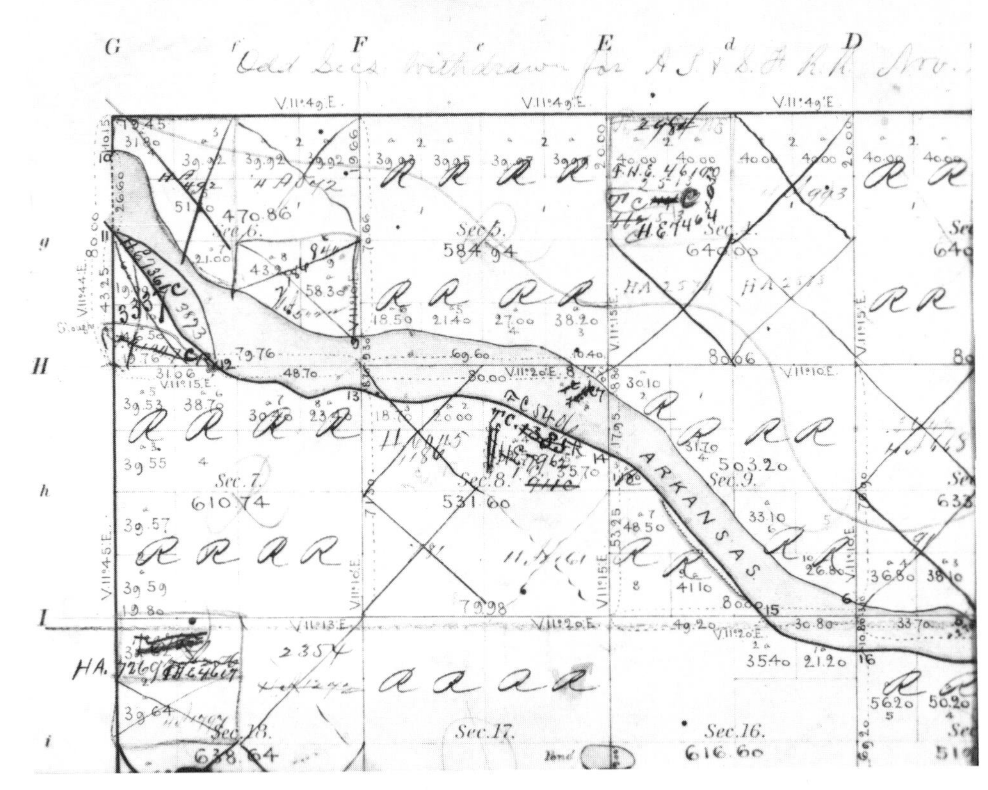

Fig. 8.7 Detail of triplicate or local office copy of plat for township 25 south, range 3 west, surveyed by Stuck and Hill, 20–24 November 1860. Annotations added in the local land office indicate the initial disposition of land: alternate sections marked with "RR" were granted to the Atchison, Topeka and Santa Fe Railroad, while those sections patented through the homesteading process are marked with an X and initialed with "HA" (homestead application), "HE" (homestead entry), or "FC" (final certificate). (Cartographic and Architectural Branch, National Archives, Record Group 49, Kansas Township Plats)

of individual parcels of land. School lands (sections 16 and 36) were left blank; alternating sections selected by the Atchison, Topeka, and Santa Fe Railroad were marked by the abbreviation "RR." The remaining lands were marked with an X and appropriate land entry or homestead final certificate numbers. A survey of the corresponding tract book indicates that most of the public lands in the northwest portion of this township were homesteaded, with initial entries made in the early 1870s and final proof presented in the late 1870s.[24] By correlating the information found on the local office plats, tract books, and land entry papers, it is possible to recreate a cadastral map showing the original landownership in this township.

Examples from two other townships better illustrate the variety of information that is portrayed on the basic township plat as well as on supplementary plats compiled for separate surveys of Indian reserves. From the plat for township 11 south, range 16 east (fig. 8.8), the location of Topeka, it is immediately evident that there is a greater variety of information depicted—woodland and prairie, numerous roads, and a few presurvey settlements, including the town of Tecumseh (lower right corner). The plats are not comprehensive, however: streams, roads, and wooded areas are not continued in the Indian reserves. In addition, several roads are marked as crossing the township and section lines in the southern tier of sections but are not continued to their completion. There are plats for the separate surveys of Wyandott Reserve No. 20 and the Kansas Half-Breed lands, in which some but not all of these physical and cultural features are continued. For example, the 1864 plat of Kansas Half-Breed Survey No. 5 shows the continuation of streams but not

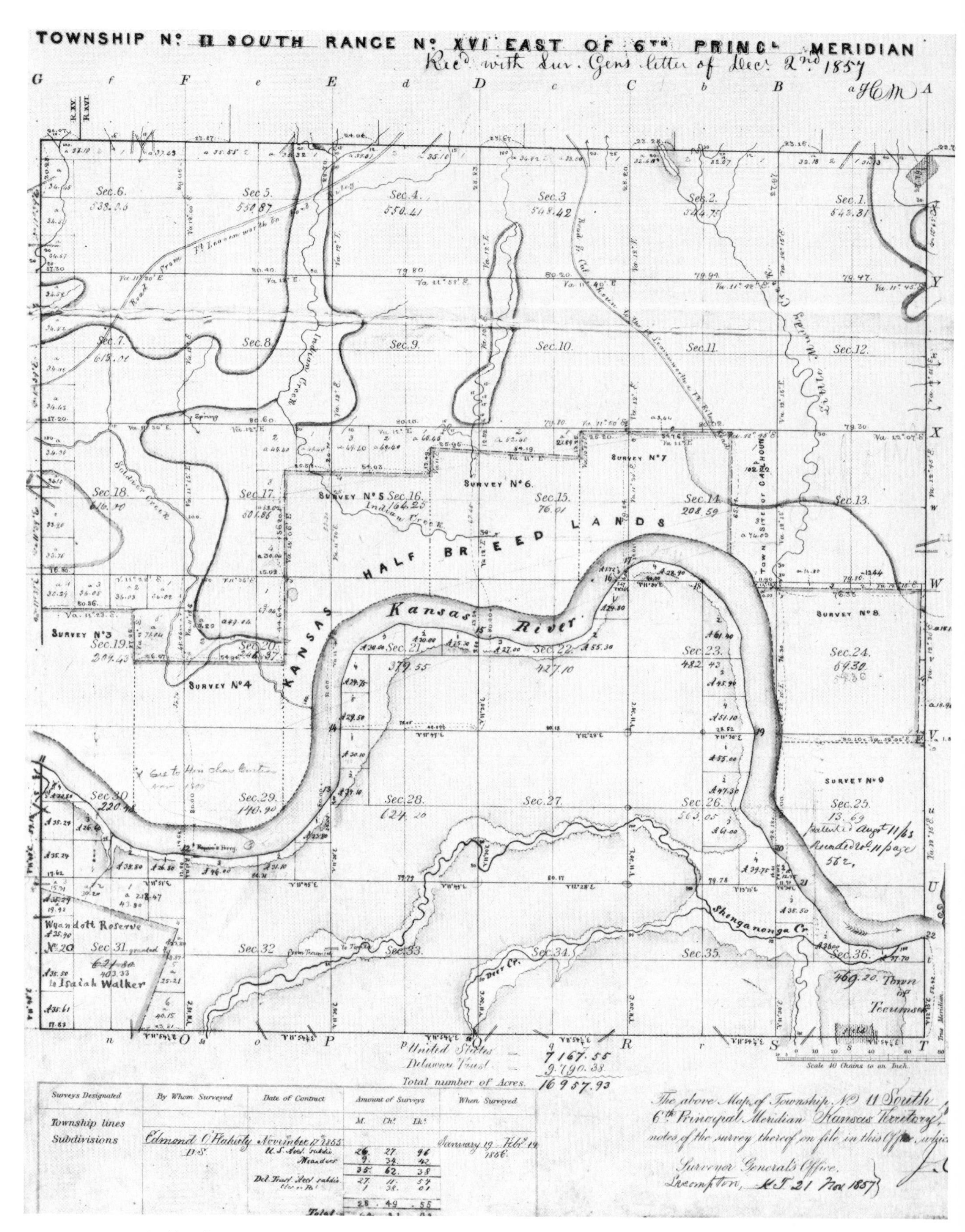

Fig. 8.8. Detail of headquarters copy of the plat for township 11 south, range 16 east, surveyed by Edmund O'Flaherty, 19 January–14 February 1856. "Wyandott Reserve No. 20" in section 31 was the eventual site of Topeka. (Cartographic and Architectural Branch, National Archives, Record Group 49, Kansas Township Plats)

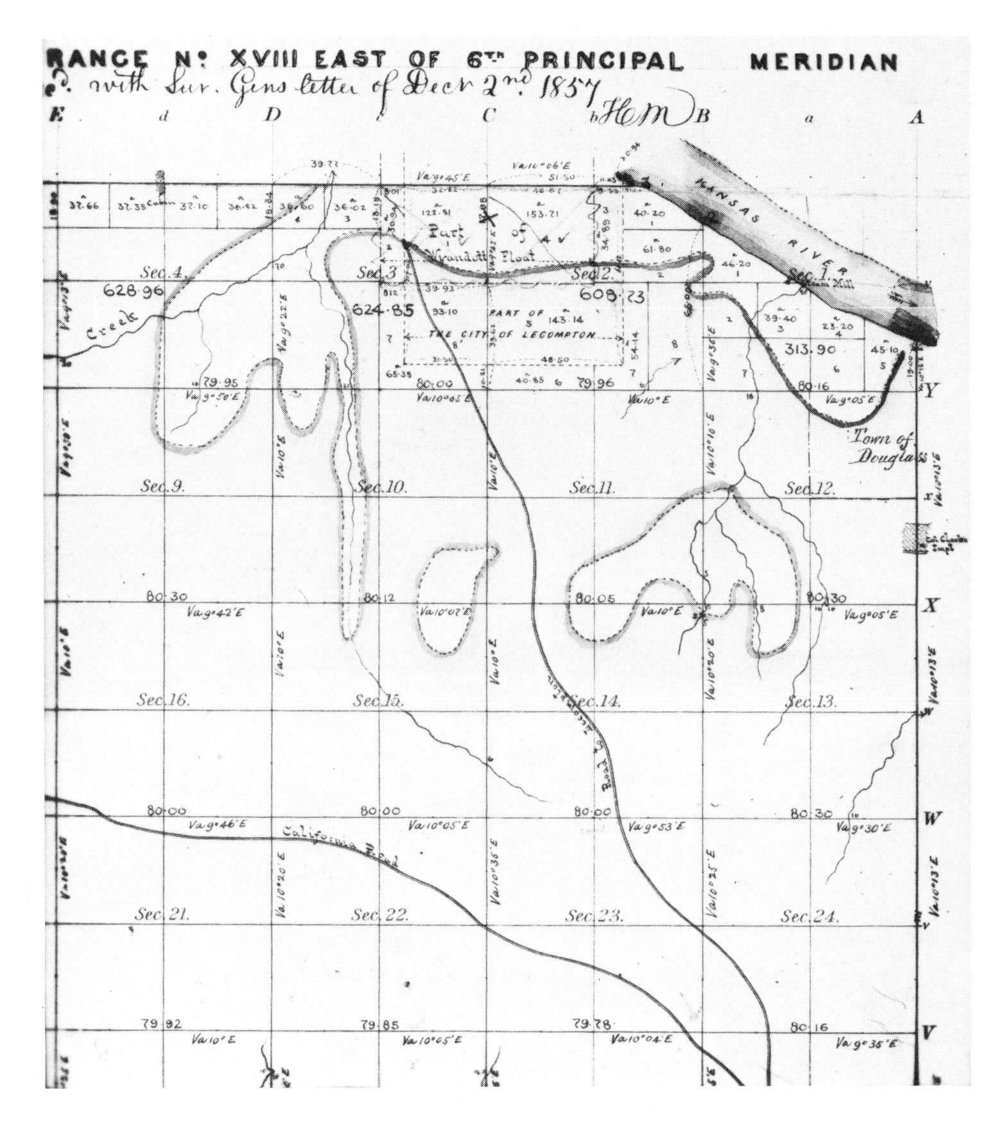

Fig. 8.9. Detail of headquarters copy of the plat for township 12 south, range 18 east, surveyed by William J. Card, 5–15 March 1856. The town of Lecompton is located in the northeastern portion of this township. (Cartographic and Architectural Branch, National Archives, Record Group 49, Kansas Township Plats)

the vegetation pattern.[25] The plat for township 12 south, range 18 east (fig. 8.9) displays similar information, but in this case, the 1858 plat for Wyandott Reserve No. 31 (fig. 8.10), which was used to locate the town of Lecompton, not only shows the continuation of the vegetation line, but also shows the location of a steam sawmill, the post office, and the surveyor general's office.[26]

Composite State Maps

Other cartographic records produced by the General Land Office during this period were composite state maps designed to show the progress or status of surveys and eventually to serve as general reference maps. The production of these maps can be divided into three chronological periods based on the frequency and source of publication and the increased amount of information depicted.

Prior to 1866, state maps showing the progress of surveys were issued with the annual report of the commissioner of the General Land Office, published as part of the Congressional Serial Set. Manuscript maps were submitted by each surveyor general to accompany his report to the commissioner. The maps were redrawn in a standard style and were

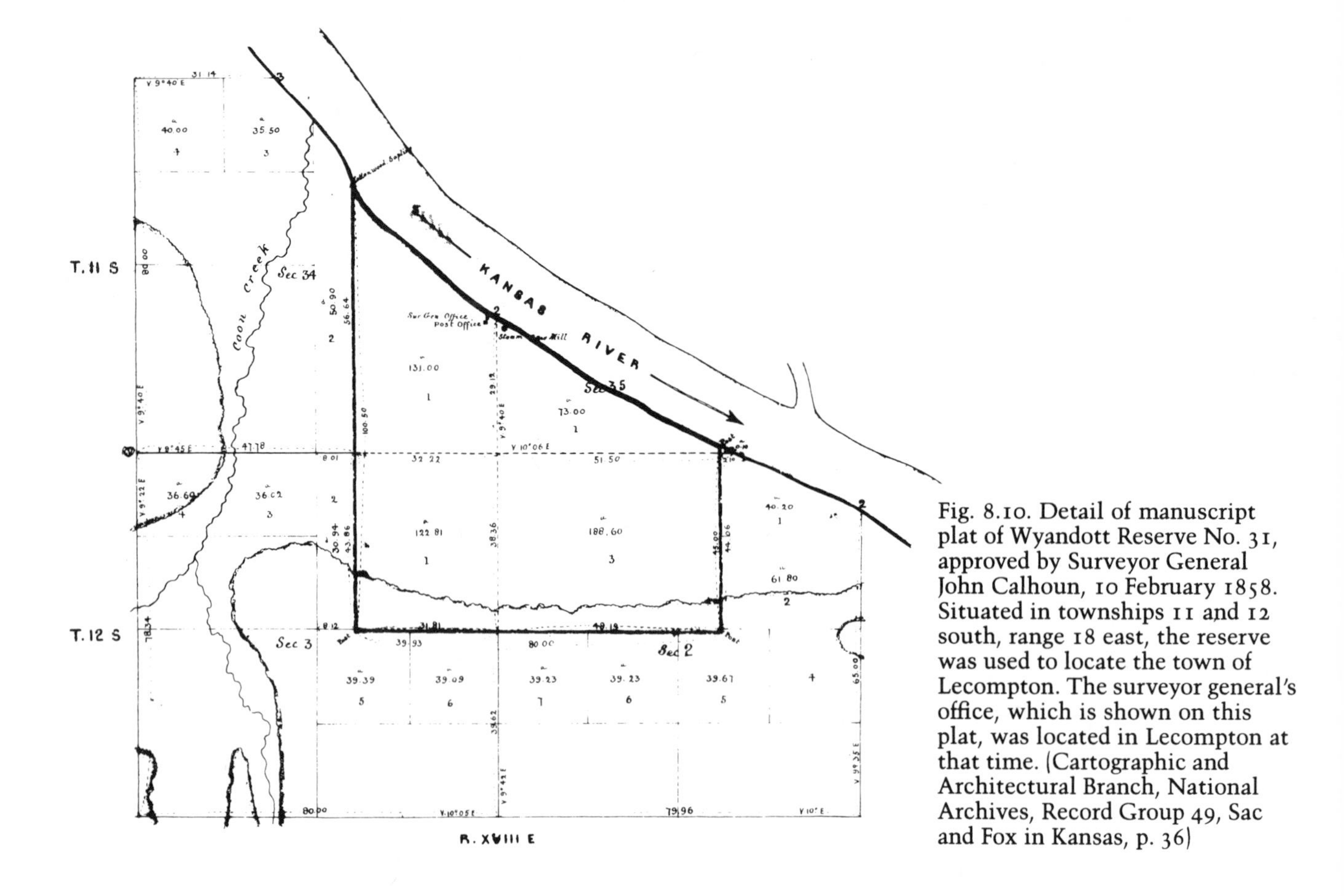

Fig. 8.10. Detail of manuscript plat of Wyandott Reserve No. 31, approved by Surveyor General John Calhoun, 10 February 1858. Situated in townships 11 and 12 south, range 18 east, the reserve was used to locate the town of Lecompton. The surveyor general's office, which is shown on this plat, was located in Lecompton at that time. (Cartographic and Architectural Branch, National Archives, Record Group 49, Sac and Fox in Kansas, p. 36)

issued in separate map volumes in the Congressional Serial Set. The surveyor general for Kansas and Nebraska submitted one map showing the status of surveys in the entire district, which encompassed both territories. Progress maps were issued for Kansas-Nebraska every year between 1855 and 1866 except for 1864 and 1865, when the publication of all General Land Office maps was suspended because of the Civil War.[27]

As the 1859 map (fig. 8.11) illustrates, the progress maps were simple diagrams, showing the base line and principal meridian, standard parallels and guide meridians, and township lines. Major rivers, a few towns, and Indian reserves were added for reference points. Completed townships were shown by solid lines and proposed surveys by broken lines. Township surveys that were completed or were in the process of being surveyed during the fiscal year were classified into several categories. For example, the 1859 map (fig. 8.11) shows a number of categories or stages of completion:

> Townships under contract and being Surveyed
> Townships, Surveys reported to the Office
> Townships Field and Office work West, complete, duplicates forwarded to Genl Land Office and triplicates to Local Land Office
> Canceled but field work wholly or partially completed . . .
> Townships under Contract but Contract Canceled . . .

Unfortunately, consistent categories were not used from year to year, making a yearly comparison difficult although not impossible. The original manuscript maps submitted by the Kansas-Nebraska surveyor general to the commissioner have survived

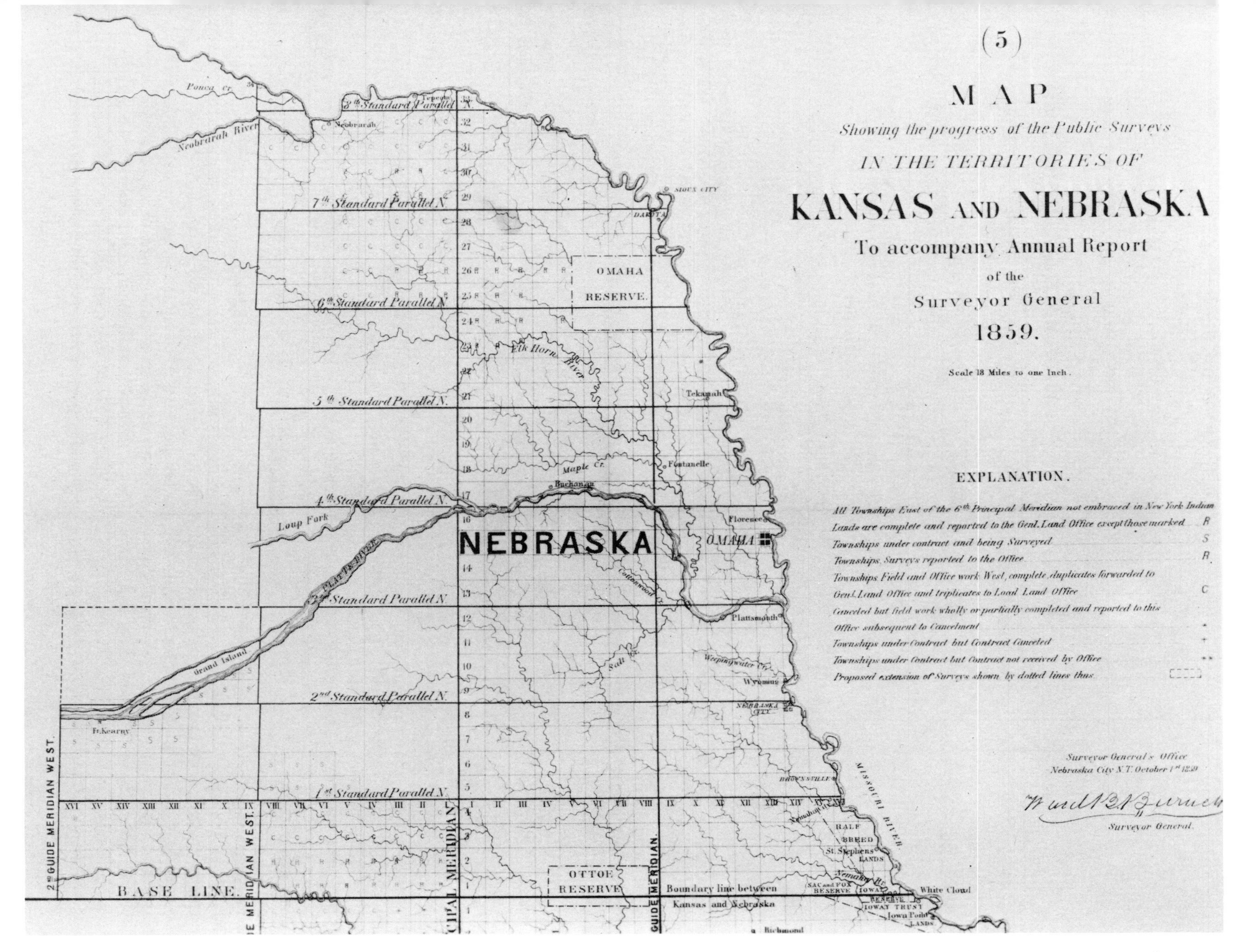

Fig. 8.11. Detail of published map showing the progress of the public surveys in the territories of Kansas and Nebraska to accompany annual report of the surveyor general, 1859. (Geography and Map Division, Single Map Collection, Library of Congress)

Fig. 8.12. Detail of manuscript map showing the progress of the public surveys in the territories of Kansas and Nebraska to accompany annual report of the surveyor general, 1859. (Cartographic and Architectural Branch, National Archives, Record Group 49, Old Map File, Kansas 8)

for 1857, 1859, 1860, 1861, and 1865 and are now filed in the National Archives.[28] The information shown on the 1859 manuscript map (fig. 8.12) is almost the same, although the lettering is more artistic. Close inspection shows a few additional place names on the manuscript map.

The publication of progress maps was not resumed after 1866, although there are occasional references in the annual reports to progress or sectional maps that were submitted with the surveyor general's annual report. Manuscript maps have survived for Kansas from the 1870, 1871, and 1873 reports, and for Nebraska from the 1868, 1869, 1870, and 1871 reports.[29] A distinct difference from the maps published before 1866 was the preparation of separate maps for each state, reflecting the creation of separate surveyor general's offices for each state in 1867. The later maps also show more detail, including complete drainage patterns, railroads, counties, and a greater number of towns resulting from the spread of settlement. Two unique maps from this group are the 1870 Kansas map and the 1869 Nebraska map (fig. VII.3). The Kansas map shows basic features as well as the extent of wooded areas, while the Nebraska map also shows soil types, coal deposits, and wooded areas.[30]

A third phase of state mapping began in 1876 with the publication of standard reference maps for each of the public land states. As the 1876 Nebraska map (fig. 8.13) illustrates, the maps still show the

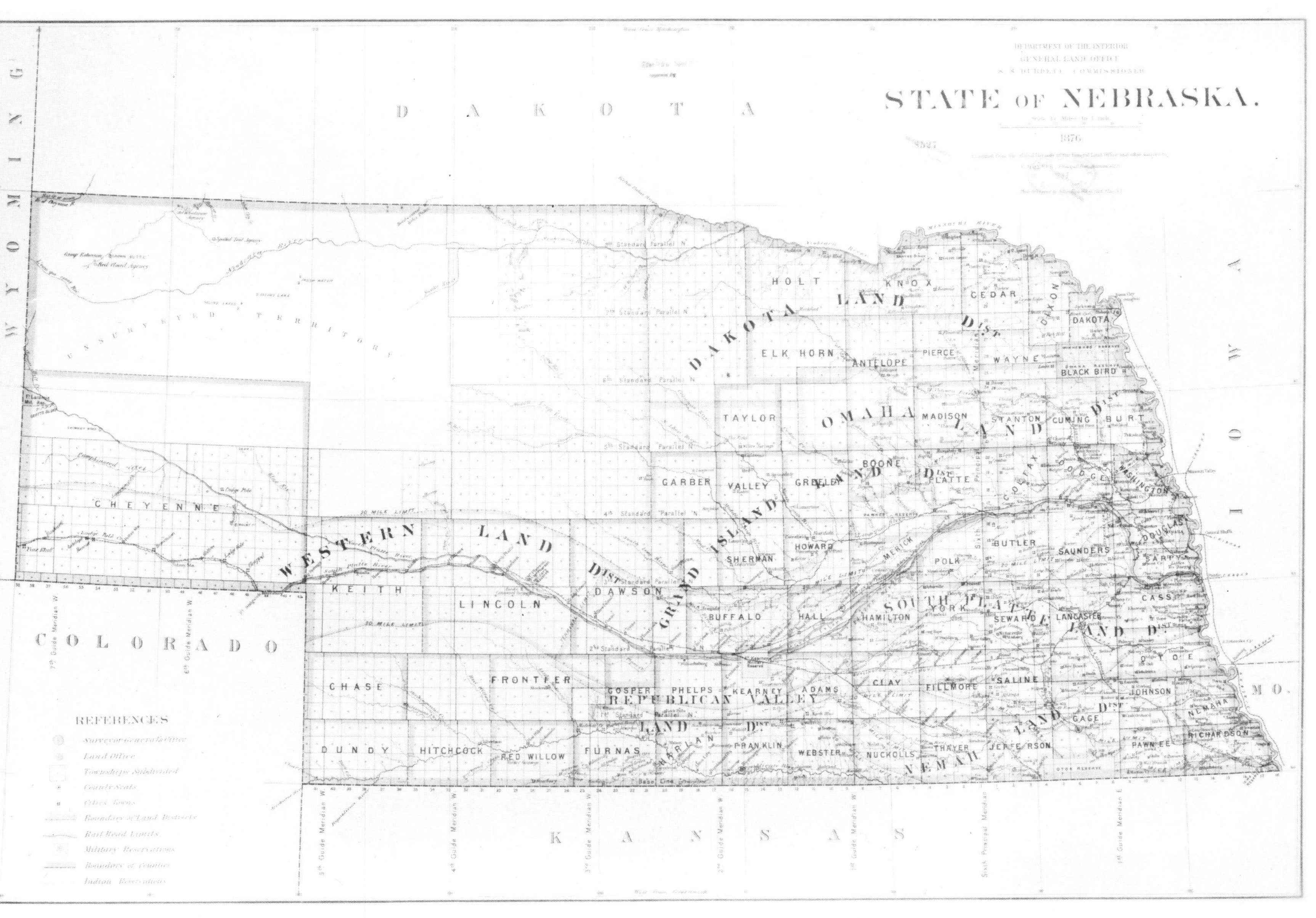

Fig. 8.13. Published General Land Office map of the state of Nebraska, compiled from official records by C. Roesser, 1876. (Geography and Map Division, Single Map Collection, Library of Congress)

progress of surveys, but this information appears to be secondary to the portrayal of counties, towns, drainage patterns, railroads and railroad land-grant limits, and Indian reservations. Separate maps were prepared for each state, but they were prepared by the "principal draughtsman" of the General Land Office rather than independently in each surveyor general's office. Six subsequent Kansas editions were published between 1879 and 1925, and five Nebraska editions between 1879 and 1922.[31]

As the example of Kansas and Nebraska shows, the General Land Office's mapping activities in the Great Plains were decentralized but resulted in a set of cartographic records that was consistent in style and type of information depicted. Supervised by a small bureaucracy in Washington, D.C., most of the cartographic work was performed by a small, transitory staff in each surveyor general's office, based on survey data gathered by contract deputy surveyors. The survey of each township was often completed as a result of two or more different contracts. While the actual surveys took only several days, the time span from the initial survey of exterior boundaries to the final subdivision of the township into sections may have extended from several months to several years. Guided by preprinted township forms and a manual of instructions based on some seventy years of experience, the draftsmen and copyists produced a series of township plats that provided the first comprehensive coverage of the two states. As the progress and composite state maps indicate, it took only twenty years to map the entire state of Kansas and three-fourths of Nebraska. Although these maps do not document a single date or even a single year, they do show a transitional phase in the region's historical geography, marking the end of the exploratory era (characterized by rudimentary exploration, fur trading, and military reconnaissance) and the beginning of permanent agricultural and urban settlement.

Notes

This essay was prepared as part of my official duties in the Library of Congress. Its contents are not subject to copyright. My thanks go to William Hawken, currently working with the National Ocean Service and formerly with the Library of Congress, for drafting fig. 8.2 and 8.3. I also acknowledge the help provided by John Dwyer, Robert Richardson, Graeme McCluggage, and Linda Cullember, National Archives staff members who assisted with the research and/or preparation of reproductions for this essay.

1. The surveying system and techniques are discussed in Lola Cazier, *Surveys and Surveyors of the Public Domain, 1785–1975* (Washington, D.C.: Government Printing Office, 1976); John G. McEntyre, *Land Survey Systems* (New York: John Wiley, 1978); Lowell O. Stewart, *Public Land Surveys: History, Instructions, Methods* (1935: reprint, Minneapolis: Myers Printing Co., 1976); and C. Albert White, *A History of the Rectangular Survey System* (Washington, D.C.: Government Printing Office, 1983). Discussions by historical geographers on the elements of the system and its effect on the landscape include Sam B. Hilliard, "An Introduction to Land Survey Systems in the Southeast," *West Georgia College Studies in the Social Sciences* 12 (June 1973): 1–15; Hildegard Binder Johnson, *Order Upon the Land: The U.S. Rectangular Land Survey and the Upper Mississippi Country* (New York: Oxford University Press, 1976); Terry G. Jordan, "Division of the Land," in *This Remarkable Continent: An Atlas of United States and Canadian Society and Culture,* eds. John F. Rooney, Wilbur Zelinsky, and Dean R. Louder (College Station: Texas A&M University Press for the Society for North American Cultural Survey, 1982), pp. 54–70; William D. Pattison, *Beginnings of the American Rectangular Land Survey System, 1784–1800* (Chicago: University of Chicago, Department of Geography, 1957); and Norman J. W. Thrower, *Original Survey and Land Subdivision: A Comparative Study of the Form and Effect of Contrasting Cadastral Surveys* (Chicago: Rand McNally, 1966).

2. The primary historical discussion of the laws pertaining to land alienation is found in Paul W. Gates, *History of Public Land Law Development* (Washington, D.C.: Government Printing Office, 1968). Gates, the preeminent historian of the public lands, has also discussed land policy in Kansas in *Fifty Million Acres: Conflicts over Kansas Land Policy, 1854–1890* (Ithaca: Cornell University Press, 1954) and "Land and Credit Problems in Underdeveloped Kansas," *Kansas Historical Quarterly* 31 (Spring 1965): 41–61. A good overview of the land alienation process in Kansas is presented in Homer E. Socolofsky, "How We Took the Land," in *Kansas: The First Century,* ed. John D. Bright (New York: Lewis Historical Publishing Co., 1956), pp. 281–306. The land alienation process in the Nebraska Sandhills is discussed by C. Barron McIntosh in "Patterns from Land Alienation Maps," *Annals of the Association of American Geographers* 66 (December 1976): 570–82; "Use and Abuse of the Timber Culture Act," *Annals of the Association of American Geographers,* 65 (September 1975): 347–62; and "Forest

Lieu Selections in the Sand Hills of Nebraska," *Annals of the Association of American Geographers* 64 (March 1974): 87–99.

The development of the administrative bureaucracy involved in the disposal of the public domain is discussed in Malcolm J. Rohrbough, *The Land Office Business: The Settlement and Administration of American Public Lands, 1784–1837* (New York: Oxford University Press, 1968). Both the surveying and land disposal records are described in Jane F. Smith, "Settlement on the Public Domain as Reflected in Federal Records: Suggested Research Approaches," in *Pattern and Process*, ed. Ralph E. Ehrenberg (Washington, D.C.: Howard University Press, 1975), pp. 290–304.

3. The General Land Office's mapping activities in two other midwestern states are mentioned in Diana J. Fox, "Iowa and Early Maps," *The Palimpsest* 59 (May–June 1978): 77–87, and LeRoy Barnett, "Milestones in Michigan Mapping," *Michigan History* 63 (November–December 1979): 29–38. The use of plats and field notes in historical research is discussed by William D. Pattison in "Use of the U. S. Public Land Survey Plats and Notes as Descriptive Sources" *Professional Geographer* 8 (January 1956): 10–14.

4. An overview of the mapping of the Great Plains is provided by John L. Allen in "Patterns of Promise: Mapping the Plains and Prairies, 1800–1860," *Great Plains Quarterly* 4 (Winter 1984): 5–28.

5. Socolofsky, "How We Took the Land," p. 288. Calhoun, who served as state surveyor in Illinois, was a friend of Abraham Lincoln, to whom he taught surveying, and Stephen Douglas, with whom he was aligned in local politics. In Kansas he was a nationally known Democrat championing the proslavery cause. See Robert W. Johannsen, "The Lecompton Constitutional Convention: An Analysis of Its Membership," *Kansas Historical Quarterly* 23 (Autumn 1957): 225–43. Sleeper, who came to Kansas from New York via Illinois, was a surveor but became involved in local politics in Kansas. See David E. Bullard, "The First State Legislature," *Kansas State Historical Society Collections*, 10 (1908): 241. The other surveyors general were either lawyers, newspaper editors, or businessmen. For example, Mark Delahay, who was trained as a lawyer, set up a newspaper when he came to Kansas. His wife was a relative of Lincoln's, and he worked actively in Lincoln's campaign before he was appointed surveyor general. See Mary E. Delahay, "Judge Mark W. Delahay," *Kansas State Historical Society Collections* 10 (1908): 638.

6. Lists of staff members appear in the annual reports of the Kansas-Nebraska surveyors general to the commissioner of the General Land Office, which were included as attachments to the latter's annual reports published in the Congressional Serial Set. The reports for 1855–65 appear in serial volumes 810, 875, 919, 974, 1023, 1078, 1117, 1157, 1185, 1220, and 1248.

7. The contract system and the problem of fraudulent surveys is discussed in Stewart, *Public Land Surveys*, pp. 59–75.

8. U.S. General Land Office, *Instructions to the Surveyors General of Public Lands of the United States for Those Surveying Districts Established in and Since the Year 1850; Containing also a Manual of Instructions to Regulate the Field Operations of Deputy Surveyors* (Washington, D.C.: A. O. P. Nicholson, 1855). Although this manual was the first general manual for all surveyors general, it was based on the 1851 instructions to the surveyor general of Oregon. A new manual was issued in 1871, but it was a copy of the 1855 edition. A revised edition was not published until 1881. The development of the instructions in this period is discussed in McEntyre, *Land Survey Systems*, pp. 94–110, while a list of instructions is found in Lane J. Bouman, "The Survey Records of the General Land Office and Where They Can Be Found Today," *Proceedings of the American Congress on Surveying and Mapping, 36th Annual Meeting* (Washington, D.C., 1976), pp. 261–71. Some of the earliest instructions are described in Thomas A. Tillman," Before Tiffin? Newfound Instructions for the Survey of the Public Lands," *Proceedings of the American Congress on Surveying and Mapping, 32nd Annual Meeting* (Washington, D.C., 1972), pp. 24–30, and "Who Wrote the Earliest Instructions?" *Our Public Lands* 24 (Spring 1974): 7–10.

9. Burt's solar compass is described in Burton H. Boyum, "The Compass That Changed Surveying," *Professional Surveyor* 2 (September–October 1982): 28–31. John Calhoun, the first surveyor general in Kansas and Nebraska, expressed reservations about the use of Burt's solar compass. See Calhoun to Hon. John Wilson, 7 October 1854, in Letters Received from Surveyors General, Kansas and Nebraska, 1854–1855, and Wilson to Calhoun, 21 October 1854, in Letters to Surveyors General, Kansas and Nebraska, 15 August 1854 to 14 February 1859, Records of the former General Land Office, Record Group 49, Scientific, Economic and Natural Resources Branch, National Archives, Washington, D.C.

10. U.S. General Land Office, *Instructions to the Surveyors General* (1855), p. 15.

11. Ibid., pp. 17–18.

12. Ibid., p. 26.

13. See note 6.

14. The surveyor general's office was subsequently moved to Leavenworth (July 1855), Wyandotte (September 1855), Lecompton (October 1856), Nebraska City (April 1858), Leavenworth (July 1861), and Lawrence (May 1869). See Socolofsky, "How We Took the Land," p. 288.

15. Surveyor General John Calhoun to Hon. John Wilson, 1 June and 31 July 1855, Letters Received from Surveyors General, Kansas and Nebraska, 1854–1855, records of former General Land Office, RG 49, National Archives.

16. This composite map was reconstructed from seven maps published or prepared by the General Land Office. The 1856 and 1866 status maps for Kansas-Nebraska and the 1876 standard published maps of Kansas and Nebraska are all found in the single-map collection in the Geography and Map Division, Library of Congress. The 1871 maps of Kansas and Nebraska are available only in manuscript form in the Cartographic and Architectural Branch, National Archives, where they are filed as Kansas 18 and Nebraska 9 in the Old Map File in the records of the former General Land Office, RG 49. The 1861 status map was not available in the Geography and Map Division's single map collection but was obtained from serial volume 1120 of the Congressional Serial Set in the Library of Congress Law Library.

17. The figures in this graph were derived from the General Land Office annual reports. In addition to the serials listed in note 6, serials 1326, 1366, 1414, 1449, 1505, 1560,

1601, 1639, and 1680 were used. From 1860 on, the yearly figure is derived from tables that list the number of acres surveyed during each fiscal year. Prior to 1860, the figures are based on the acreages reported to individual land offices. Since there are inconsistencies in these figures, the graph represents general patterns rather than precise measurements.

18. The current disposition of these records is an archivist's nightmare. For Kansas, the surveyor general's copy is now in the custody of the Kansas secretary of state, while the duplicate and triplicate copies are in the Cartographic and Architectural Branch, National Archives. The original field notes and a microfilm copy of the original plats are in the Kansas State Historical Society, Topeka. For Nebraska, the original plats and field notes are in the custody of the Nebraska state surveyor; the duplicate plats and field notes are in the custody of the Bureau of Land Management, Eastern States Office, Alexandria, Virginia; and the triplicate copy is held by the Nebraska State Historical Society.

19. Kansas Exterior Boundaries, vol. 1, p. 129 (for fifth standard parallel south), and vol. 2, p. 2 (for exterior boundaries), RG 49, Cartographic and Architectural Branch, National Archives.

20. Fig. 8.6 is reproduced from the duplicate or headquarters copy of the plat for township 25 south, range 3 west. The Kansas headquarters plats are filed among the records of the former General Land Office, RG 49, Cartographic and Architectural Branch, National Archives. Kansas plats are available on microfilm publication T1234, rolls 23–31.

21. Kansas Field Notes, vol. 48, p. 1165, RG 49, Cartographic and Architectural Branch, National Archives. Kansas field notes are available on microfilm publication T1240, rolls 117–81.

22. Ibid., p. 1177.

23. In the National Archives, the triplicate copy is known as the local office plat. The local office plats have been integrated in one series with the headquarters plats. See note 20.

24. The tract books and corresponding land entry papers are filed among the records of the former General Land Office, RG 49, General Archives Division, National Archives. The land entry papers are listed in Harry P. Yoshpe and Philip P. Brower, comps., *Preliminary Inventory of the Land-Entry Papers of the General Land Office,* Preliminary Inventory no. 22 (Washington, D.C.: National Archives, 1949). The research potential of these records is discussed in Richard S. Maxwell, *Public Land Records of the Federal Government, 1800–1950, and Their Statistical Significance,* Reference Information Paper no. 57 (Washington, D.C.: National Archives and Records Service, 1973).

25. "Kansas Half-Breed Indian Lands," p. 17, RG 49, Cartographic and Architectural Branch, National Archives.

26. "Sac and Fox in Kansas, Omaha in Nebraska, and Wyandotte Reserves in Kansas," pp. 36–38, RG 49, Cartographic and Architectural Branch, National Archives. The background of the Wyandotte reserves is discussed in Homer E. Socolofsky, "Wyandot Floats," *Kansas Historical Quarterly* 36 (Autumn 1970): 241–304.

27. All but the 1861 map are listed in the Newberry Library's *Checklist of Printed Maps of the Middle West to 1900,* vols. 12 and 13, *Kansas and Nebraska,* comp. Helen Brooks and Ann Hagedorn (Boston: G. K. Hall, 1981). Only the 1855, 1858, and 1861 maps are not in the Library of Congress's single map collection. These maps were located in the Congressional Serial Set, serials 813 and 843 (1855), 978 and 1001 (1858), and 1120 (1861).

28. These maps are filed as Kansas 4, 8, 10, 11, and 12 in the Old Map File, RG 49, Cartographic and Architectural Branch, National Archives. They are listed in Laura E. Kelsay, comp., *List of Cartographic Records in the General Land Office,* Special List no. 19 (Washington, D.C.: National Archives and Records Service, 1964), pp. 40–41.

29. Kelsay, *List of Cartographic Records,* pp. 41 and 58. These maps are filed in the Old Map File as Kansas 17, 18, and 19, and Nebraska 6, 7, 8, and 9, RG 49. Cartographic and Architectural Branch, National Archives.

30. These maps are not listed in August Wilhelm Küchler, "The Vegetation of Kansas on Maps," *Transactions of the Kansas Academy of Science* 72 (Summer 1969): 141–66. The use of General Land Office surveys in reconstructing native vegetation is discussed in Eric A. Bourdo, Jr., "A Review of the General Land Office Survey and of Its Use in Quantitative Studies of Former Forests," *Ecology* 37 (October 1956): 754–68. Examples of the use of General Land Office surveys in reconstructing native vegetation are found in Paul Bigelow Sears, "The Native Vegetation of Ohio," *Ohio Journal of Science* 25 (1925): 139–49; ibid., 26 (1926): 128–46, 213–31; and Walter A. Schroeder, *Presettlement Prairie of Missouri,* Natural History Series, no. 2 (Missouri Department of Conservation, 1981). The evaluation of woodland and prairie for settlement is discussed in Brian P. Birch, "The Environment and Settlement of the Prairie-Woodland Transition Belt—A Case Study of Edwards County, Illinois," *Southhampton Research Series in Geography* 6 (1971): 3–31; Terry G. Jordan, "Between the Forest and the Prairie," *Agricultural History* 38 (October 1964): 205–16; and Douglas R. McManis, *The Initial Evaluation and Utilization of the Illinois Prairies, 1815–1840* (Chicago: University of Chicago, Department of Geography, 1964).

31. Kelsay, *List of Cartographic Records,* pp. 154–55. These maps are also available in the single map collection of the Library of Congress.

9. MAPPING THE INTERIOR PLAINS OF RUPERT'S LAND BY THE HUDSON'S BAY COMPANY TO 1870

Richard I. Ruggles

BY royal charter, Charles II in 1670 granted to a small coterie of London entrepreneurs, united in a joint stock company, exclusive trading privileges in a vast territory of then unknown dimensions. The group was the "Company of Adventurers of England tradeing into Hudson's Bay," the Hudson's Bay Company. The territory was Rupert's Land, named for Prince Rupert, cousin of the monarch, who graciously consented to act as the first governor of the company. By charter, Rupert's Land included "all the Landes Countryes and Territoryes upon the Coastes and Confynes of the Seas" lying within Hudson Strait, that is, the area drained by waters flowing into Hudson and James bays and Hudson Strait.

The new enterprise erected trading factories at the mouths of several of the large rivers, Rupert, Moose, Albany, and Nelson-Hayes, and established a trading system based on the annual journeying of Indian customers to these export posts. The executive committee of Hudson's Bay Company urged employees to accompany Indian groups inland from the factories at the bay shore to winter among the tribes and to encourage them at river break-up time to return to the factories with their furs and other trade items. Not only would this policy allow the company winterers to recruit customers, but it would also develop a cadre of experienced travelers. For many years, no one accepted this challenge, except for Henry Kelsey—a young scamp to some, a young hero to others—who undertook a lone journey onto the Saskatchewan plains between 1690 and 1692.

Kelsey, who eventually became a senior trader in the company, operating mainly out of the York and Churchill factories, was certainly the company's first winterer and the first European to journey onto the northern plains of North America. Regrettably, he did not draw a map depicting his route or the extent of his penetration of the plains. Therefore, there is no cartographic memorial to the commencement of the Hudson's Bay Company's long involvement with the Canadian western interior. After Kelsey's voyage, the vast region was not intruded upon again by British traders for over sixty years. For some thirty years, the company was deeply embroiled in defending the Hudson and James Bay littoral against incursion by the French, whose forces occupied several of the chief Hudson's Bay Company factories during Anglo-French wars. The company's explorers struggled to extend their knowledge of the northwest shore of Hudson Bay, investigating inlets for a possible opening to a Northwest Passage. Success in this venture could give the company great advantage in the extension of trade. Forays onto the plains by wintering company employees were not resumed until 1754, when Anthony Henday reached nearly to the foothills of the Rocky Mountains. Perennial occupation of the Great Plains, with the erection of trading houses and the posting of complements of officers and servants, was not initiated until 1774.

Hudson's Bay Company mapping of the plains began in 1755, when Anthony Henday arrived back at York Fort with a party of Plains Indians. Henday had made a sketch of his river and overland track to within sight of the Rockies, and he turned it over to his immediate superior. The company terminated its cartographic endeavors in 1870, when it surrendered its territorial rights to Rupert's Land to the British crown.

Rupert's Land

The region concerned in this analysis is that of the Great Plains lying essentially within Rupert's Land. More specifically, it is the drainage basin of the Nelson River, comprising in the plains the Saskatchewan and the Red-Assiniboine river networks; Lakes Winnipeg, Manitoba, and Winnipegosis and associated lesser lakes; and a small part of the upper Churchill River basin, especially the Beaver River valley. A few maps, however, depicted territory beyond Rupert's Land: to the north into the Mackenzie watershed, and to the south into the Missouri. The vast domain is almost eight hundred miles wide along the international border but tapers toward the Mackenzie basin. From the higher, more dissected tracts to the southwest and west, the plains slope north to the Arctic, northeast to the Laurentian Shield, and east to the extensive flat lowlands of the large lakes of Manitoba. Although many of the earlier commentators were overawed by the immensity and levelness of the vistas that confronted them, the plains are more commonly undulating to rolling in form, rising into hills. The few hundreds of feet of relief occasioned by river valley wall or hill front were sufficiently salient to attract the traveler's attention, especially because they were often more wooded than the level ground.

Early occupancy of the Rupert's Land plains was mainly in the crescent of boreal-mixed forest and aspen grove parkland that frames the grassland core of the plains. This more wooded, transitional zone was the paramount habitat of the beaver and other main peltry of the fur trade. Even though some earlier explorers and traders penetrated the open plains, more frequent passage into the drier grasslands did not prevail until the nineteenth century. Cumberland, the company's first post on the forested plains, was erected in 1774, and the first establishment in the park belt, Hudson House, followed in 1780, but it was not until 1800 that the first grasslands fort, Chesterfield House, was built, far out at the junction of the South Saskatchewan with the Red Deer River.

Maps as Business Records

The Hudson's Bay Company used maps, charts, and plans for business purposes from the inception of its activities in 1670. Of the company's total archival holdings, which would have amounted to about 4,800 items if all were still available, the most significant for cartography are approximately 800 manuscript maps and charts prepared from 1670 to 1870.[1] Two-thirds are still extant in the collection; a one-third attrition has occurred.[2] In addition, there are some 557 segmental sketches of certain waterways in the journals of two of the company cartographers, Peter Fidler and George Taylor, Jr. In all, 160 men have been identified as having been involved, among them about 50 Indian and Inuit persons.[3]

The mapping of the plains represented only a small segment of the total cartographic effort of the company, whose maps flowed into Hudson's Bay House in London from sea to sea, and from northern California to the Arctic archipelago. From 1755 to 1870 about ninety maps were drafted that delineated some portion or all of the interior plains—about 11 percent of the manuscript total. Of these, one-quarter have not survived in the archives of the company. During this period, several hundred segmental sketches of western waterways were produced, involving at least thirty-three persons, among whom were seven Plains Indians named as primary informants and providers of original sketches. Few of the exploration maps concerned only the plains area. Most were of the forest and park belt, and extend across into the Laurentian Shield. Only a small number were focused on the grasslands and the plains alone.

Visitors to Hudson's Bay House in London during these years could not have viewed a busy map-drafting office nor discussed maps with a chief company cartographer, for neither of these existed. Nevertheless, the company was map-conscious. From the beginning of operations, officials proclaimed the necessity of encouraging employees to travel inland and of hiring individuals who had the ability to observe the record. Such persons were even more valuable if they could use instruments to measure distance and direction and to determine their astronomical location, especially if they could also sketch, or compile a map and draft it. The company also hired young apprentices who had been trained in mathematical classes at Christ's Hospital and Grey Coat Hospital in London, and who, in their marine service, or as explorer-surveyors, such as David Thompson, could prepare maps for the use of the factors and the executive committee in England.[4]

In 1778, the company hired a full-time inland surveyor, Philip Turnor, but once his major tasks were completed he was turned increasingly to trading duties. The head office did not establish and maintain in the field a surveying-drafting section. If they had, and if they had appointed outstanding young men to such posts, undoubtedly the company could have kept Thompson in its employ. Instead, the geographers had to subordinate their mapmaking interests to the regular duties of clerks and traders. Thompson, probably one of the greatest practical geographers of all time, left the Hudson's Bay Company for the North West Company in 1797 and stayed with the rival group until he retired to Montreal in 1812. The larger number of company personnel involved in cartography had not originally come to North America to make measurements, to engage in geographical investigations, or to map; nevertheless, they became entailed in such pursuits in the course of their various careers with the Hudson's Bay Company.

The executive committee of the company persistently requested that sketches, charts, drafts, plans, and maps be sent by the factors back to the main office in London for the use of company officials. They were examined there when the packets of official correspondence from the chief factors in America were opened and the letters, journals, and reports were read and discussed at executive committee meetings. Usually maps were commented upon for their usefulness or their inadequacies. The company requested maps for a variety of purposes. On the east and west coasts, they needed charts for sea navigation and coastal and rivermouth charts. Inland, the main concern was to be able to visualize the details of river and lake networks; their interconnections; their relationship to major terrain features; hazards due to waterfalls and rapids; the numbers, locations, and difficulties of portages; and the locations of company posts and those of their competitors.

The major role of maps was to provide locations and other spatial information for company officers who were developing trading strategy, transport routing, and an understanding of their entire territory. Officials requested post layouts, the characteristics of fur post locales, and the pattern of land use on their properties. The officers also expected district masters to provide maps of their regions, but not all of them were adept enough at mapping to do this, and some ignored the directive. Some maps were made for special purposes—for example, those that indicated property holdings, such as in the Red River settlement, or a map that delinated the route of a proposed telegraph line across the plains.

Exploration and mapping under the conditions prevailing on the plains in the eighteenth and nineteenth centuries could not be portrayed as a lighthearted occupation. Hunger, or near-starvation, was a present specter. Miserable weather conditions made traveling and camping disagreeable and often dangerous. There was the risk of being overtaken by summer fires, both in the forest and out on the grasslands. Mosquitoes and black flies in the still forests, away from clearing breezes, were the bane of existence.

The specific problems and exigencies of field observation were manifold. Instruments were lost or left behind, were broken, could not be properly calibrated, or had not reached the explorer in time because of the great distances from London or from the Bayside port. Delicate instruments were difficult to transport, and were especially at risk during loading or unloading from canoes and boats; while packed on horses, dogs, or sleds; and during transport across rocky portages. Dangerous to life, as well as to instruments, was the overturning of canoes in storms, stoving in on rocks, and spilling in white water. Such mishaps could lose an observer's vital sextant or compass, or wash away his sketch or records. Low temperatures affected observers on the plains in many ways and caused special problems with the use of instruments. Holding metal instruments and putting them to the eye could incur pain and injury at extreme temperatures. Quite often, intense cold caused the liquid in bulbs and tubes to expand and burst the glass. Grey, overcast skies, dense clouds, or the "smoky exhalations" of grass or forest fires often obscured the view of sun, moon, or stars.

Slow transportation led to serious delays in field observation. The waiting period between the breakage or loss of a vital instrument and its replacement could extend to two years. Ships normally left London in late May or early June, arrived at Bay ports in late August or early September, and turned around as quickly as possible to avoid being caught in the forming ice cover. Equipment losses had to be reported to York or another factory to await the early autumn ship, which would return the following spring. Even the time that elapsed between the executive committee's request for a map and its receipt, without any other form of delay or evasion, was lengthy.

As could be expected, the larger share of Hudson's Bay Company maps of the interior plains—two-thirds—were exploratory sketches of river and lake networks. Most were simple in drafting style, with black india ink line work, or with grey ink wash to outline waterways. Little color was used; one or two colored inks or color washes might have been added. These maps were rarely ornamental, and only a few cartographers used a decorative cartouche. Legends were rare, since the map-makers made little use of symbols. The most significant elements represented with symbols beyond the hydrography were fur trading posts, locations of waterfalls and rapids, portages, and perhaps trails or certain hill-lands. The depiction of terrain was of the simplest form.

Winterers As Mapmakers

The history of the Hudson's Bay Company's mapping of the interior plains separates tidily into two main periods. The exploratory period, from 1755 to 1815, encompasses almost all of the maps derived from primary exploration, sketches based on native people's knowledge and concepts, and composite regional maps illustrating the growth of geographical information. The second period, from 1815 to 1870, was one of greater diversity of subjects. Cadastral maps (recording property boundaries and other details of settlement), district maps, and maps made for special purposes were concentrated in this span of years. During the same period the company received many maps in correspondence with other businesses.

The pattern of geographical discovery of the West in 1690, when the Hudson's Bay Company entered the northern plains, was generally unchanged from that of twenty years before. Coastal knowledge still dominated, for none of the company's servants ever traveled more than a few miles up the rivers, usually in search of firewood or game. The extent of involvement of the wintering servants from 1754 to 1774 is difficult to describe since most were illiterate, and few left accounts of their travels. The only map produced by a winterer has not survived. Six different men appear to have traveled inland between 1754 and 1763, on nine different journeys. Living with Indian groups along the line of the Saskatchewan and North Saskatchewan rivers, these company explorers essentially defined the arc of mixed forest, park belt, and northern grasslands, in concert with the French, who were operating mainly from the Manitoba plains. The pace increased from 1763 to 1774, the period of dominant wintering, when twelve servants were involved.

In 1755, when Henday gave his sketch to James Isham, the York factor, Isham made a fair copy and sent it to London.[5] Isham probably threw away the original, for it must have been very rough. In any case, neither has survived; nor did Isham's 1757 rendition of Smith and Waggoner's wintering itinerary northwest of Lake Winnipeg and in the upper Assiniboine valley.[6] Furthermore, the draft of William Tomison's 1767–68 trip from Severn Fort to the Lake Winnipeg region, drawn by William Falconer, the sloop master at Severn, is no longer available in the collection.[7] To compound the loss, Falconer's 1774 map of the "interior part of the Country" has also disappeared.[8]

Only two maps of this early period relating to the inland winterers are extant (fig. 9.1). Both maps, made by Andrew Graham at York in 1772 and 1774, are successive versions of some of the full data available.[9] The most significant inclusions are the routes followed by Tomison from Severn to the grassland plains, including the mention for the first time on an English map of the dry grasslands, "Barren Ground. Buffalo plenty in winter," and the earliest map of Indian tribal regions in the West, inserting

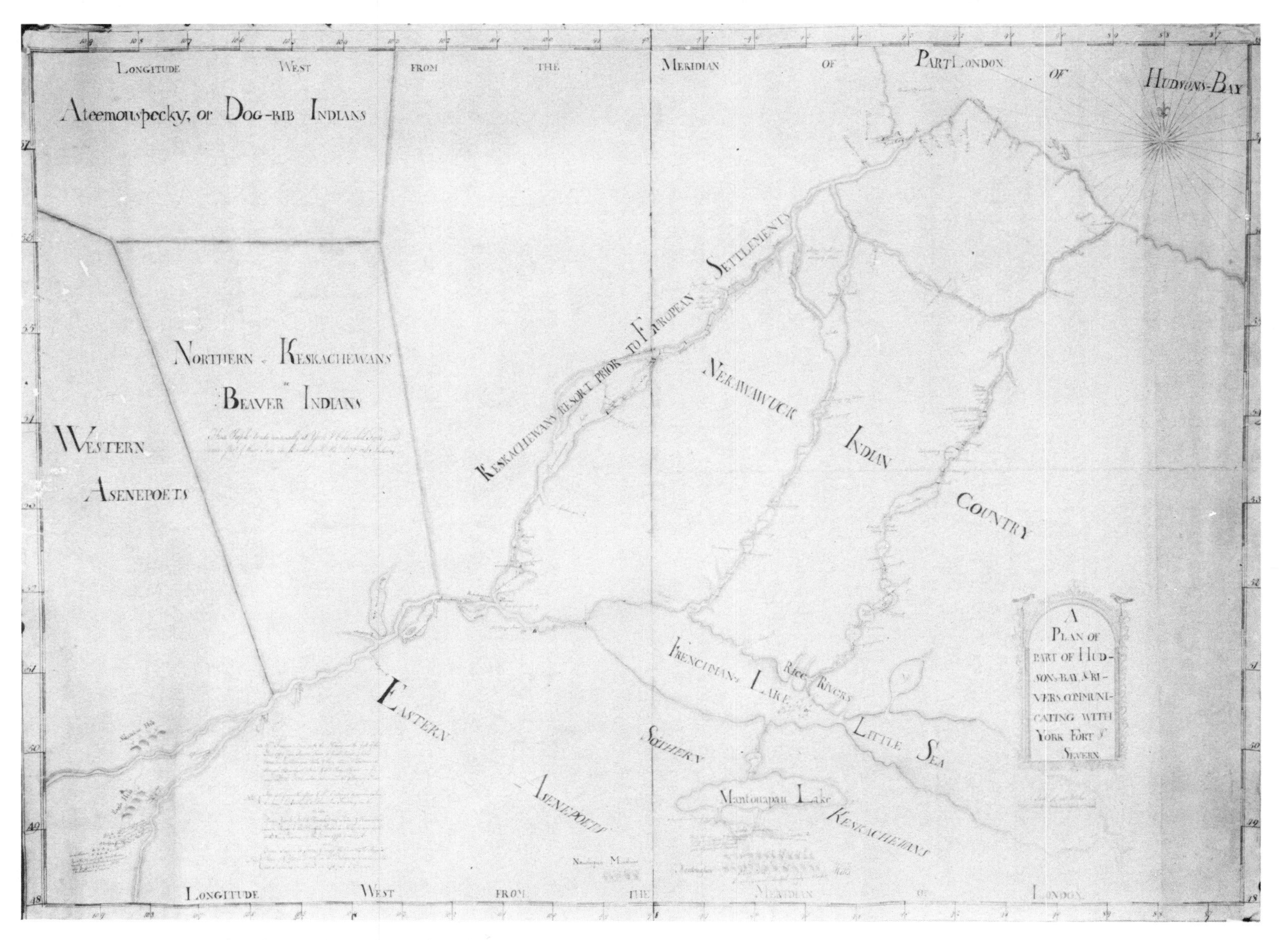

Fig. 9.1. [Andrew Graham], "A Plan of Part of Hudson's Bay and Rivers, Communicating with York Fort and Severn, 1774." (Hudson's Bay Company Archives, Provincial Archives of Manitoba, Winnipeg, G. 2/17; hereafter cited as HBCA, PAM)

the boundaries in a rather geometric fashion. The geographic detail is selective and covers only Cocking's journey near the Saskatchewan River forks. Graham's orientation, scale, and shape are very much off for Lake Winnipeg. He enjoyed preparing his personal cartouches and used color coding.

The Saskatchewan and the Mackenzie

In the next twenty years, the Hudson's Bay Company was forced to change its tactics on the plains, since the wintering program had proved to be absolutely inadequate for the task that confronted the company. Independent Canadian traders had occupied many of the strategic trading sites on the plains. With the erection of Cumberland House in 1774, the Hudson's Bay Company became inland traders also, and began the fascinating but cutthroat chess game of post construction. The pattern that developed was peripheral, that is, along the northern and eastern fringes of the grasslands, in mixed forest and park belt, from which traders moved west and south onto the open grassland, and farther north and west on the forested plains. For the first time, the company also passed out of Rupert's Land into the Mackenzie basin, along the eastern edge of the interior lowlands.

The cartographic record for these two decades is more extensive than for the wintering era. Yet, unfortunately, about half of the eighteen maps relating to the region are no longer extant. Nine of the maps record the accumulating knowledge and more precise locating of the topographic features of the Nelson-Saskatchewan line and northern Manitoba lakes, but only one of these maps depicts the plains region alone. One map is a witness to the founding of Cumberland House, drawn by the expedition leader, Samuel Hearne, on his return to York in 1775.[10] It is a classic map of its genre, similar in style to many drafted later, although less precise in its locational framework than the map of 1779 made by Philip Turnor (fig. 9.2). Turnor had been hired full-time to take careful observations of latitudes and longitudes on the established trade routes in the west and north.[11] This map, the first in a series by Turnor, was the result of his journey to the plains from York and up the Saskatchewan River as far as the new cabin that was opened that winter above the forks of the Saskatchewan.

David Thompson came onto the mapping scene during this period. A Grey Coat Hospital apprentice being groomed by the company to join Turnor's expedition into the Athabasca country, Thompson had the great misfortune to crush his leg and to be replaced by Peter Fidler. Both Thompson and Fidler had been tutored in surveying methods and computation at Cumberland House by Turnor. When his leg had nearly mended, Thompson used his revived training to make observations and then to draft his first map, showing the route from Cumberland, through Lake Winnipeg, to York Fort.[12] This map is not in the collection; however, Thompson's developing style may be observed in a large compilation of 1794, which extended his mapping on the Nelson River and particularly up the Saskatchewan to Buckingham House in the forest-parkland transition zone.[13]

Peter Fidler: Company Cartographer

Turnor's teaching was not misplaced in Peter Fidler, who became the preeminent explorer-surveyor-cartographer of the Hudson's Bay Company during the next several decades. On his first posting in 1790 to 1792, across Methye Portage and on to Great Slave Lake, Fidler began to develop his careful observational and mapping methods, making continuous and sequential annotated sketches with full journal notes, which distinguished him from other employees engaged in mapping. Fidler drafted detailed, larger-scale maps of the entire journey, while Turnor produced a sixteen-sheet map of the shield portion of the track. The series is not now in the company archives, but smaller-scale renditions appeared later.

After his Athabasca tour, Fidler was sent to Buckingham House, and almost immediately was assigned to wander southwest through the parklands and grasslands to the upper Bow River. Remaining there until early winter, he drifted back northeast across the Red Deer River to the fort. Later, at York Factory, he put together all of his sketches and notes into a map of a "Journey to the Stony Mountains." In addition, for one of the factors, he compiled a map of the normally used waterways from York Fort up to Edmonton House. Neither of these useful studies is in the collection today.

Three regional maps (1791, 1792, and 1794) re-

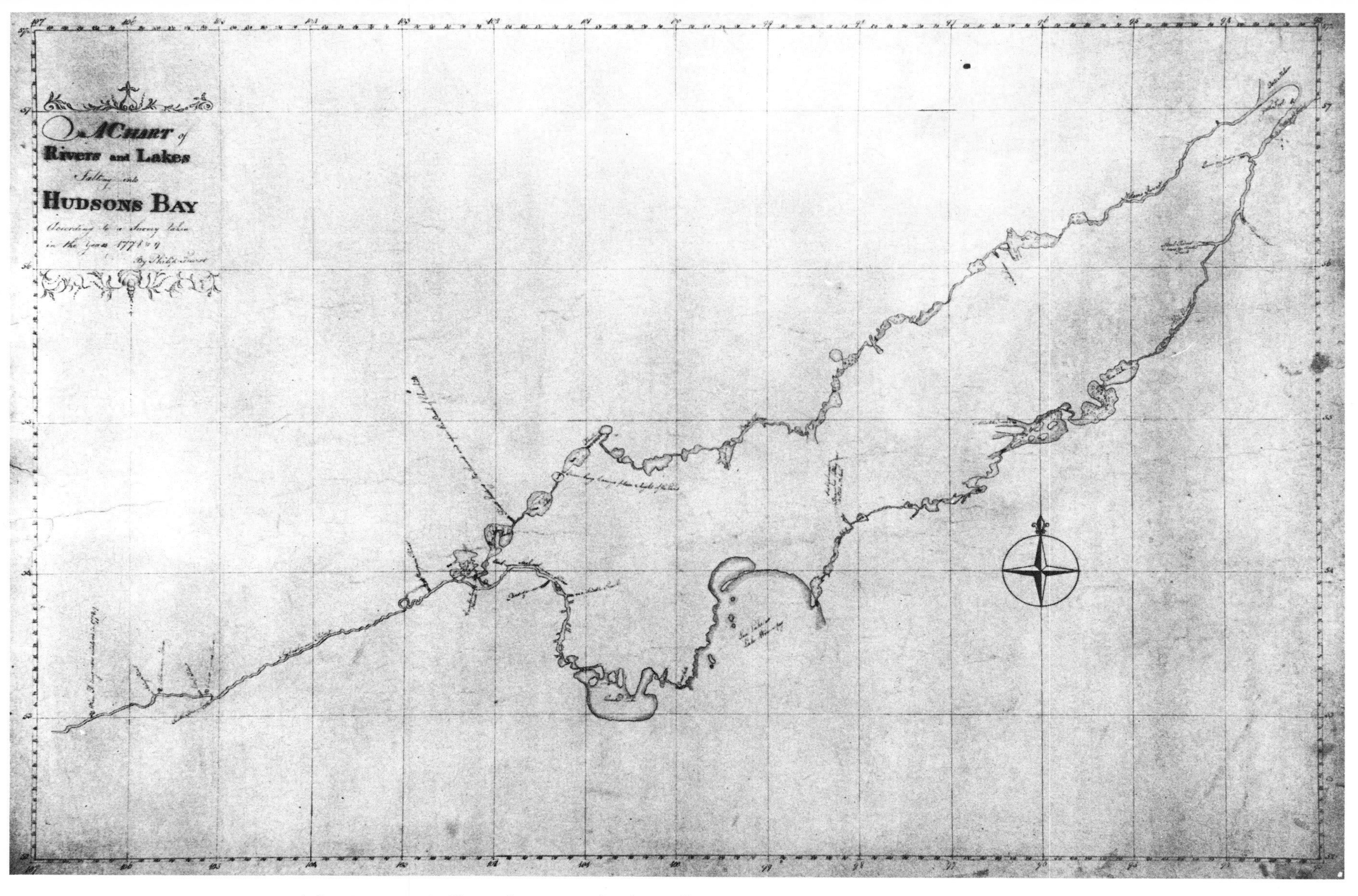

Fig. 9.2. Philip Turnor, "A Chart of Rivers and Lakes Falling into Hudsons Bay, 1779." (HBCA, PAM, G. 1/21)

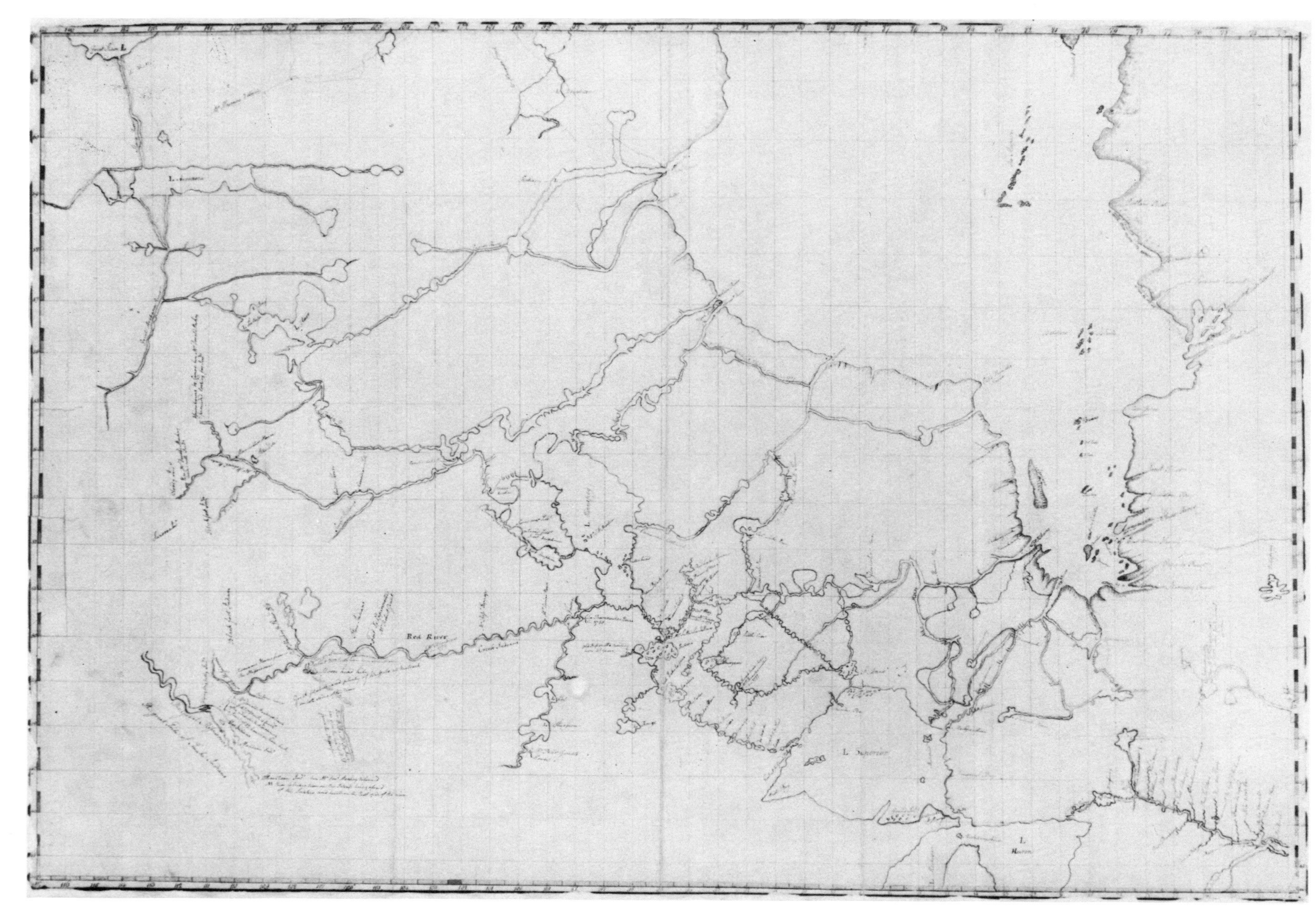

Fig. 9.3. Edward Jarvis and Donald McKay, "A Map of Hudsons Bay and interior Westerly particularly above Albany, 1791." (HBCA, PAM, G. 1/13)

capitulate the company's knowledge of the Great Plains and illustrate the effect of its developing trading strategy. The first, undoubtedly drawn in 1791 by Donald McKay and Edward Jarvis of Albany, indicates the areas of experience of McKay, formerly a North West Company trader, who came over to the Hudson's Bay Company with a plan to have the company cut west across the main track of the Canadian traders from the Albany system (fig. 9.3).[14] His map lays out the strategy, which was to establish company posts from the Winnipeg River into the Red and Assiniboine country, a direct thrust west from Albany River headwaters. The Assiniboine River, labeled the Red, stretches west to within a short overland walk to the Missouri.

None of McKay's cartographic forms appear on Turnor's 1792 map.[15] It shows the detail of the Athabasca journey, which extended along the east side of the plains, and displays (most intriguingly, in red ink) configurations of the Beaver and Peace river areas, obtained from Indians and various Canadian traders. A compendium of company activity and cartographic production is memorialized on Philip Turnor's map of 1794.[16] Drafted in Britain, this magnum opus is a large work, six by nine feet in dimension. Prepared at the end of Turnor's career with the company, the map reflects, especially for the plains region, the significance of the mapping by Fidler. It also demonstrates that few of the inland employees located at the Saskatchewan posts had been encouraged sufficiently to observe and record their wide-ranging travels.

The fact that the plains were still largely terra incognita is quite apparent also on the historic map of 1795 prepared by Aaron Arrowsmith. This young cartographer had been chosen by the company to be given access to all its maps and travel records when the company changed its policy from secrecy to a more public disclosure of its activities. Arrowsmith and his cartographic descendants became unofficial cartographers of the Hudson's Bay Company until the demise of the business. Turnor's maps of 1792 and 1794, and the McKay-Jarvis map of 1791, were the crucial elements of Arrowsmith's plains configurations.

During the last twenty years of this major exploratory period, from 1795 to 1815, Peter Fidler was the dominant cartographer of the interior plains, either reworking his many exploratory sketches, compiling them into synthetic maps, or transcribing the sketches provided by Indian customers or fellow traders into ink versions in his private journals. In all he drew some thirty-two maps at various scales, and made at least eighty segmental sketches. These maps included the lakes and rivers of the Beaver River area west to Lesser Slave Lake—the region that Turnor had first depicted and Arrowsmith copied—based on rough detail from Fidler's informants.

Fidler was suitably chosen by the company to expand its collecting of pemmican and furs far south into the heart of the dry plains at the turn of the century. No company employee had followed the South Saskatchewan far upstream from the forks, and except for the upper courses of its tributaries, the river was unmapped. From 1800 to 1802 Fidler operated Chesterfield House, which he had had built, and there, far out in the plains in the midst of a congeries of Indian tribes who visited by the hundreds, he fashioned his concepts of the western high plains and mountain ranges. Fidler perfected his technique of questioning the Indians, inducing some of them to sketch out their geographical understanding of their territories, and inserting Indian place-names.

Fidler produced nine maps based on Indian data and one large composite map of the western plains. The Ackomokki, Akkoweeak, and Kioocus maps have become well known. The Ackomokki map of 1801, oriented to the west, depicts the vast extent of the Rocky Mountains south into the great fan of Missouri River tributaries.[17] A second version from 1802 gives greater attention to the South Saskatchewan and its tributaries.[18] Akkoweeak's map of 1802, similarly, is more concerned with the Canadian plains rivers.[19] The Kioocus contribution, though not as comprehensive, provides a more accurate picture of the vegetative and terrain pattern of this immense land.[20] The map depicts the "woods edge" somewhere in the lower course of the Battle River, reaching southwest to the foothills and stretching unbroken to clothe the bow of the Absaroka Range and Big Horn Mountains. Across the plains, Kioocus traced out various Indian trails, with distances depicted by a symbol for each night's sleep (fig. 9.4).

Fidler's compilation was a six-sheet series extend-

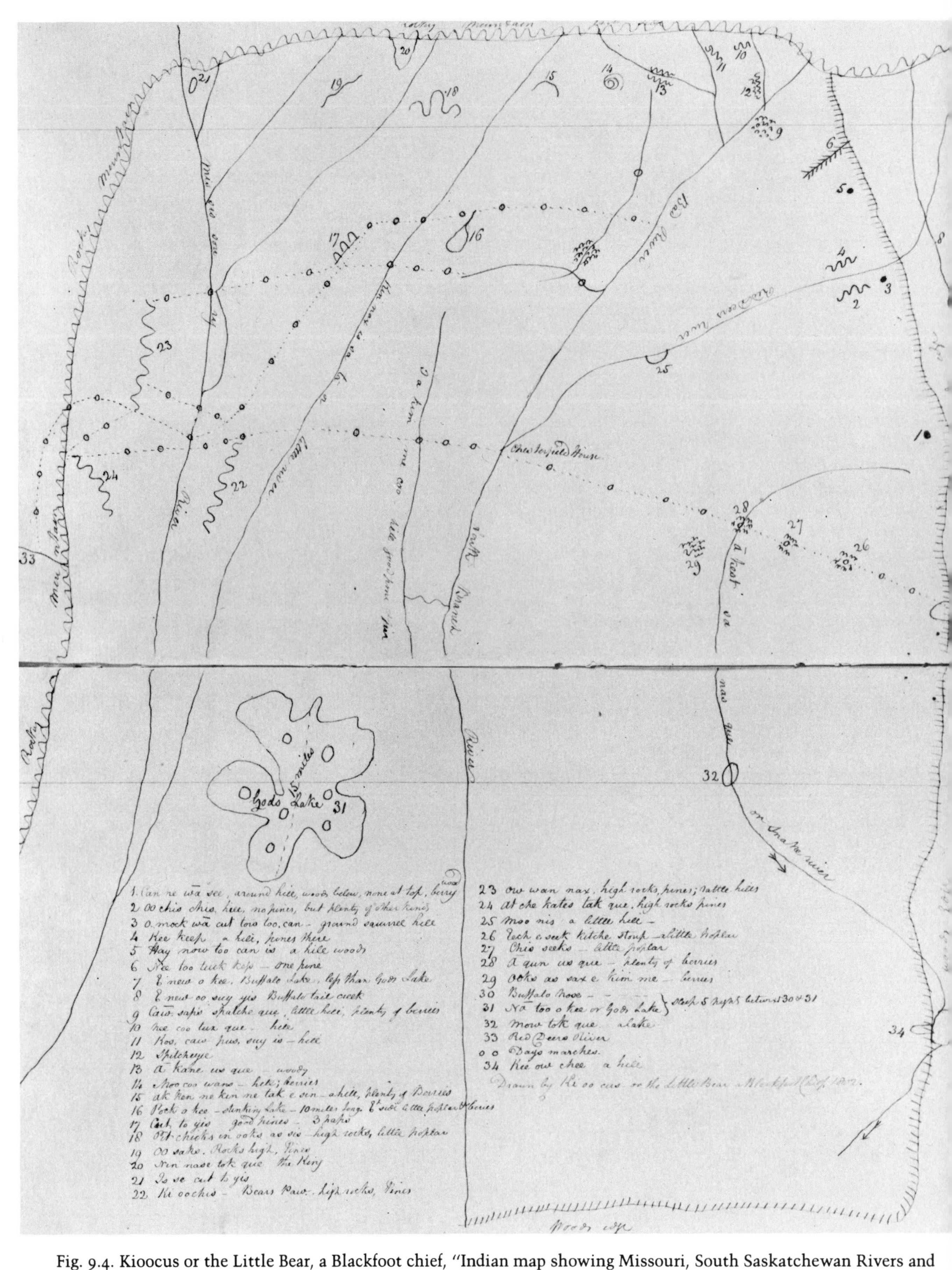

Fig. 9.4. Kioocus or the Little Bear, a Blackfoot chief, "Indian map showing Missouri, South Saskatchewan Rivers and Northwards, drawn by Peter Fidler in his journal, 1802." (HBCA, PAM, E. 3/2, fols. 104d–105)

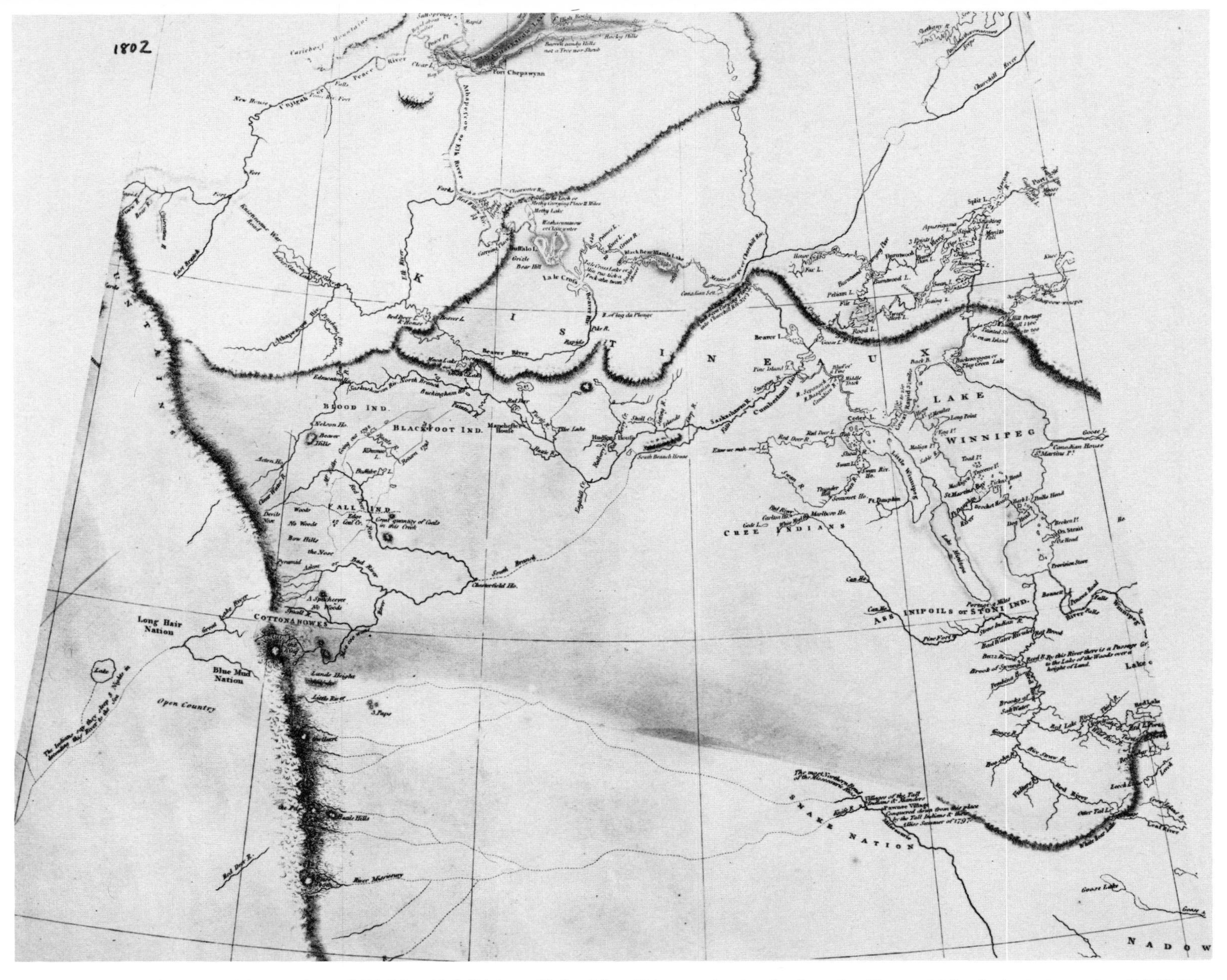

Fig. 9.5. Aaron Arrowsmith, portion of "A Map Exhibiting all the New Discoveries in the Interior Parts of North America, 1802." (HBCA, PAM, G. 3/672)

ing from Buckingham House to the Bow River, with an appended one-and-one-half sheets extending south into the Missouri country, based on information from his Blackfoot sources.[21] The map reached London and was turned over to the cartographer Aaron Arrowsmith. The executive committee also notified Alexander Dalrymple and Sir Joseph Banks of the Royal Society because "these Discoveries should be of sufficient Importance to attract . . . [their] notice."[22] The original series was apparently not returned to the company's office—a tragic loss for the history of western cartography. Fortunately, the details from Fidler's lost map appeared on Arrowsmith's map of 1802 and on several later editions of these famous maps based on Indian sources, because they were the only up-to-date printed representations of this area (fig. 9.5).

Fidler's cartographic renditions of interior plains locales are diverse in area and in form. They range from the 1796 map of the Swan River and upper Assiniboine, based on information gathered during his posting there, through various transcriptions of Indian and traders' sketches, to the detailed contour of Lake Winnipeg, drawn from his measurements in the field and his own sequential drawings.[23] Careful study of a page of Fidler's sketches, based on some sketches by a Canadian trader who conversed with Fidler in the Swan River area in 1808, shows the care he took before he himself mapped the Assiniboine and Red rivers.[24]

Another mapmaker who made significant contributions to Hudson's Bay Company cartography during the early exploration period was Joseph Howse. The first company man to cross the Rocky Mountains and the first to open company trade in the Columbia region, Howse produced a map of the South Saskatchewan River in 1809 and another, in 1812, of both sides of the Rockies from the Athabasca River to the Missouri.[25]

Mapping Tasks, 1815 to 1870

The second major period of Hudson's Bay Company mapping, from 1815 to 1870, witnessed not only a reduction in the amount that was undertaken but also a change in the company's situation. For one thing, topographical mapping was greatly reduced. The elimination of the most difficult portages in order to expedite boat transport was the main concern of waterway mapping. For this purpose, William Kempt and George Taylor, Jr., prepared maps of the York Fort to Red River route. Taylor also mapped the Saskatchewan-Athabasca network through Fort Edmonton to aid the main supply brigades. At the request of Chief Factor McTavish of York, probably between 1824 and 1827, Taylor used Fidler's maps and records to piece together a "Fidler" map of the West, but it never reached the head office in London. The map is now located in the National Map Collection, Ottawa.

Among the thirty or so maps attributable to this second period are five district maps. At the time of the reorganization of the company into departments and districts, the district masters were instructed to write detailed reports on their regions, with illustrative maps. Some factors found the cartography too difficult to undertake, and a number just ignored the directive. The maps that came in varied in quality and usefulness. Fidler provided two, one of the Red River (fig. 9.6), and one of the Manitoba district, his final assignment.[26] Robert Kennedy sent a map from the Lesser Slave Lake district, reaching from the Beaver River to the Peace River.[27] The scale is distorted somewhat, but he provided enough detail to be useful to the chief factors and executive committee members. James Bird's lack of drafting skill is apparent in his map of the Fort Carlton district at the forks of the Saskatchewan, but he provided valuable topographic detail, including rivers, creeks, lakes, sloughs, hills, and grassland plains.[28]

The largest corpus of maps of this period are cadastral, showing details related to settlement. Eight of the twelve cadastral maps depict the growth of the surveyed plan of the Selkirk colony at Red River. Fidler was the first to aid the settlers in laying out some of their lot lines, and he was followed by Kempt and Taylor from 1823 to 1838. Taylor was seconded to the Selkirk colony to develop a census, to act as surveyor, and to prepare an official plan for the settlement. He brought out an earlier version, but from 1836 to 1838, Taylor laid out the lines on the ground and drew a full plan of the village. The quality of Taylor's work in drafting the plan was unmatched by any other company employee in the interior during this period (fig. 9.7).[29]

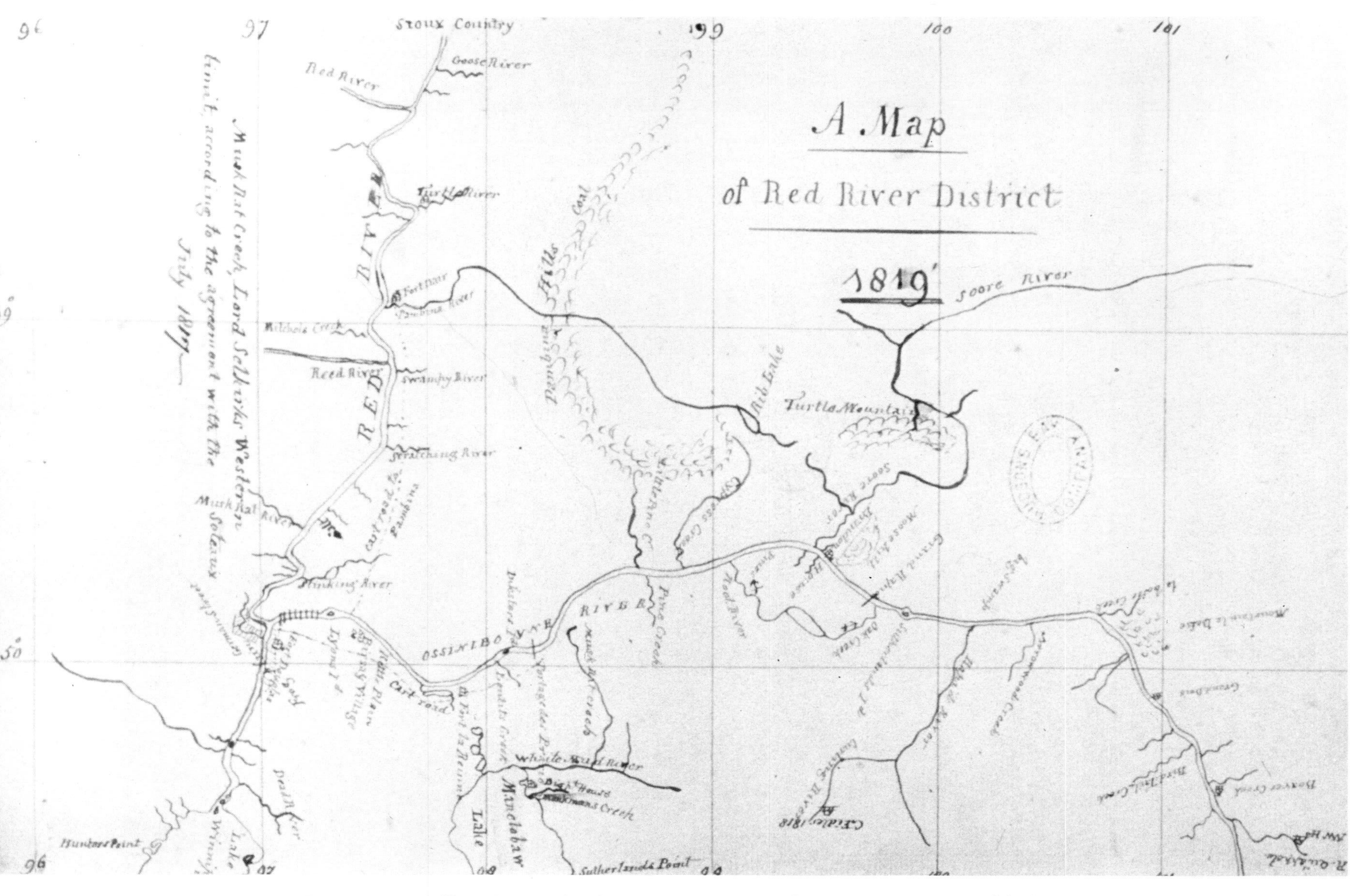

Fig. 9.6. Peter Fidler, "A Map of Red River District, 1819." (HBCA, PAM, B22/e/1, fol. 1)

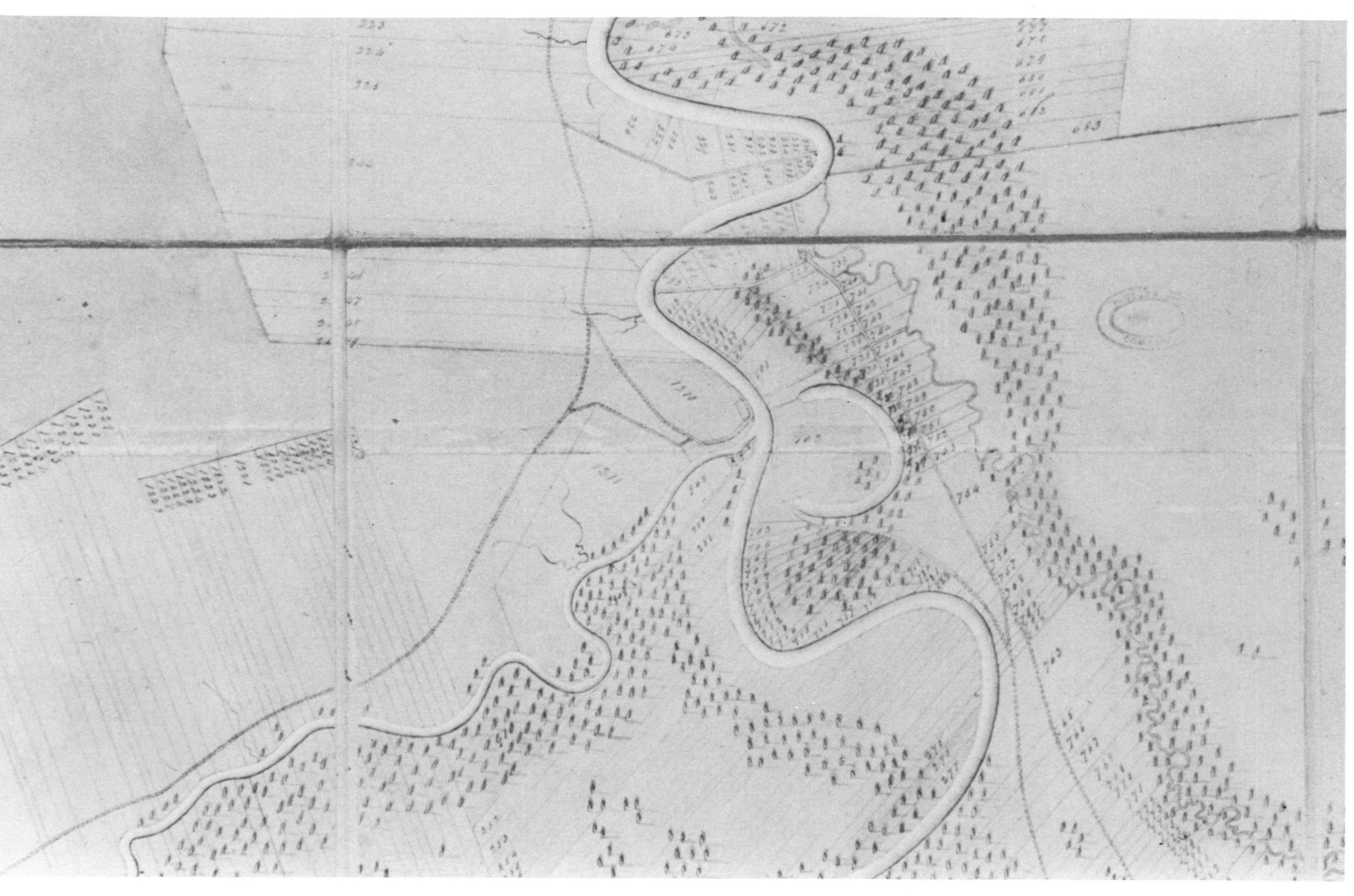

Fig. 9.7. George Taylor, Jr., "Part of Plan of Red River Colony Surveyed in 1836, 1836–1838." The Red River runs north and the Assiniboine tributary flows in from the west. (HBCA, PAM, E.6/14)

In the collection are a suite of maps and plans drawn by Mervin Vavasour. Along with Henry Warre, a fellow British army officer at Quebec, Vavasour was sent surreptitiously and incognito across the plains to the Columbia in 1845 and 1846. This was a period of border and territorial stress, and the company was anxious to have a professional assessment of the situation and advice on the defensive possibilities of specific posts and sites, in case of American incursion. Acting as wealthy young British travelers, sportsmen, and sketchers, Vavasour and Warre provided maps and sketches on the plains and across the mountains, and drew plans of Fort Ellice, Fort Carlton, and Fort Edmonton, showing military capabilities. The company was given a copy of their report and a set of all their maps and plans.

Another group of maps prepared on behalf of the Hudson's Bay Company in 1864 were five dealing with the examination of a route for a proposed telegraph line, to be financed by the company and built across the plains from Red River to the Yellowhead Pass, and eventually on to Victoria. The company chose Dr. John Rae, the distinguished Arctic explorer and a former chief factor, to carry out the investigation, along with a young assistant, A. W. Schwieger, and a small crew. They traversed the country estimating timber resources, the amount of filling and cutting required, the difficulties of the terrain, and the distances between segments of the proposed course. Rae worked on four field maps of sections of the route, including one of the entire proposal. Not all four have remained in the collection, but Schwieger's final map of the route across the plains is available.[30]

For eight decades after 1755, the Hudson's Bay Company was particularly concerned with the cartographic delineation of the Precambrian shield and the northern plains west of Hudson Bay. After this, it's mapping efforts were concentrated in the cordillera, on the Pacific coast, in the Arctic, and in interior Quebec. The earlier phase witnessed the work of many individuals and of some small groups, searching for easier and more efficient routes, seeking out customers, and engaging in trade with them. Their maps illuminated the major lineaments of this immense region, the plains of Rupert's Land. In concert, these company travelers provided the cartographic foundation for the work of scientific expeditions that assessed the natural environment later in the nineteenth century, and for public surveyors who parceled out the land. In 1870 the company relinquished its charter rights and obligations on the Rupert's Land plains to the British crown. In cartographic terms, the Hudson's Bay Company had played a significant role as Canada's first "national" mapping agency.

Notes

The author gratefully acknowledges the Hudson's Bay Company Archives for use of their map records.

1. These figures include manuscript and printed maps and were obtained through a recent detailed preliminary inventory by the author.
2. Some of the maps are in other collections, but most have not been found.
3. From the author's detailed study of the history of the cartography of the Hudson's Bay Company between 1670 and 1870.
4. Richard I. Ruggles, "Hospital Boys of the Bay: The Hudson's Bay Company Surveying and Mapping Apprentices," *The Beaver*, Outfit 308 (Autumn 1977): 4–11.
5. Hudson's Bay Company Archives, Provincial Archives of Manitoba, Winnipeg (hereafter HBCA, PAM), A11/144, fol. 197.
6. HBCA, PAM, A11/115, fol. 10d.
7. HBCA, PAM, A64/45, no. 11.
8. HBCA, PAM, A5/1, fol. 169.
9. HBCA, PAM, G2/15 and G2/17.
10. HBCA, PAM, G1/20.
11. There are two copies of this map, HBCA, PAM, G1/21 and G1/22.
12. HBCA, PAM, A11/117, fols. 54, 109d, and A5/3, fol. 64d.
13. HBCA, PAM, G2/18.
14. HBCA, PAM, G1/13.
15. HBCA, PAM, G2/13.
16. HBCA, PAM, G2/32.
17. HBCA, PAM, E3/2, fols. 106d–107.
18. HBCA, PAM, E3/2, fol. 104.
19. HBCA, PAM, E3/2, fol. 103d.
20. HBCA, PAM, E3/2, fols. 104d–105.
21. HBCA, PAM, A11/52, fols. 1, 2d; B39/a/2, fols. 22, 23d.

22. HBCA, PAM, A5/4, fol. 103d.
23. HBCA, PAM, G2/19; HBCA, PAM, G1/28b.
24. HBCA, PAM, E3/4, fol. 18.
25. HBCA, PAM, A64/52, no. 73; HBCA, PAM, A64/52, no. 58.
26. HBCA, PAM, B22/e/1, fol. 1d; HBCA, PAM, B51/e/1, fols. 1d–2.
27. HBCA, PAM, B115/e/1, fol. 1d.
28. HBCA, PAM, G1/27.
29. HBCA, PAM, E6/14.
30. HBCA, PAM, G1/327.

10. MAPPING THE QUALITY OF LAND FOR AGRICULTURE IN WESTERN CANADA

James M. Richtik

THE original impetus that brought explorers and settlers to the East Coast of North America had, at least as early as the eighteenth century, evolved into, among other things, an interest in the potential of the Canadian West for European types of agriculture. As settlement spread across the continent, the perceived value of the West changed from fur hinterland to possible agricultural empire. With this shift in interest there was a change in the purpose of exploration, and, as features such as rivers, lakes, and mountains became known, assessing and mapping the agricultural potential of the land began. Cartographers would henceforward record soil types, rainfall and drainage patterns, and the types of vegetation existing upon the land. This paper looks at the process of evaluating and mapping the agricultural potential of the Canadian prairies, the area that now comprises the provinces of Manitoba, Saskatchewan, and Alberta (fig. 10.1).

Presettlement Maps

The first explorers were fur traders who, understandably, showed only passing interest in a region's agricultural potential. Their maps, reflecting an interest in routes and Indian tribes, located physical features and sometimes described vegetative cover or locations of gardens and crops, but they carried no clear implications about the value of the land for agricultural production in general. Even after the establishment in 1812 of the Selkirk Settlement at the forks of the Red and Assiniboine rivers, there was no effort to map the agricultural potential of the rest of the Canadian West, perhaps because the banks of the rivers provided far more land than the small number of farmers could possibly use. Not until the 1850s, when the American frontier had spread into Minnesota and the Hudson's Bay Company's mandate for the Canadian northwest was being reexamined, were the first attempts made specifically to chart the potential of the Canadian prairies for settlement and agriculture. Then both the British and Canadian governments sent out expeditions, the former under Capt. John Palliser and the latter eventually under Henry Youle Hind.

The Palliser expedition, though multifaceted, always considered the examination of agricultural possibilities as a major responsibility. In January 1857, the president of the Royal Geographical Society wrote to the British secretary of state for the colonies, Henry Labouchere, proposing an exploring expedition to the Canadian West to find all-Canadian access routes and to examine "the general capability of the country." He pointed out that the Americans had just finished exploring their Great Plains, whereas the Canadian West was barely explored but was "said to be well fitted for agriculture." Labouchere recommended the plan to the Treasury as deserving of public support, "considering that the region referred to is supposed to contain a considerable extent of fertile soil, and that in the rapid progress of British N. America and the United States public attention is beginning to be di-

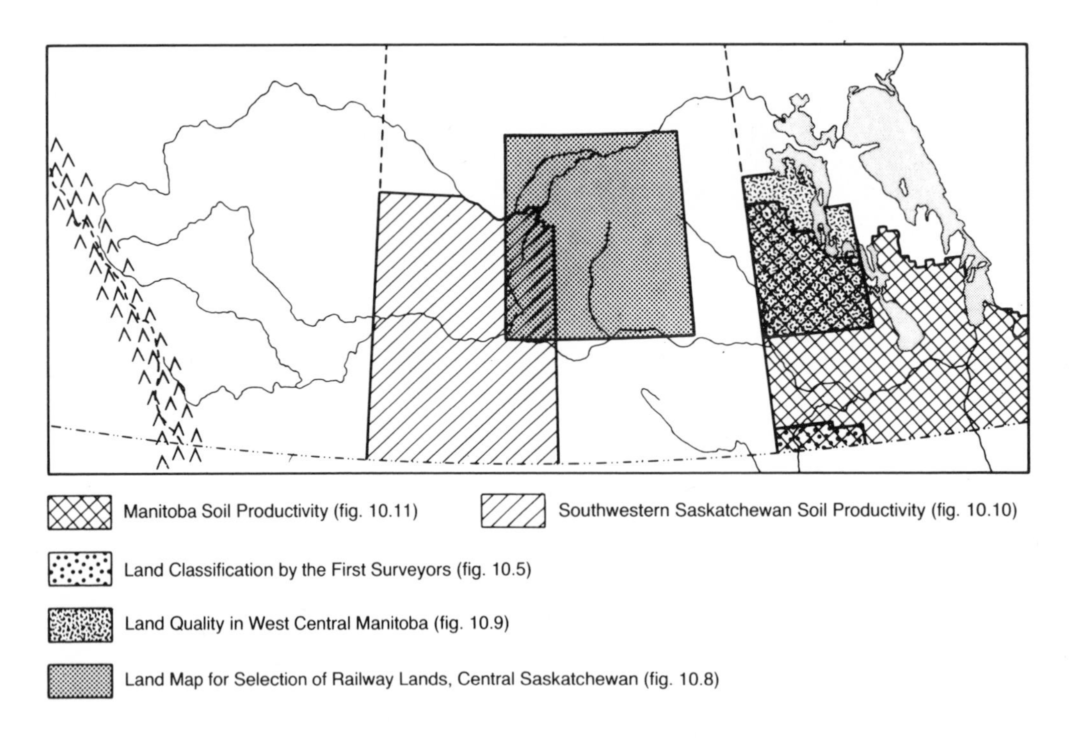

Fig. 10.1. Location of study area and detailed maps.

rected towards it." In his instructions to Palliser, sent in March of that year, Labouchere stressed the importance of the agricultural mission: "I have to impress upon you the importance . . . of regularly recording the physical features . . . the nature of the soil, its capability for agriculture, the quantity and quality of its timber."[1]

In keeping with these instructions, Palliser produced a map of the Canadian West dividing the territory into three categories: (1) "The true forests where spruce and pine predominate," (2) "the fertile belt," and (3) "the great plains with poor soil, scanty herbage, and no wood except on moist northern exposures." Palliser said little about the true forest belt, but seemed to imply at least that it was not very suitable for agriculture. In the fertile belt he found the soil "abounds in vegetable matter" and that "a sufficiency of good soil is everywhere to be found." The land had previously been forested but was partially wooded with willow and poplar and elsewhere had been denuded by fire; thus, incoming settlers would not have "to encounter the formidable labour of clearing the land." Palliser's account noted equally good land and agricultural potential—a "superior class of soil"—along the eastern foothills of the Rockies, although this does not appear on his 1863 map. There can be little doubt that Palliser's negative reaction to the plains was a result of his awareness of the "Great American Desert" to the south—an awareness that, according to John Warkentin, he got from Hind. However, Palliser explained that even "the most arid plains" in Canada did not include "the great expanses of true desert country that exist further to the south." He found the Canadian plains generally "sterile" and "sandy," and his geologist, James Hector, blamed the soil more than the climate for the lack of vegetation.[2] Furthermore, the 1859, 1860, and 1863 maps show different boundaries for the three soil zones (fig. 10.2).

The Canadian expedition under Henry Youle Hind was similar to the British one in purpose and timing. It too was at least partly motivated by an awareness of American expansion and exploration to the south, and it too was organized to look for a better water route from Lake Superior to Red River "and ultimately to the great tracts of cultivable land beyond them." In 1858 Hind was instructed to ex-

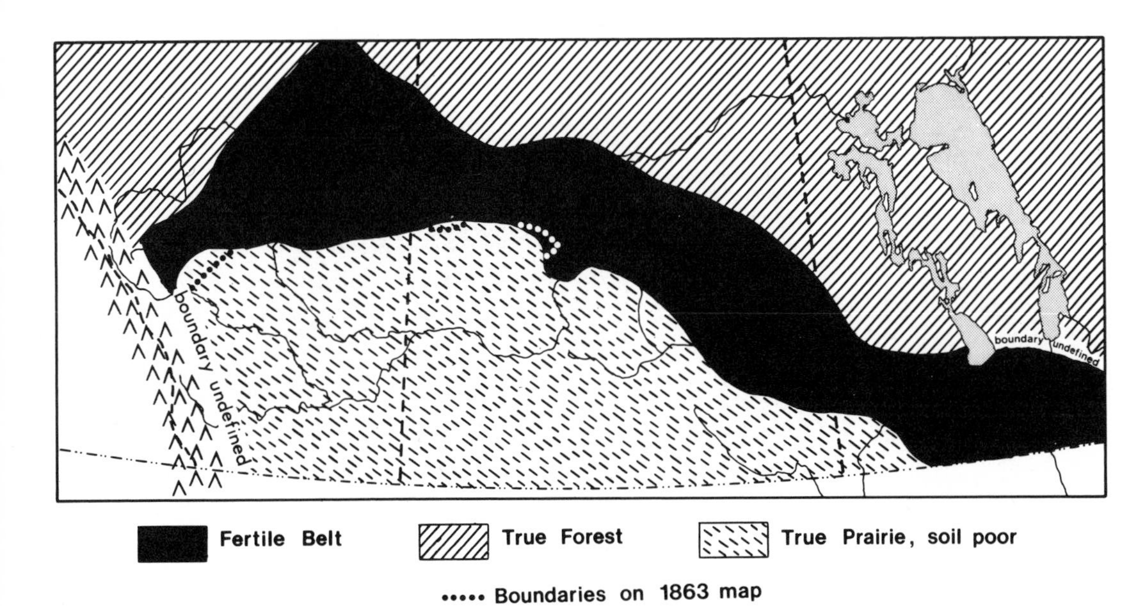

Fig. 10.2. Capt. Henry Palliser's map of land types. (after Palliser)

plore west to the South Saskatchewan River and to "endeavour to procure all the information in your power." He was to describe "the general aspect of the whole region," observe "the character of the timber and soil," and ascertain "the general fitness of the latter for agricultural purposes . . . as far as may be from observation and inquiry."[3]

S. J. Dawson, leader of a portion of the Hind party, had only praise for the Canadian prairies. Although aware of the "Great American Desert," he quoted Lorin Blodgett's claim that the dry areas "*are not found above the 47th parallel in fact.*" He also declared that "the plains . . . present a soil apparently of as great fertility" as the Red River Valley. Dawson quoted extensively from A. J. Russell, who implied that the whole of the prairie territory would make excellent farmland because of the ease of cultivation. He can be faulted for making his judgment on the basis of second-hand information, but James A. Dickinson, the biologist of the group, seemed to agree with him. From a hill on the edge of what later was mapped as desert, he looked "across the boundless plains, no living thing in view," and "thought of the time to come" when the railway would create a thriving agricultural area. This vision of the West appeared on the map that accompanied Dawson's version of the joint report with Hind, and an area that Palliser included in his prairie or poor soil zone was identified as "good clay soil" on Dawson's map.[4]

In his own report, Hind identified an extension of the "Great American Desert" into Canada. He described the area southwest of a line from the Great Bend of the Souris River through the Qu'Appelle mission to the Moose Woods on the South Saskatchewan River as a "treeless plain with a light and sometimes drifting soil . . . and not, in its present condition fitted for the permanent habitation of civilized man." Hind referred to good soil only in the wooded or partly wooded areas and felt the "sterility of the Great Prairie" was "owing to the small quantity of dew and rain, and the occurrence of fires." He further explained that part of this desert "does not appear necessarily sterile from aridity, or poverty of soil." However, he added that "from the character of its soil and the aridity of its climate, the Grand Coteau is permanently sterile and unfit for the abode of civilized man." Hind included a map showing an approximation of Palliser's three belts, but with a larger area of aridity and poor soil than Palliser found. In fact, in the London edition of his report he included a more detailed map of the area his expedition covered, west to the elbow of the South Saskatchewan.[5] Much of the southern portion of the area west of the Assiniboine River that he and Palliser elsewhere called the fertile belt is

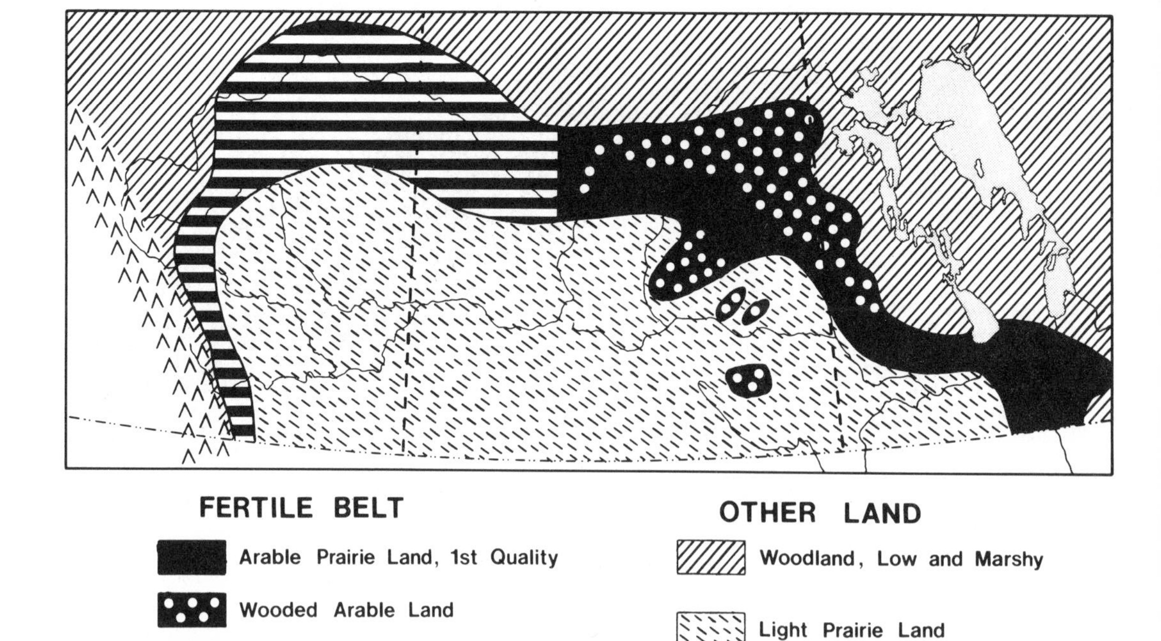

Fig. 10.3. Henry Youle Hind's map of land types. (After Hind)

shown as "arable prairie land 1st quality." The rest of their fertile belt appears as "wooded arable land" (fig. 10.3).

The maps by Palliser and Hind were important because they established a popular image of a huge triangle of infertile land unfit for settlement—the same image that was still being perpetuated by government agencies. G. M. Dawson, exploring the Canadian-American boundary line in 1872–75, raised questions about the aridity of the desert, but he attributed the lack of vegetation to poorer soil because of the "special quality of the Cretaceous shale parent material." However, he was not specifically mapping soil quality and confined his work to areas along the border. In 1878, the Department of the Interior produced a "Map of Part of the North-west Territory, including the Province of Manitoba showing an Approximate Classification of the Lands" (fig. 10.4). On this map, Palliser's triangle appeared mostly as "open plains; poor soil; possessing occasional tracts fit for settlement; extensive pasturage," but a smaller core of the triangle was shown as "principally barren lands; excellent pasture." The equivalent of Hind's "Arable Prairie Land 1st Quality," but extending through an area at least as far south as Palliser's fertile belt and reaching westward into that belt, was shown as "mixed prairie & timber—soils rather light—but produces fine crops—good grazing lands." Hind's "wooded arable land" and a considerable extension northward into his "swampy wooded land" was labeled as a "vast region—generally excellent soil with abundance of wood & water proved to be admirably adapted for the growth of cereals—especially wheat." The same area was shown extending into the Peace River country, which was described as a "tract of extraordinary fertility." That region included a clearly defined foothills extension, described as "superior grazing and grain growing country." A small area along the foothills west of Edmonton, which had not been differentiated by Palliser and Hind, was shown as "swampy but well timbered country." The 1878 map did show an enormous increase in fertile lands over Palliser's version. Palliser's triangle had encompassed 80,000 square miles of infertile land fringed by 65,000 square miles of fertile belt. The 1878 map showed only 64,000 square miles of poor soil and barren lands and 219,000 square miles of good agricultural land.[6] The expansion of the perceived good land area mostly occurred in the north, but there was a preliminary shrinking of the triangle.

The real shrinkage of the triangle occurred in the

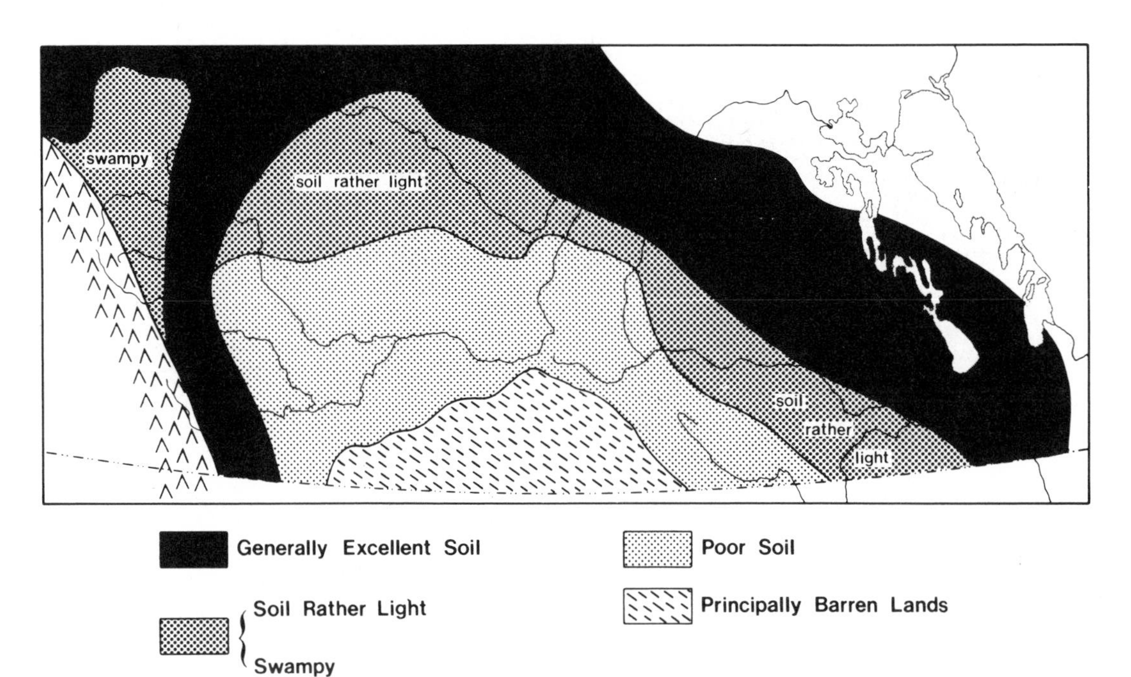

Fig. 10.4. Department of the Interior map showing land quality, 1878. (National Map Collection, Public Archives of Canada)

late 1870s, especially after 1879, when John Macoun began to insist that the entire area of the prairies was suitable for agriculture, climatically and pedologically. Macoun found "excellent soil" between Old Wives Lakes and Cypress Hills in the "barren lands" (see fig. 10.4). However, he did not produce maps of land quality.[7]

Late Nineteenth-Century Maps

Whereas Palliser, the Dawsons, and Hind mapped soil on a small scale, the Dominion Land Surveyors (DLS.) worked only on a large scale. In 1871 they began the arduous task of dividing the newly opened West into the grid pattern of roads with the enclosed square-mile sections of land on the American plan. Their primary task was to have the land accurately surveyed and available for selection when the first settlers arrived. They were also instructed to record terrain and vegetative cover. The surveyors' manual also required them to indicate "the nature of the soil" and classify it "according to its fitness for agriculture, as first, second, third or fourth rate" but gave no instructions on how they should arrive at the ratings.[8] As John Tyman has pointed out in his assessment of their work, "Certain surveyors . . . were unable to bring themselves to grade any land as 4th class," and there was considerable variability in the ratings that different surveyors gave to identical areas. In another analysis, T. R. Weir used a map to illustrate the same point (fig. 10.5). He showed that DLS. classifications did not correlate with either the more recent Manitoba Soils Survey classifications of land quality or the timing of settlement. Tyman also showed lack of correlation of DLS. classifications with the Canada Land Inventory classification, with vegetation, or with time between survey and settlement. Some surveyors were still working in the snow, when "good opportunity for observing the soil seldom occurred," so it is not surprising that their ratings are at variance. Furthermore, individual surveyors gave higher ratings for similar prairie soils after a year or two of surveying. At any rate, this early classification was mapped only in the field notes, and has never appeared in maps, except those published in recent academic studies.[9] The field notes were available for purchase by intending settlers, but there is no evidence that any great number used them. By 1880 the surveyor general could report that the surveys showed much more first-class land than earlier estimated and that the fertile areas "have their limits

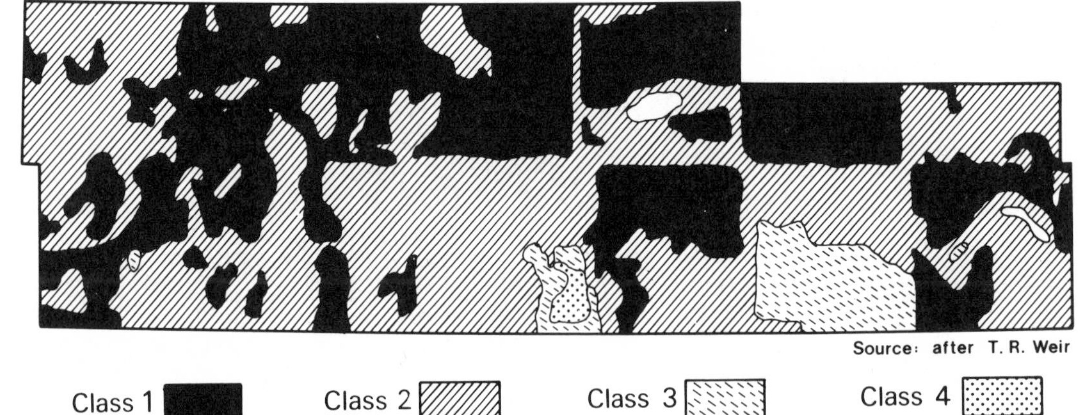

Fig. 10.5. Land classification by the first surveyors in southwest Manitoba. (After T. R. Weir)

extended the fuller our information becomes."[10] This detailed soil information had the effect, along with Macoun's propagandizing, of expanding the perceived fertile belt.

During 1879 the Hudson's Bay Company also began evaluating land quality throughout the Canadian West. That year, as surveying proceeded, the Company put up for sale the lands it was receiving as its one-twentieth share of the "fertile belt." Charles J. Brydges, the Company's land commissioner, complained that "there is no information" on any of the Company's lands, "which makes it very difficult to talk about price." He therefore hired a surveyor "to go over some of the lands and report as to wood on them, whether they are wet or dry." The next year he arranged for three surveyors to report on the same characteristics and also the "nature of soil." He was very concerned about the accuracy of the information because "a good many complaints" had been made about the claims of "speculators" and he wished to preserve the Company's good name. The classification of soil during the first two years was unsystematic, so a four-class grading system was introduced. Montague Aldous, the chief surveyor, explained the grades:

> 1st class indicates a country with rich loam soil from 12 to 20 inches or more in depth, with a good clay or sandy clay subsoil, it may be prairie or partially wooded, but in any event it is what was considered by our inspectors as 1st class agricultural country; 2nd class is land well adapted for settlement, but having the drawbacks of being broken with hills and ponds or possibly consisting of open prairie land where the soil is somewhat light, say from 10 to 15 inches of loam or sandy loam; 3rd class is what I call the plains as distinguished from prairie lands, the soil is light and shallow, often of a very fair quality but always having a large percentage of sand or gravel, the ground is dry and baked and the ponds for the most part alkaline or brackish, 4th class means the worst grade of 3rd class sand hills without any soil overlying them, and is generally termed worthless by those inspecting it.

The same system was being used in 1888 when the "general inspection of the whole country was completed.[11] The resultant map (fig. 10.6) shows enough correlation with Weir's and Tyman's maps of the DLS. classifications to suggest the DLS. may have used much the same system.

Both Brydges and Aldous found the land in Manitoba to be "very superior," especially below the Manitoba escarpment, where most land was designated as class 1. Above the escarpment and west to the Missouri Couteau, they found mostly class 2 land and west of there they found mostly class 3 and 4 land. They also questioned whether it was possible to grow wheat west of the Missouri Couteau because the elevation was too high, and there was not a month without "more or less frost." Only Brydges found this area to be "clearly the northern apex of the Great American Desert," an opinion he expressed in both 1883 and 1888. Much of what Brydges called desert was considered suitable for cultivation by one of the surveyors if it got "abundant rainfall" from May to July. Aldous's soils map showed some class 2 land west of the Couteau and even some class 2 land west of Moose Jaw Creek, which Brydges identified as the start of the desert.[12]

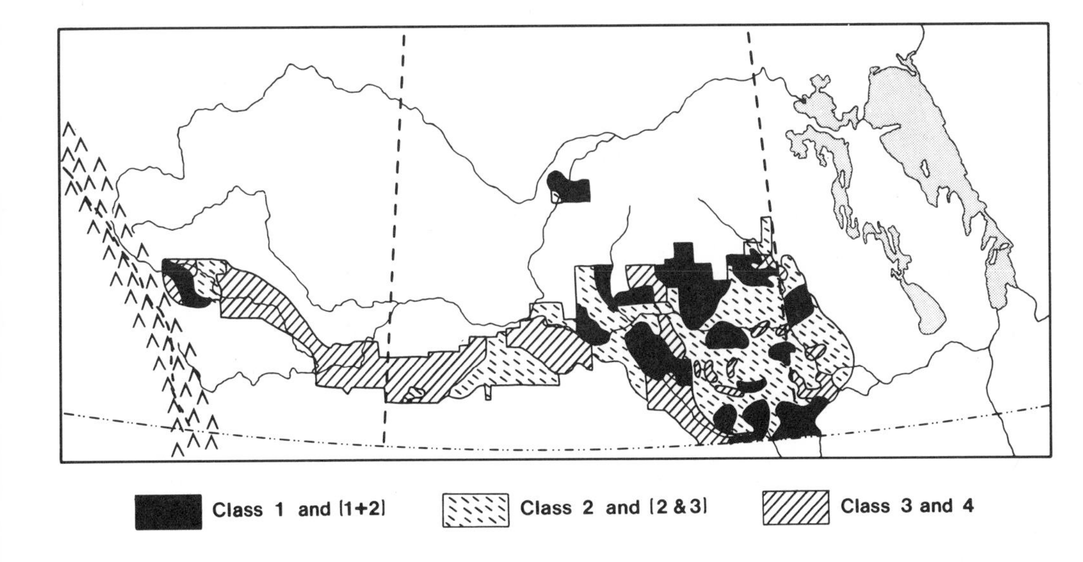

Fig. 10.6. The Hudson's Bay Company's land classification. (Hudson's Bay Company Archives, Public Archives of Manitoba)

Finally, it seems clear that the Hudson's Bay Company's land evaluation had direct results. The values placed on lands for sale were partly based on soil quality, and Brydges found the Company reports "of the greatest use in the sale of lands" because incoming settlers "*all* want to have particulars." In fact, he complained in early 1882 that he was having trouble selling land because all the evaluated or graded lands had been sold.[13] The collapse of sales in 1883–84 made the surveys largely superfluous for almost two decades because so little land was being sold.

Like the Hudson's Bay Company, the Canadian Pacific Railway (CPR) was entitled to huge acreages of western land. The CPR had been promised twenty-five million acres of land "fairly fit for settlement." As early as 1882, Brydges reported that the railway would reject everything from just west of Moose Jaw to near Calgary, a prediction that proved accurate in the event (fig. 10.7). Like the Hudson's Bay Company, the CPR had surveyors assessing land quality. The rejection of most land was largely on the basis of climate as the drought of the mid-1880s produced crop failure throughout most of Palliser's

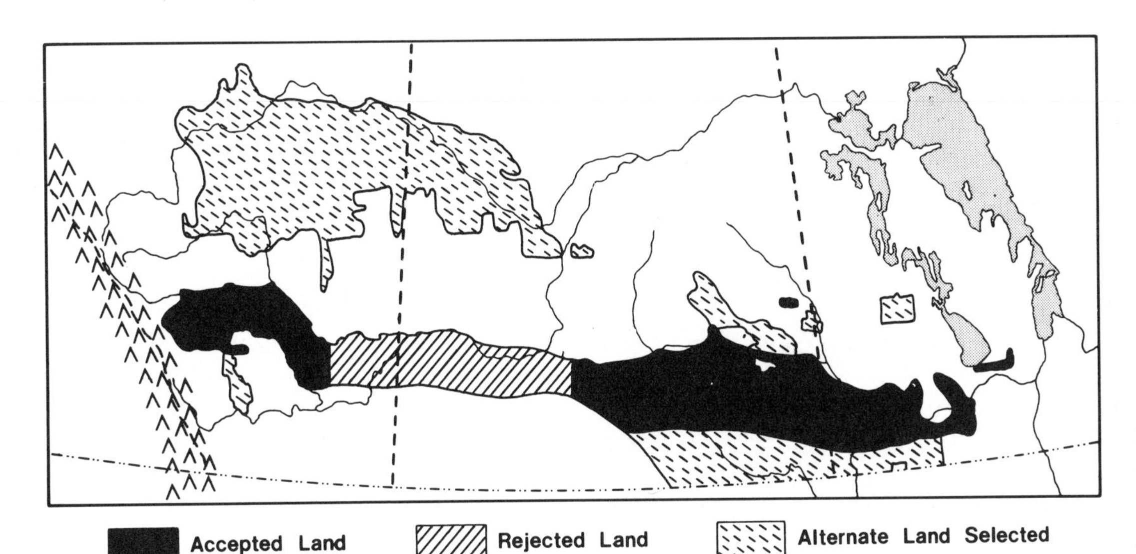

Fig. 10.7. Lands accepted and rejected by the CPR. (Public Archives of Canada)

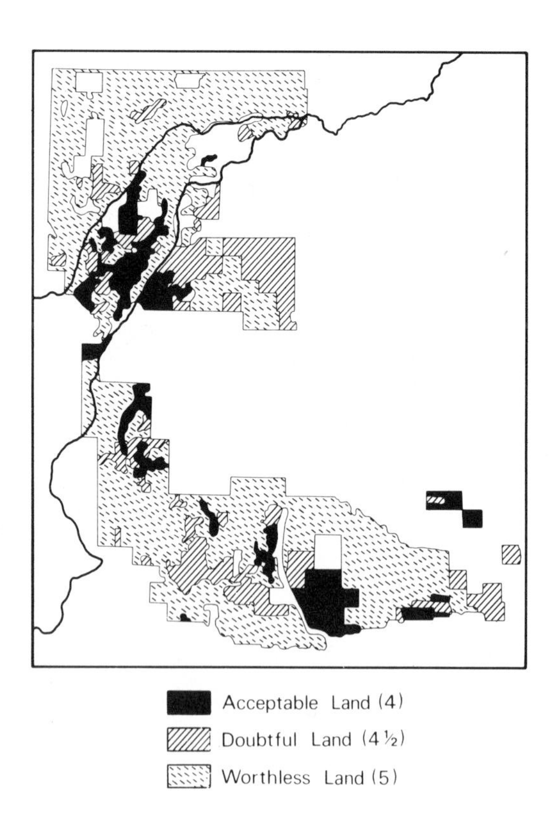

Fig. 10.8. Land map for selection of railway lands in central Saskatchewan. (Public Archives of Canada)

triangle. In Manitoba, the CPR's rejection was more selective and was mostly of the excessively sandy areas in the Carberry and Oak Lake districts. As Tyman pointed out, much land ultimately rejected as unfit for settlement had been offered for sale in the 1881–82 boom and some had even been sold but had reverted to the company for nonpayment before its rejection by surveyors.[14] The CPR's selection of alternate lands was generally within Palliser's fertile belt. Using Macoun's claims of the fertility of the entire prairie area as a rationale, the syndicate running the CPR had chosen a southern route for the line. However, when it came time to select land, the vision of the Great American Desert canceled that choice.

Slightly more detailed land quality maps were produced in 1893 for the Qu'Appelle-to-Prince Albert area for railway land grants (fig. 10.8). Part of the grant was for construction of a branch line from Regina to Prince Albert, including land to be substituted for unacceptable or unavailable sections, part was for land owed for the Manitoba and Southwestern Railways land grant, and part was for the CPR mainline grant in lieu of land rejected elsewhere. The categories used in these maps were "worthless" (class 5), "doubtful" (class 4 1/2) and "accepted" (class 4 and better). Again, the basis for selection was not identified, but there was a fairly good correlation with the ratings shown by the Hudson's Bay Company.

Beginning in 1883, as part of the "better terms" for the province of Manitoba, the federal government began deeding swamplands to it. At first the province received only alternate sections of land actually drained, but later, after swampland inspectors examined the land, it received all except four sections per township. The first maps of the swamplands appeared about 1900, shortly before the transferring was discontinued. These maps showed only one of the features of land quality and made no reference to soil materials. Such land transfers never occurred after Saskatchewan and Alberta became provinces in 1905, so the phenomenon was unique to Manitoba.[15]

Twentieth-Century Maps

More ephemeral were the manuscript ratings on the 1906 maps of the northern Interlake and Duck Mountain areas of Manitoba. The categories were "good land" and "poor land and swamp" (fig. 10.9). The area mapped included the present Duck Mountain forest reserve and Riding Mountain National Park. The almost barren high lime soils of the Interlake were designated as "good" and the land near Arborg (presently considered excellent) was shown as "swamp," which indeed it then was. The maps appear never to have been used.

The expectation that returning soldiers would take up land after World War I led the dominion government to undertake land classification to help guide the potential settlers. The topographical surveys branch produced maps showing "classification of land by quarter sections for settlement purposes." It actually mapped land quality only on areas still available for homesteading.[16] Like the 1906 map, these maps give a very optimistic view of land that

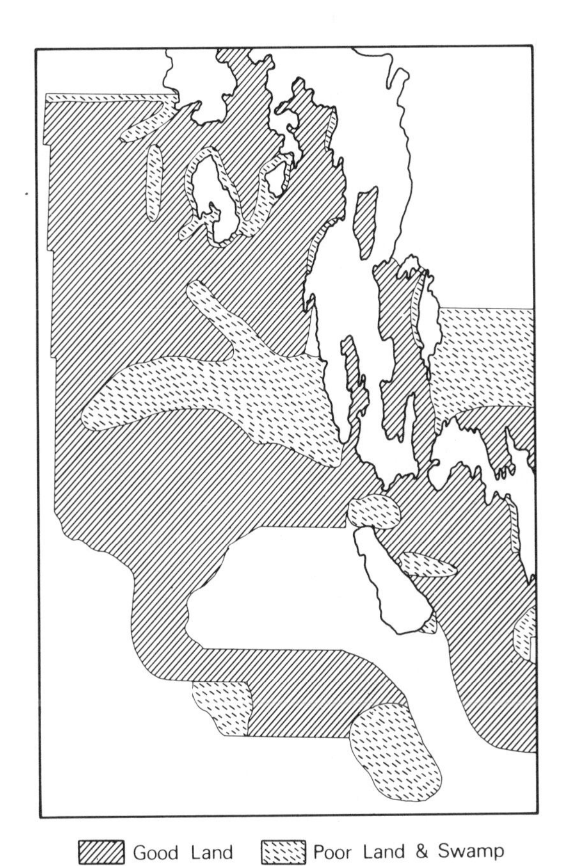

Fig. 10.9. Map of land quality in west central Manitoba based on a 1906 manuscript map. (Public Archives of Canada)

is now abandoned and considered of no future value. Their primary purpose was to save the settler time by directing him to the more promising areas, but they reminded the settler that the final choice was his. Considering the optimistic view represented by these maps, the caution was certainly in order.

In the 1930s the Manitoba Soil Survey began reconnaissance soils surveys. Their object was to obtain "the essential facts about the soils" in order to ascertain "the characteristics, the possibilities and the problems of the respective soils." The writers of the reports hoped to improve conservation by providing a useful guide to land use policy. They did not claim to evaluate land quality but rather to describe the soil profiles. A table identified the suitability of the various soils for particular crops, but generally as a range. It was only by working from this table that T. R. Weir was able to construct his maps of land quality for southwestern Manitoba.[17] These maps were of significant help to government agencies advising farms, but the farmers themselves were generally unaware of their existence.

In 1953 the Canadian Department of Agriculture prepared a map of part of Saskatchewan showing potential wheat production per quarter-section based on long-term average wheat yields (fig. 10.10). Production on Class 1 soils was less than 350 bushels per quarter-section because of "poor soil texture or low arability or both." Class 2 soils had similar lim-

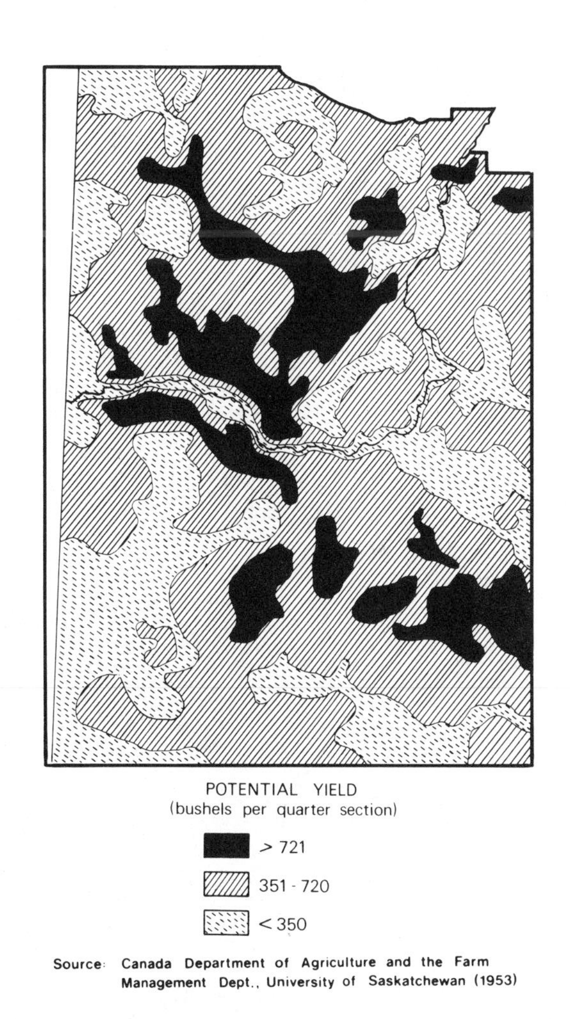

Fig. 10.10. Map of soil productivity in southwestern Saskatchewan.

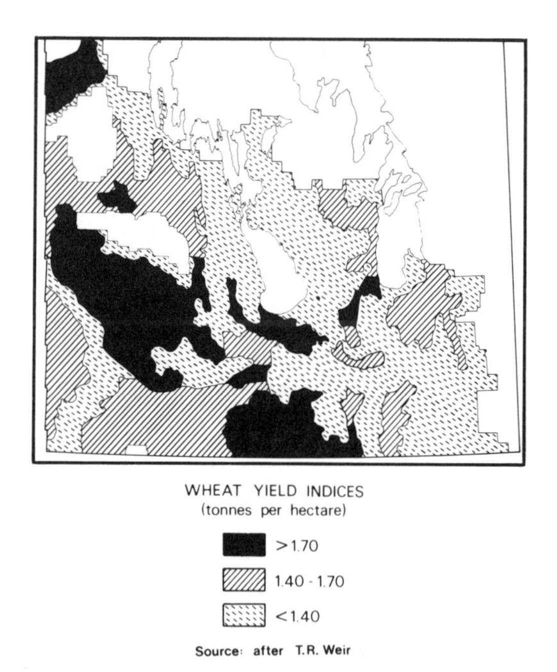

Fig. 10.11. Crop insurance map of soil productivity in Manitoba. (After T. R. Weir)

itations but produced 357 to 475 bushels. Class 3 soils produced 476 to 720 bushels on loam soils with level to rolling topography. Class 4 soils produced 721 to 900 bushels on "superior loams to clay . . . with very few stones." Class 5 produced over 900 bushels per quarter section on soils "of a heavy clay texture . . . usually well drained and stone free." Figure 10.10 shows that some of the highest yielding land was in areas rejected by Palliser, Hind, and the CPR, and the actual production correlates poorly with the map of land quality for the railway grant lands north of Qu'Appelle. Only the Hudson's Bay Company surveyors produced a similar map of land potential.

A more recent example of maps based on soil productivity is that by T. R. Weir (fig. 10.11). It is based on indices of soil and climate together, which were calibrated by actual twenty-five-year yields of red spring wheat. The Manitoba Crop Insurance Corporation produced the yield estimates for their own purposes, but the resulting map is a visual expression of one aspect of soil quality.[18] Correlations with other land quality maps are not great.

The most recent effort at fairly complete mapping of soil quality was that carried out by the Canada Land Inventory (fig. 10.12). It was based on limitations for agriculture, recognizing that these might be climate, erosion, flooding, stoniness, topography, or high water tables. Class 1 soils have no significant limitations for agriculture. Class 2 soils have moderate, and Class 3 have moderately severe limitations. Land falling in these three classes is considered good for agriculture. Class 4 land, with severe limitations, and Class 5, with very severe limitations, are marginal to submarginal. Class 6 land is only marginally suitable for forage crops,

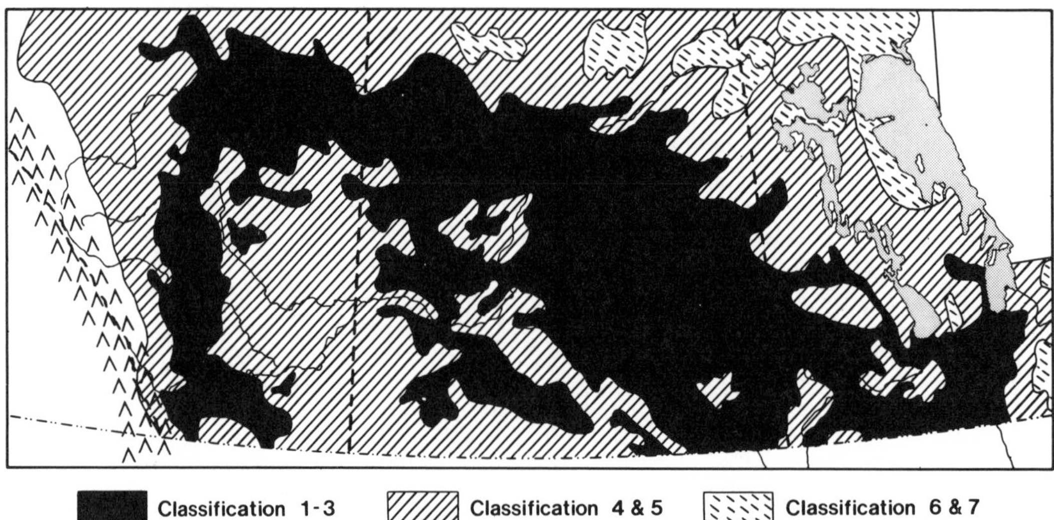

Fig. 10.12. CLI map of soil capability for agriculture.

and class 7 is unsuitable for any agricultural use.[19] The correlation with the Saskatchewan crop yield map is, not surprisingly, very high. More surprising is the correlation of the CLI maps with Palliser's and Hind's maps, although the CLI maps show a much smaller area of submarginal land.

Conclusion

The mapping of land quality in western Canada began early and has continued to be an important scientific objective. The maps produced by the many evaluators have varied enormously, reflecting the different purposes involved in the mapping, the degree of information available, and the perceptions of the evaluators. Although it is easy to see, in retrospect, that Palliser's triangle and its equivalent shown by Hind included enormous areas of what is now prime farmland, the CLI maps show at least a remnant of it, and other early maps are at least partly reflected in more recent land quality maps. No doubt future maps of agricultural land quality will differ in detail as scientific knowledge increases and perceptions of value continue to evolve.

Notes

1. Irene M. Spry, ed., *The Papers of the Palliser Expedition, 1857–1860,* (Toronto: Champlain Society, 1968), pp. 5, 495–502.
2. Ibid., pp. 22, cx, map in pocket.
3. Henry Youle Hind, *Narrative of the Canadian Red River Expedition of 1857* (1860 reprint, New York: Greenwood Press, 1969), 1:4, 269–270; S. J. Dawson, *Report of the Exploration of the Country Between Lake Superior and the Red River Settlment* (1859; reprint, New York: Greenwood Press, 1968), p. 2.
4. Dawson, *Report;* Hind, *Narrative,* 1:373.
5. Hind, *Narative,* 1: 317, 348, 351; Dawson, *Report,* p. 2; Henry Youle Hind, *Reports of Progress Together with a Preliminary and General Report* (London: Houses of Parliament, August 1860).
6. George M. Dawson, *Report on the Geology and Resources of the Region in the Vicinity of the Forty-ninth Parallel* (Montreal: Dawson Bros., 1875), pp. 144–45, 299; Map 0008438, National Map Collection, Public Archives of Canada.
7. John Macoun, *Manitoba and the Great North-West,* (Guelph, Ontario: World Publishing Co., 1882), especially pp. 60–79.
8. Canada, Department of the Interior, *Manual Showing the System of Survey of the Dominion Lands with Instructions to Surveyors* (Ottawa: 1881), p. 17.
9. John L. Tyman, "Subjective Surveyors: The Appraisal of Farm Lands in Western Canada, 1870–1930," in Brian W. Blouet and Merlin P. Lawson, eds., *Images of the Plains: The Role of Human Nature in Settlement* (Lincoln: University of Nebraska Press, 1975), pp. 75–89; Thomas R. Weir, "Pioneer Settlement of Southwest Manitoba: 1879 to 1901," *Canadian Geographer* 8, no. 2 (1964): 64–71; Canada, *Sessional Papers: 1883,* no. 23, "Report of the Department of the Interior," pt. 1, pp. 8–9; Original Surveyors' Plans, Provincial Archives of Manitoba, Winnipeg (hereafter PAM).
10. Canada, *Sessional Papers: 1881,* no. 12, "Report of the Department of the Interior," pt. 1, p. 9.
11. Brydges to Armit, 9 June 1879, A12/8, fol. 31–34; 4 May 1880,A12/19, fol. 200; 11 June 1880, A12/19, fol. 258; 5 November 1888, A12/26, fol. 47; Aldous to Brydges, 5 December 1883, A12/22, fol. 515, in Hudson's Bay Company Archives (hereafter HBCA), PAM.
12. Brydges to Armit, 28 September 1882, A12/21, fol. 289–91; 4 October 1883, A12/22, fol. 383; 5 November 1888, A12/26, fol. 471–72; Aldous to Brydges, 5 December 1883, A12/22, fol. 515–16; W. J. Riley to Aldous, 10 November 1884, A12/24, fol. 20, in HBCA, PAM.
13. Brydges to Armit, 4 May 1880, A21/19, fol. 200–201; 1 June 1881, A12/20, fol. 153–54; 29 May 1882, A12/21, fol. 164; 15 February 1886, A12/25, fol. 42, in HBCA, PAM.
14. Canada, *Statutes: 1881* (Ottawa: 44 Vic, C. 1, S. 11); Brydges to Armit, 6 October 1882, A12/21, fol. 312 in HBCA, PAM; Tyman, "Subjective Surveyors," pp. 75–76; T. P. Shaughnessey, CPR president, to L. A. Hamilton, CPR land commissioner, 25 Janaury 1900, and attachments, Shaughnessey Letter Bank 69, file 67821, Canadian Pacific Corporate Archives, Montreal.
15. Chester Martin, *Dominion Lands Policy* (1938; reprint, Toronto: Carleton Library, McLelland and Stewart, 1973), pp. 175–78.
16. Tyman, "Subjective Surveyors," p. 78; "Map of the District Adjacent to Lake Winnipeg and Manitoba" prepared by the Topographical Surveys Office, 1921 edition.
17. J. H. Ellis and W. H. Shafer, *Reconnaissance Soil Survey of South-Western Manitoba,* Soils Report no. 3 (1943; reprint, Winnipeg Manitoba Soil Survey, 1974), p. 5; Weir, "Pioneer Settlement," pp. 64–71.
18. Thomas R. Weir, ed., *Atlas of Manitoba* (Winnipeg Surveys and Mapping Branch, Department of Natural Resources, Province of Manitoba, 1983), pp. 14–15.
19. Canada, Department of Regional Economic Expansion, *The Canada Land Inventory: Objectives, Scope and Organization,* Report no. 1 (Ottawa: 1970), pp. 22–25.

11. MAPPING THE NORTH AMERICAN PLAINS

A CATALOG OF THE EXHIBITION

Ralph E. Ehrenberg

THE EXHIBITION "Mapping the North American Plains" was assembled to serve two basic purposes: first, to enrich the symposium by assembling more than seventy historical maps of the plains, dating from the sixteenth century to the beginning of the twentieth, and second, to commemorate the 150th anniversary of the 1833 expedition conducted by Maximilian, Prince of Wied-Neuwied, to the upper Missouri River.

The exhibition presented the major historic phases in mapping the North American plains. It included intricate and decorative maps by sixteenth-century European cosmographers and geographers; French, English, and American maps derived by explorers from native accounts and personal observation; maps prepared by Indian, British, and American soldiers; representative maps from major series initiated by federal mapping agencies in the United States and Canada; and maps prepared for promotional and commercial purposes.

The catalog is arranged according to the following categories:

I. Sixteenth-century images
II. French and Spanish exploratory maps
III. British exploratory maps
IV. American exploratory maps
V. Military maps
VI. Jurisdictional maps
VII. Geological and topographical maps
VIII. Commercial maps

More than one-third of the maps included in the exhibition are illustrated here. Selections were made on the basis of such considerations as importance, representativeness, and whether or not a given map has been reproduced elsewhere.

Uniform information is provided for each map in the catalog accounts. A summary of technical information includes title, cartographer, and place and date of production. Each map is then described; its uniqueness, importance, or other special quality is identified. Each entry concludes with the map's provenance and references to it in published books and articles. Since the exhibitions in Lincoln and Washington were not identical, differences in the two are noted. The catalog ends with a bibliography of the references included in the individual entries.

I. Sixteenth-Century Images

The general configuration of North America gradually emerged on maps during the sixteenth century. Coastal features of the New World were charted by navigators from many nations. French and Spanish explorers seeking overland routes to the mythical cities of gold in the Southwest and sea passages to the riches of eastern Asia provided information on inland features near the coasts and rivers. Jacques Cartier sailed up the St. Lawrence in 1535, reaching present-day Montreal. Hernando de Soto, with six hundred soldiers, journeyed overland as far as eastern Oklahoma. Francisco Vásquez de Coronado discovered the high plains of Kansas in the early 1540s. Despite these initial journeys to the fringe of the vast interior of the North American continent, the plains region remained unknown to European mapmakers. As a consequence, the interior of North America as shown on sixteenth-century printed maps was either left blank, filled with mythical cities and fanciful hills and rivers, or decorated with ornate cartouches.

I.1. *Tabula nouarum insnlarum, quas diuersis respectibus Occidentales & Indianas uocant*

Sebastian Münster
Woodcut, with place names set from metal type; 27 × 35 cm
Basle [1554]

This early European view of the New World first appeared in Sebastian Münster's 1540 edition of Claudius Ptolemy's *Geographia* and was subsequently published in numerous editions of Münster's *Cosmographia.* It portrays North America on the eve of Spanish exploration of the western interior. While the coastline from Newfoundland to Florida and the Gulf Coast is recognizable but distorted because of inaccurate astronomical information, the interior of North America has been left blank except for generalized symbols representing mountains and forests. The outline of South America is based on information obtained from Ferdinand Magellan's voyage, which is symbolized by his ship pictured in the Pacific Ocean.

G3290 1554 .P Vault
Geography and Map Division
Library of Congress

References: Skelton, 38–40; Klemp, 9.

I.2. *Il Disegno del discoperto della noua Franza, ilquale s'e hauuto ultimamente dalla nouissima nauigatione de Franzesi in quel luogo: Nel quale si uedono tutti l'Isole, Porti, Capi, et luoghi fra terra chein quella sono* (Illustrated)

Bolognini Zaltieri
Engraving; 26 × 39 cm
Venice, 1566

Although this example of Italian cartography is of interest primarily because it is possibly the first to portray North America as a continent separate from Asia, it incorporates important Spanish discoveries as far north as the Great Plains within a framework of fanciful geography current in its day. Bolognini Zaltieri shows *Quivira,* an Indian tribe in south central Kansas reached by Francisco Vásquez de Coronado in 1541. *Quivira* was the name of the mythical kingdom of gold sought by Coronado but became the Spanish word for Wichita, the Indian tribe that Coronado found instead of gold. Because Coronado's maps were apparently not available to European cartographers, *Quivira* is depicted too far westward on the *Tigna* (Colorado River), since it was incorrectly believed that Coronado's expedition had at one point reached the Pacific Ocean. Zaltieri portrays *Quivira* and *Quivira Pro* (Quivira Province or country) as situated in a broad lowland between two mythical east-west mountain chains. The Rocky Mountains are not shown. Other evidence of Spanish association with the western interior plains is found in the place name *Civola Hora,* a variant spelling for *ciuola,* the Spanish word for buffalo.

National Map Collection
Public Archives of Canada
Ottawa, Canada

References: Wheat, 1: 21–22; Swanton, 306.

I.3. *Americae sive Novi Orbis, Nova Descriptio*

Abraham Ortelius
Engraving; 39 × 49 cm
Antwerp, 1570

Derived largely from Gerard Mercator's great world map of 1569, this map, from Abraham Ortelius's *Theatrum Orbis Terrarum,* depicts the continental interior as largely unknown. Except for Francisco Vásquez de Coronado's *Quivira,* incorrectly located on the western edge of the continent, there are no references to the North American Plains. The St. Lawrence River, a feature that was to play an important role in the exploration and mapping of the western interior, is fairly accurately represented. Discovered and explored in 1535 by Jacques Cartier, this river later served as a major route to the North American Plains. Ortelius also followed Mercator in depicting a freshwater lake in the vicinity of the Great Lakes. Ortelius, Mercator, and other cartographers of the Netherlands and Rhineland during the last half of the sixteenth century based their work more on scientific method and systematic presentation than did the Italian map printing establishments. Nevertheless, in order to appeal to the burgeoning European map trade, their maps were decorated with ornate strapwork cartouches and finely drawn ships and animals by Franz Hogenberg and other artist-engravers in the style of the Flemish Renaissance.

Lowery Collection 59
Geography and Map Division
Library of Congress

Reference: Skelton, 6.

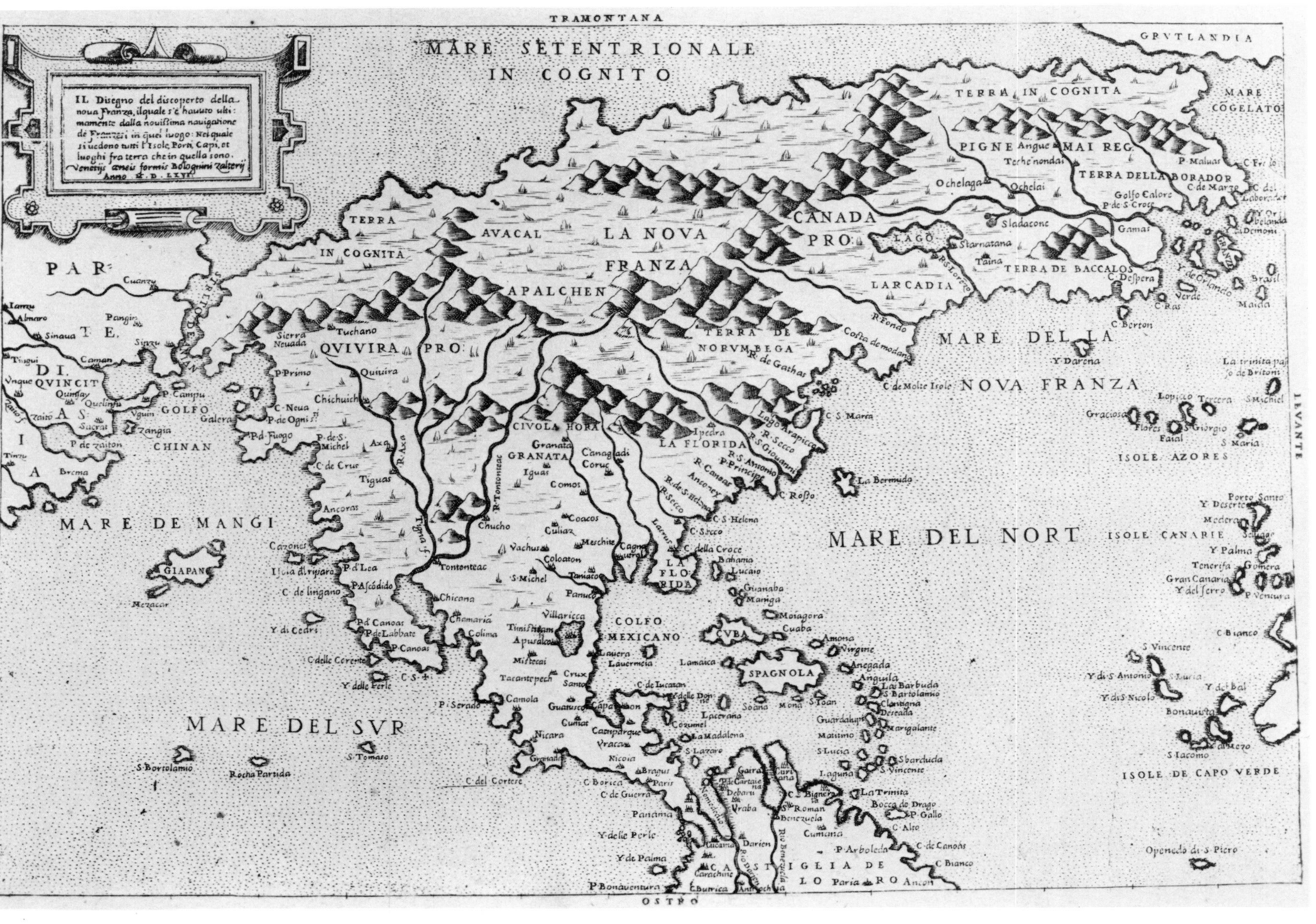

I.2. "Il Disegno del discoperto della noua Franza, ilquale s'e hauuto ultimamente dalla nouissima nauigatione de Franzesi in quel luogo. . . ." (Public Archives of Canada)

I.4. *Americae Pars Borealis, Florida, Baccalaos, Canada, Corterealis*

Cornelis de Jode
Engraving; 36 × 50 cm
Antwerp, 1593

Sixteenth-century cartographers often decorated unexplored regions with hypothetical or conjectural features. An interesting example is this map of North America published by Cornelis de Jode in his *Specvlvm Orbis Terrae* (Antwerp, 1613). De Jode closely followed the coastal delineations found on maps of North America by Gerard Mercator and Abraham Ortelius but added symbols for numerous imaginary mountains and rivers supposedly located in the western interior. Flowing across the top of the map is a large river representing the fabled northwest passage to India. In the general area of the Great Plains, De Jode depicted *Ceuola* (Cibola), a cluster of Zuñi pueblos in western New Mexico discovered and named by the Franciscan Fray Marcos de Niza in 1539. Cibola became associated with the mythical Seven Cities of Gold sought by Francisco Vásquez de Coronado on his expedition to the Great Plains. Another significant feature of this fanciful map is the location of the headwaters of the St. Lawrence in the continental interior near Cibola.

University of Nebraska Exhibition
Yale University
Library of Congress Exhibition
G1015.J6 1613 Vault
Geography and Map Division
Library of Congress

II. French and Spanish Exploratory Maps

In the contrast to the mythical representations of the North American interior that appeared previously, seventeenth- and eighteenth-century maps portrayed information on the Great Plains gathered from direct observation and reliable Indian accounts.

Traders and missionaries from New France were the first to map the Great Plains. French mapmaking can be divided into three distinct phases. In the 1670s French military campaigns against the Iroquois and Mohawk Indians opened direct routes to Lake Superior. This enabled exploration of the area near Lake Superior and along the Mississippi valley. A number of manuscript maps of the eastern edge of the Great Plains were based on these explorations. The second period of mapmaking coincided with the expeditions of the Sieur de la Vérendrye and his sons between 1731 and 1749. For the first time they explored and mapped a section of the northern plains from Lake Winnipeg to the Missouri River. With the fall of New France in 1760, official French exploratory mapping of the northern plains ceased, but French, as well as American, traders employed by Spanish authorities made important contributions to the mapping of the trans-Missouri region.

II.1. *Ameriqve Septentrionale* (Illustrated)

Nicolas Sanson
Engraving; 39 × 56 cm
Paris, 1650

Nicolas Sanson's map of 1650, based on the latest geographic information from French, British, and Spanish sources, represents the European view of the North American interior just before French penetration of the region. Sanson was the royal geographer and the leading French cartographer in the last half of the seventeenth century. With this map he introduced the high quality of French mapmaking which led to the shift of the center of mapmaking from Holland to France before the end of the seventeenth century.

Sanson's map is an intriguing mixture of fact and fancy. The western interior is undefined in the north, but in the south a large area surrounded by mountain ranges is portrayed, suggestive of a great interior plains. With this region, Sanson depicted *Quivira*, the Wichita Indian country visited by Francisco Vásquez de Coronado in 1541 and Don Juan de Oñate in 1601. Also in this region he placed the *Apaches Vaqueros* (cowboy Apaches), a nomadic Plains tribe first described by Coronado as the *Querechos* (Buffalo Eaters) and then by Oñate.

The Great Lakes, which later served as the major eastern water route to the western interior, are shown for the first time with some accuracy. The configuration of the individual lakes, conforming to a remarkable degree to their correct position and size, was derived from an unidentified Jesuit map now lost and from descriptions found in the *Jesuit Relations*, a series of reports sent each year to France with news of religious activities as well as information about the country. Lake Superior and Green Bay (*Lac des Puans*) are shown without their western shorelines, since the Jesuits had not yet explored the western end of the Great Lakes. Sanson probably based his depiction of Lake Superior on Father Ragueneau's *Relations* of 1647, in which he described "a superior lake . . . larger than the fresh-water sea [Lake Huron], into which it discharges . . . [that] extends toward the northwest."

Sanson's unique configuration of Hudson Bay was derived from combining Samuel Champlain's map of 1632 and the maps of the English explorers Luke Fox and Thomas James, who explored the bay in 1631 and 1632, respectively. In relying on Champlain, Sanson perpetuated two

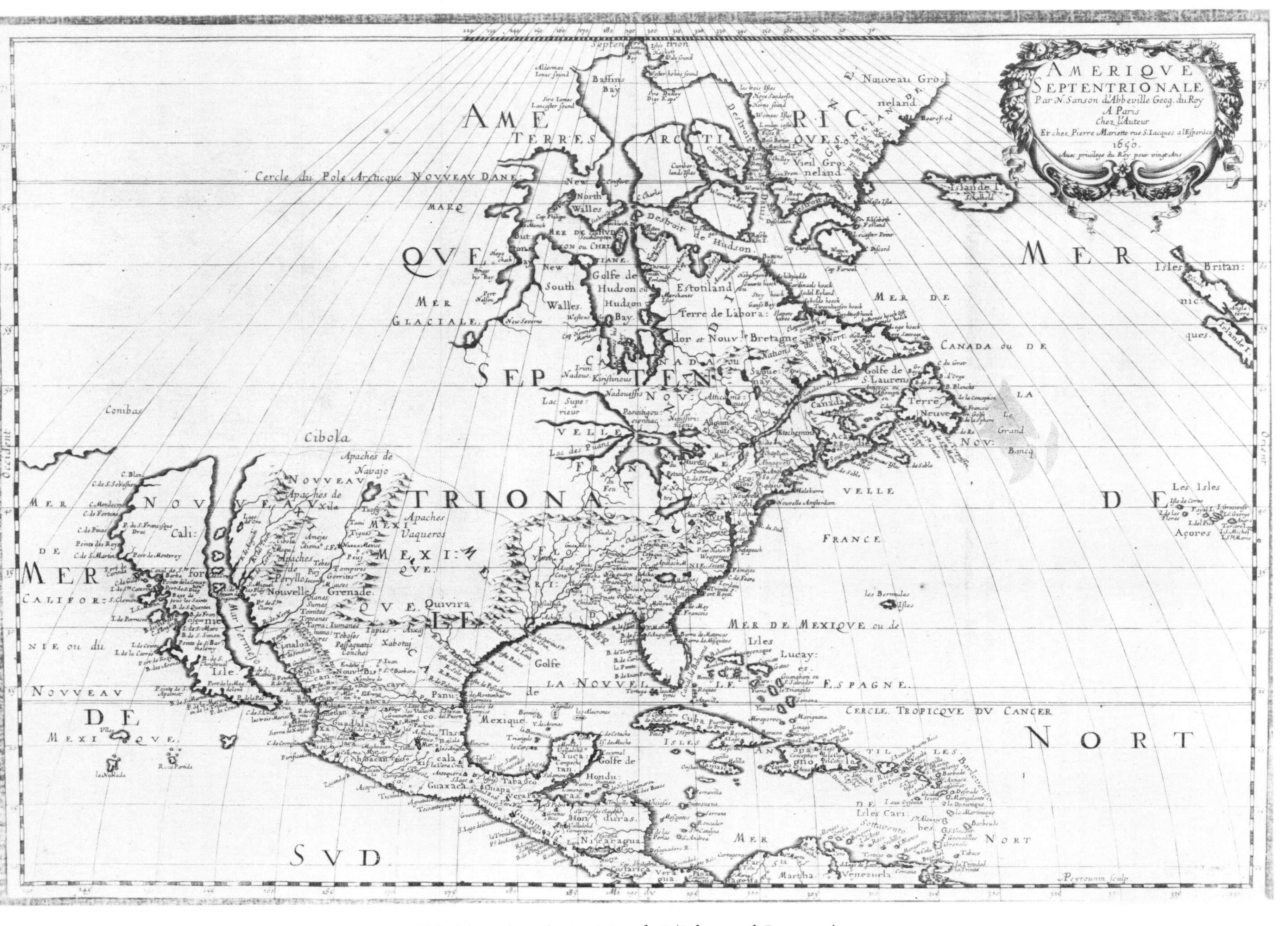

II.1. "Ameriqve Septentrionale." (Library of Congress)

geographical misconceptions. Champlain introduced the concept of a large triangular peninsula of land between the mouth of the Nelson River and James Bay. Also, he adopted the rectangular configuration of James Bay, first delineated on Hessel Gerritz's prototype map of the bay in 1612, which recorded the discoveries of Henry Hudson's voyage of 1610. English officials later used Sanson's map as documentary evidence to support their territorial claims to Hudson Bay, contending that Sanson's use of the reports and maps based on Hudson, Fox, and James implied tacit French recognition of the English claim.

On the southeastern periphery of the western interior, Sanson portrayed a lengthy mountain range encircling a great river basin leading to the Gulf of Mexico. This section of Sanson's map was derived from Gerónimo de Chaves's map of Florida depicting the discoveries of Hernando de Soto in 1539–43. It was published by Abraham Ortelius in 1584. As the royal Spanish cosmographer, Chaves had direct access to the expedition's records. The concept of the encircling mountain range was based on De Soto's accounts of the foothills of the Appalachian mountains. It led later seventeenth-century geographers and cartographers to the incorrect assumption that the length of the Mississippi River, discovered by Spanish explorers in the 1520s and named the *Rio del Espiritu Santo*, was restricted to the coastal region by this mountain barrier.

The region of New Mexico, located south and west of the western interior, was taken from the latest Spanish sources available to Sanson, for he included new place names not found on other printed maps of the period. Despite this new knowledge, Sanson continued to portray two popular geographical misconceptions. One is the delineation of California as an island. This concept originated in 1620 when Father Antonio Ascension's map of the region was captured at sea by Dutch seafarers from Antwerp. The other is the misplacement of the *Rio del Norte* (Rio Grande), which Sanson shows as flowing from a large lake in a southwesterly direction into the Pacific Ocean, rather than into the Gulf of Mexico.

G3300 1650.S Vault
Geography and Map Division
Library of Congress

References: Swanton, 1; Crouse, 131; Warkentin and Ruggles, 10–11; Delanglez (1944), 229; Cumming, 11.

II.2. *Carte de la decouverte faite l'an 1663 dans l'Amerique Septentrionale*

Melchisédech Thévenot
Engraving by Liebaux; 16 × 39 cm
Paris, 1681

The first printed map to depict the eastern fringe of the Great Plains based on direct observation was Louis Jolliet's map of his discovery of the Mississippi River, which appeared in Melchisédech Thévenot's *Recueil de voyages.* It was published in Paris in 1681. Although Jolliet and his companion, Father Jacques Marquette, did not reach the Gulf of Mexico during their expedition of 1673, Thévenot's map shows the entire length of the Mississippi River. In addition, it shows for the first time two large unnamed rivers entering the Mississippi from the west which were to serve as major water routes to the western interior plains. These were the Arkansas and the Missouri.

Jolliet and Marquette did not venture west of the Mississippi River, but from discussions with Indians they learned of several tribes that inhabited the eastern portion of the Great Plains. Thévenot introduced these tribal names for the first time on a printed map. Later, these names became well-known place names for western states, cities, towns, and rivers. Four plains tribes are shown along the Missouri River. These include, from west to east, the *Paniassa* (French name for Wichita), first referred to by Francisco Vásquez de Coronado in 1541 by their Spanish name, the *Quivira;* the *Kamissi* (Kansas), located above the mouth of the Kansas River; the *autrechaha* (Osage); and the *8missouri* (Missouri). To the north of the Missouri River, Thévenot denoted the *Otontanta* (Oto), a tribe generally located in the historic period on the lower course of the Platte River but which may have been situated on the headwaters of the Des Moines River during the 1670s.

Thévenot's map was derived from a map drawn by Jolliet from memory in 1674 for the governor of New France, Louis de Buade, Comte de Frontenac. Jolliet's original map was lost when his canoe overturned. Redrawn by Father Claude Dablon in Quebec, it became known as the Manitoumie map because the Jesuit included a small drawing of a man in the center of the map. This drawing represented a stone outcrop shaped in the form of a human figure that the Indians of the region worshiped as a Manitou or spirit. To the Jesuits, the placing of this figure on the map symbolized the idolatrous nature of the inhabitants of that region. Dablon's copy was sent to Paris in 1678 along with his narrative of the expedition, which he had based on Marquette's notes and a copy of Jolliet's journal. Dablon was the director of the Jesuit mission in New France. Intensely interested in the geography of North America, he probably also compiled the first map of Lake Superior, which appeared in the *Jesuit Relations* in 1670–71.

The incorrect date of 1663, cited in the title of the first printing of the map, was corrected in the second printing.

National Map Collection
Public Archives of Canada
Ottawa, Canada

References: Delanglez (1948), 72–77; Swanton, 270, 271, 287, 294, 305.

II.3. *America Settentrionale Colle Nuoue Scoperte fin all'Anno 1688*

Vincenzo Coronelli
Engraving; 61 × 45 cm
Venice, 1688

Vincenzo Coronelli's decorative map of 1688 was issued in two sections for inclusion in his atlas *Atlante Veneto* (Venice, 1690–96). The western section is described here. It was derived from his globe of 1683, the first printed work to show the entire length of the Mississippi River following La Salle's discovery in 1682.

The delineation of the upper course of the Mississippi was based on the model provided by explorer Daniel Greysolon Duluth and the Jesuit mapmaker Claude Bernou. Two parallel rivers, one of which is shown emanating from *Lago Buado* (Lake Mille Lacs), enter the main course of the Mississippi from the northeast, while the main course flows from the northwest. At the apex of this triangle, the map depicts in perspective, with four larger towers, the Dakota Sioux village (*Issatis Populi*) that Duluth visited in July 1679. Halfway between this village and Hudson Bay, and to the northeast of *Lago Buado,* the cartographer depicted a large distinctive lake with four short rivers flowing from a small mountain range. It is named *Lage du Nadouessans* (Lake of the Sioux) and was apparently derived from Jean-Baptiste Franquelin's map of 1686. On the latter map, Franquelin had enlarged and tranformed Duluth's *Lac Buade* into *Lac des Sioux.* Coronelli's misinterpretation of Franquelin's work, therefore, led him to include two renditions of Lake Mille Lacs.

The lower course of the Mississippi River was derived from the La Salle-Franquelin model, which located the mouth of the Mississippi on the far western shore of the Gulf of Mexico, near the Rio Grande and some six hundred miles west of its actual position. This location was from La Salle, who explored the lower course of the Mississippi in 1682. By reporting that the Mississippi was located further to the west, he had hoped to mislead French officials into believing that the Spanish mines in New Mexico were closer to French territory than was originally believed to be the case.

The region north and west of the upper course of the Mississippi remains unfinished for lack of information. The only suggestion of a great interior plains is found in the portrayal of pictorial tree symbols scattered throughout this area and a note near the western headwaters of the Mississippi indicating the existence of the *Tinthonha o' Gens des Prairies* (People of the Prairies), in reference to the western Dakota Sioux. The southwestern shore of Hudson Bay, which later served as a base for English penetration of the northern prairies, is shown with a note respecting Thomas Button, the English navigator who was the first to explore that region in 1612.

The southern region west of the Mississippi is depicted more carefully than on earlier maps. It is bounded on the southwest by the Rio Grande, which is depicted correctly with its lower course directed eastward to the Gulf of Mexico rather than the Pacific Ocean. The headwaters of the Rio Grande are located at approximately 40° north latitude in present Colorado. Near its source we find the mythical *Quivira,* but to the south toward the Gulf of Mexico are Taos and Santa Fe, twin destinations of later French traders who explored the southern plains.
The *Apaches Vaqueros* (Cowboy Apache) are shown to inhabit the region between the Mississippi and the Rio Grande. The source for this new information was Diego de Peñalosa, whose map of the Rio Grande was one of the first Spanish maps of the western periphery of the southern plains to become known to French and Italian mapmakers.

An Italian monk who worked in Paris and Venice, Coronelli served as cartographer to Louis XIV, which allowed him access to the manuscript maps submitted to the king of France by French explorers. Coronelli's maps and globes are distinguished by their fine engraving. The pictorial renderings that adorn this map were derived primarily from the works of Theodore de Bry. In addition to his official position in France, Coronelli was cosmographer to the Republic of Venice and founder of the first geographical society, the Cosmographical Academy of the Argonauts.

University of Nebraska Exhibition
Yale University

Library of Congress Exhibition
G1015.C64 1695
Geography of Map Division
Library of Congress

References: Wheat, 1: 44–45; Devoto, 138.

II.4. *A Map of ye Long River and of some others that fall into the small part of ye Great River of Mississippi which is here laid down*

Louis-Armand de Lom d'Arce, Baron de Lahontan
Engraving
London, 1703

During the Great Age of Discovery, geographical lore was sometimes transformed into geographical fact by unscrupulous adventurers who wished to take advantage of a gullible public. One of the most notorious examples relating to the Great Plains region was the creation of the Long River by Baron de Lahontan. A lieutenant in the French Army, serving on the western frontier at the post of Michilimackinac, Lahontan compiled a fictitious map to support his claim that he led an exploratory expedition in 1688 which discovered a river to the western sea called the "Riviere Longue." This "product of imagination and speculation," as one historian described it, was first published in Amsterdam in 1703 along with an account of his travels in New France between 1683 and 1693. Extremely popular, the book and map were quickly translated into English and German. This copy is from the London edition entitled *New Voyages to North-America.*

Lahontan's map depicts a great river tributary of the upper Mississippi that flows directly westward across open country from the general vicinity of the Minnesota River to a mountain chain that separates the headwaters of the Long River from those of another which flows westward into the Pacific Ocean. Along the banks of Long River, Lahontan listed the names of several spurious Indian tribes, the *Eokoros, Esanapes, Gnacsitares,* and *Mozeemleks.*

Lahontan claimed that his information of the Long River was obtained from a Gnacsitares Indian who represented the geography of the region "by way of a map upon a Deer's Skin; as you see it drawn in this map." The Indians, he later noted, in an accurate analysis of their value as guides and mapmakers, "are as ignorant of *Geography* as of other *Sciences,* yet they draw the most exact Maps imaginable of the Countries they're acquainted with, for there's nothering wanting in them but the Longitude and Latitude of Places. . . . These *Chorographical Maps* are drawn upon the Rind of your *Birch Tree;* and when the Old Men hold a Council about War or Hunting, they're always sure to consult them."

Lahontan embellished the account of his pretended travels and accompanying map with enough accurate information obtained directly from Perrot, Duluth, Tonti, and various Indians, and from the published works of Membré, La Salle, Marquette, and Hennepin, to give just enough credence to his alleged tale of discovery to induce many European geographers to incorporate the Long River on their maps. The Long River persisted on maps in various forms late into the eighteenth century, particularly in England, where R. W. Seale included it on his *Map of North American . . . from the latest and best observations,* which was published in 1785.

Newberry Library
Not in the Library of Congress Exhibition

References: Thwaites; Verner and Stuart-Stubbs, 92.

II.5. *Carte du Canada ou de la Nouvelle France et des Decouvertes qui y ont été faites* (Illustrated)

Guillaume Delisle
Engraving; 50 × 65 cm
Paris, 1703

Carte de Mexique et de la Floride des Terres Angloises et des Isles Antilles du Cours et des Environs de la Rivière de Mississipi

Guillaume Delisle
Engraved by C. Simonneau; 47 × 64 cm
Paris, 1703

These two maps represent important advances in the cartography of the Great Plains region. Although the name Guillaume Delisle appears in the title of both maps, the major part of the design and compilation was done by his father, Claude. A series of some 120 large-scale regional sketch maps of the western interior in the hand of Claude Delisle still survive in the Archives du Service Hydrographique, and were apparently used in the construction of these maps. The sketch maps were based on numerous printed and manuscript maps and reports relating to North America that were collected by Claude Delisle before 1703.

The region northwest of Lake Superior depicted on the *Carte du Canada* was derived from Jean-Baptiste Franquelin's manuscript maps showing an immense *Lac Des Assiniboüels* covering much of central Manitoba and connected to Hudson Bay by the *Bourbon* (Nelson) River. The delineation of the upper Mississippi and the nomenclature of the Indian tribes were also based on Franquelin and the work of Father Hennepin. A new western feature is the *R[ivière] S[t]. Pierre* (Minnesota River), which is shown entering the Mississippi River from the west, in its approximate location just south of St. Anthony Falls. The source of the Minnesota River is shown as *Lac des Tintons* (Big Stone Lake), which today divides South Dakota and Minnesota. The Tintons or Teton Sioux were located on the shores of Big Stone Lake when the French arrived, having been pushed westward onto the Dakota Plains by Chippewa who were armed with European weapons. The depiction of the Minnesota River was based on information obtained from Pierre-Charles le Sueur, who entered the lower Minnesota in April 1700 in search of copper deposits. It first appeared on Franquelin's manuscript of 1699 from information received from Le Sueur as he was preparing to leave France for Canada in 1697, but its course is grossly generalized.

For the lower portion of the Mississippi River, Delisle relied heavily upon La Salle's *Carte de Mississippi;* the reports of La Salle's Italian lieutenant, Henry de Tonti, who explored the river four times between 1682 and 1700; Pierre le Moyne, Sieur d'Iberville's letters, journals, and maps relating to his explorations of the lower Mississippi, 1699–1700; and Le Sueur's memoirs, from which Delisle constructed a five-part map showing the course of the Mississippi River from its headwaters at 49° north latitude to the Gulf of Mexico. Based on information from Iberville, whom Guillaume Delisle acknowledged in the title block of his manuscript *Carte des Environs du Missisipi,* dated 1701, Delisle moved the mouth of the Mississippi eastward to its approximately true position, correcting a geographical misconception, initiated by La Salle and perpetuated by Franquelin and Coronelli, which showed the Mississippi as entering the Gulf six hundred miles to the west. Delisle represents the mouths of the Arkansas and Red rivers in their approximate positions, but the courses of these major waterways to the southern plains are incorrectly aligned in a northwesterly direction.

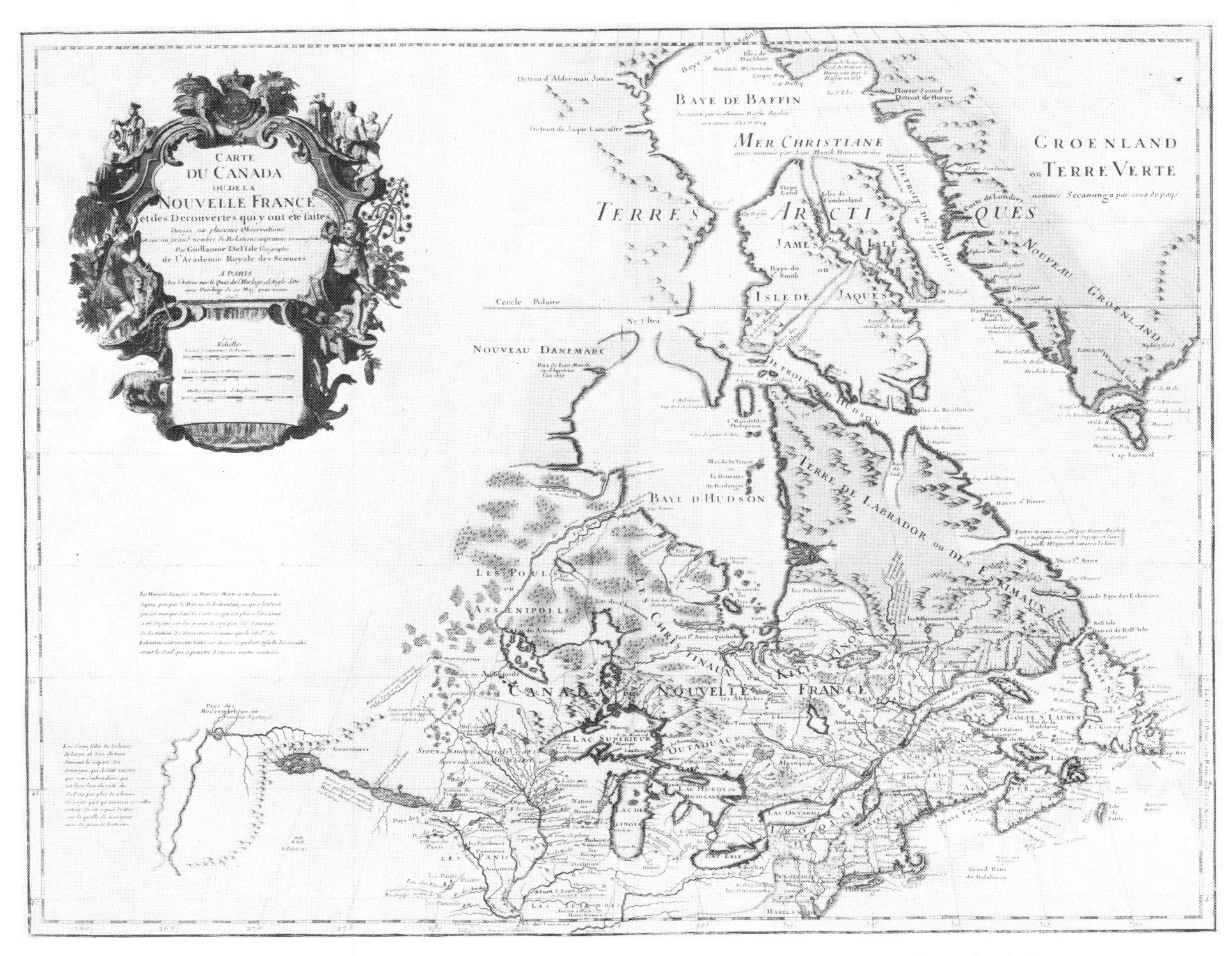

II.5. "Carte du Canada ou de la Nouvelle France. . . ." (Don Forke Collection, Lincoln, Nebraska)

The Rio Grande, the southwestern boundary of the southern plains, was derived from the Peñalosa-Coronelli model (Map II.3).

Delisle paid special attention to the Missouri valley, incorporating new information from Gabriel Marest, the Jesuit missionary who established the first settlement in the valley in 1700. The main course of the Missouri River is delineated for the first time in its more familiar alignment, curving in a northwesterly direction as far north as present-day Sioux City, rather than due west as portrayed on Delisle's map of 1700. Although depicted earlier by Franquelin, the Platte River is not shown. To the west of the Missouri, Delisle notes the existence of *Les Prairie.*

Despite Delisle's careful analysis of explorer's reports and maps, the rivers emanating from the plains were highly generalized, and in two instances completely mythical. In his desire to portray the River of the West, a venerable concept in French geography, Delisle fell victim to the fanciful notions of Baron de Lahontan, a French soldier who claimed to have explored the region, and to the testimony of Pierre-Charles le Sueur, who based his account on legendary Sioux lore. Both rivers are marked on the 1703 map. Lahontan's *Riviere Longue* is depicted as extending from the Mississippi northwestward to a chain of mountains separating it from another river leading to the Pacific Ocean (Map II.4). Its lower course, called the *Moingona,* parallels the lower course of the Missouri. Le Sueur's concept of the *Meschasipi ou Grande R[ivière]* is depicted to the southwest of the Missouri, flowing southwestward from a fictitious *Lac des Panis.* Although the Delisles may have been skeptical of Lahontan's Long River, noting on the map that the "Baron" may have "invented these things," it continued to appear on maps during the rest of the eighteenth century despite the fact that the Delisles' map for 1718 does not include it (Map II.7).

Map 1
Don Forke Collection
Lincoln, Nebraska
Not in the Library of Congress Exhibition

Map 2 Lowery Collection 256
Geography and Map Division
Library of Congress
Not in the University of Nebraska Exhibition

References: Delanglez (1939), 229; Delanglez (1943), 275–98; Hamilton, 655–56.

II.6. *Hudson's-Bay Country*

Attributed to a member of Pierre Gaultier de Varennes, Sieur de la Vérendrye's exploring expeditions; copied by Johann Georg Kohl
Manuscript; 53 × 86 cm
Original ca. 1740; copy, 1855–59

The original of this map, drawn about 1740, was based on the memoirs of Sieur de la Vérendrye. It was presented in 1750 to the Dépôt des Cartes et Plans de la Marine by Roland-Michel Barrin, Marquis de la Gallissonniere, commandant general of New France from 1747 to 1749. Beginning in 1731, La Vérendrye, France's leading western explorer, pushed beyond Lake Superior in search of a route to the "Western Sea." He eventually reached the Canadian prairies and in 1738 the Mandan villages on the great bend of the upper Missouri River. Maps such as this one were compiled by natives and French members of his expeditions and then sent back to Paris by way of Quebec to be incorporated in printed maps. On this map, the border lakes from Lake Superior to *Lac Gouinipigue* (Lake Winnipeg) are displayed with some accuracy, and reference to the Canadian prairie is found in the name of *Lac des Prairies* (Lake Manitoba). In the southwest, just below *Riviere des Assiniboilles* (Assiniboine River), is one of the earliest representations of the great bend of the Missouri River. It was also called the *Riviere des Mantanes* in recognition of the Mandan Indians, whose villages are symbolized by small circles.

This copy of La Vérendrye's map was made from the original by Johann Georg Kohl. A German geographer, Kohl copied some 750 maps from German, French, and English collections relating to the history of North American discoveries. While Kohl was employed by the U.S. Coast Survey from 1854 to 1858, many of his maps were recopied for American scholars as a result of a $6,000 grant from the U.S. Congress.

Kohl Collection, No. 130
Geography and Map Division
Library of Congress
Not in the University of Nebraska Exhibition

References: Kavanagh, 91; Wolter, 10.

II.7. *Carte de la Louisiane et des pays voisins* (Illustrated)

Jacques-Nicolas Bellin
Engraving; 48 × 61 cm
Paris, 1750; Revised 1755

The dominance of the Delisle family in French mapping is illustrated by this map published under the direction of Jacques-Nicolas Bellin, the senior hydrographic engineer of the French Dépôt des Cartes et Plans de la Marine from 1721 to 1772. Despite the fact that French explorers had reached the Mandan villages on the Great Bend of the upper Missouri in 1738, the western portion of Bellin's map of 1750 was copied directly from Guillaume Delisle's map of 1718.

It depicts the western interior framed on the east by the Mississippi River, on the north by the Minnesota and the upper course of the Missouri River, on the south by the Gulf Coast, and on the west by the Rio Grande and a chain of mountains stretching to 45° north latitude,

II.7. "Carte de la Louisiane et des pays voisins." (Library of Congress)

where it almost touches the upper course of the Missouri River.

The region south of the Missouri appears to have been taken from Father Le Maire's map *Carte nouvelle de la Louisiane,* drawn in 1716, and Vermale's manuscript map entitled *Carte Generale de la Louisiane ou du Miciscipi,* compiled in 1717. Both of these maps are now preserved in the library of the Hydrographic Service in Paris. Vermale, a French cavalry officer, appears to have introduced the notion of the Red and Arkansas rivers protruding in a northwesterly direction deep into the Great Plains, with the Red River almost reaching the point where the headwaters of the Rio Grande and the course of the Missouri River nearly meet. As shown on this map, the headwaters of the Red River are located in present central Nebraska, some 490 miles north of their correct position in the rolling plains of northwestern Texas.

The courses of two major western affluents of the Missouri are shown for the first time. These are the *Grande Riviere des Cansez* (Kansas) and the Panis (Platte) River, derived from the description of Étienne Veniard, Sieur de Bourgmont, of his expedition up the Missouri as far north as the Platte River in 1714. While the mouths of both rivers are placed near their proper latitudes, the Kansas is incorrectly depicted as being much longer than the Platte, extending northwesterly almost to 44° north latitude in the vicinity of the Black Hills. Bourgmont's exploration of the Osage River also contributed to a more accurate rendition of this region.

An important feature of Delisle's map of 1718, which was followed by Bellin, was his attempt to locate more accurately the Indian tribes that occupied the river valleys and prairies west of the Mississippi. These groups included the Teton Sioux, Omaha, Pawnee, Loup (Panimaha), Iowa, Kansas, Missouri, and Osage. On the Delisle-Bellin map, the Kansas are located farther up the Missouri River, suggesting that a northern migration occurred between the publications of Delisle's 1703 and 1718 maps. The Delisle-Bellin map also depicts villages of Comanche (*Padoucas*) on the headwaters of the Kansas River and the Rio Grande, which were shown earlier on Vermale's map.

An earlier version of Bellin's map appeared in Father Pierre F. T. de Charlevoix's *Histoire et Description Generale de la Nouvelle France* (Paris, 1744), under the title *Carte de la Louisiane Cours du Mississipi et Pais Voisins.*

Lowery Collection 406
Geography and Map Division
Library of Congress
Not in the Library of Congress Exhibition

II.8. *A Topogra[phic] Sketch of the Missouri and Upper Missisippi: Exhibiting The various Nations and Tribes of Indians who inhabit the Country: Copied from the Original Spanish MS. Map.*

Attributed to Antoine Soulard; annotated by William Clark
Manuscript; 42 × 53 cm
St. Louis, 1795

This map was probably drawn for Jean-Baptiste Truteau by Antoine Soulard just before Truteau's exploration up the Missouri River. Although the original map has been lost, several copies survive. The one displayed in the exhibit was obtained by Meriwether Lewis and used by Lewis and Clark in planning their expedition. It contains annotations by William Clark, principally on the lower portion labeled "Note." Another manuscript in French is preserved in the Bibliothèque du Service Hydrographique in Paris. A note in the cartouche indicates that this latter map was drawn for Francisco Bouligny, a colonel of the Permanent Regiment of Louisiana, from "information given by various traders." Soulard, who may have been trained in surveying and mapmaking during service with the French Royal Navy, held the office of surveyor general for Spanish Louisiana from shortly after his arrival in St. Louis in 1794 until the transfer of the territory to the United States in 1804.

Soulard's representation of the Missouri River is elongated in an east-west direction because of inaccurate measurements for longitude, which plagued early surveyors and mapmakers. The Great Bend and the Grand Detour, noted features of the Missouri, are merged and grossly exaggerated. Despite these distortions, this map depicts the entire region of the North American plains in more detail than did previous maps. In addition to the upper course of the Missouri, the Platte River is portrayed more realistically than on earlier maps, and the names of several Indian tribes to the west of the Mandan villages are shown for the first time. The region of the northern plains contains information about the Saskatchewan and Assiniboine rivers obtained from James Mackay, an English trader who explored this region in the 1780s and who collaborated with Soulard in the preparation of this map. The regions surrounding the North American plains were derived from Jonathan Carver's map of 1781. A dotted line marks Carver's route of exploration.

Western Americana Collection
Beinecke Rare Book and Manuscript Library
Yale University
New Haven, Connecticut

Reference: Diller (1955), 175–80.

II.9. *[Route Map along the Missouri River from Fort Charles to the Mandan-Hidatsa Villages, sheet 2]*

John Evans
Manuscript copy of original; 20 × 15 cm
1796–97
John Evans came to North America from Wales in search

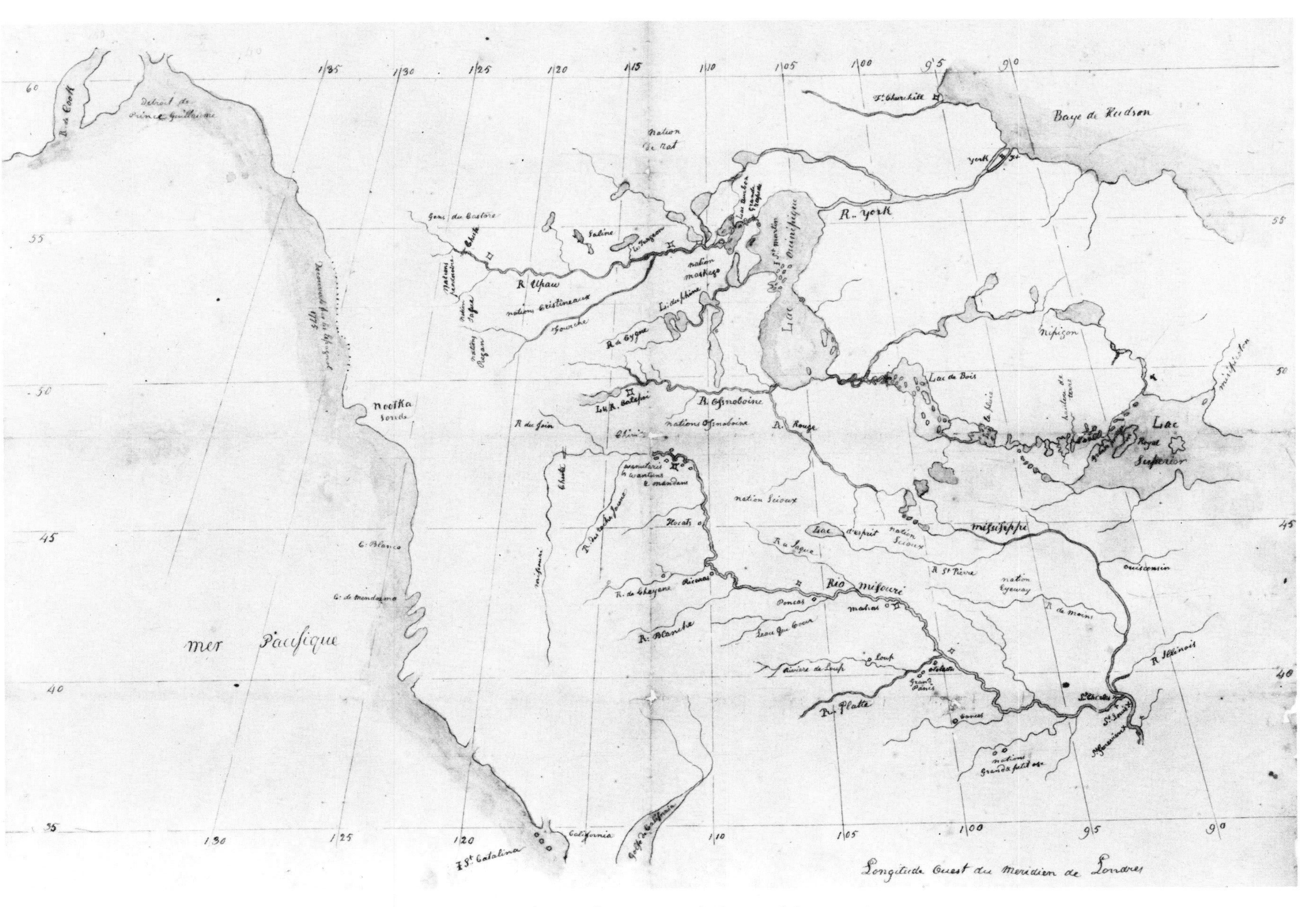

II.10. Map of Spanish Louisiana. (Library of Congress)

of the illusory "Welsh Indians" believed to be found on the Missouri River. This chart traces Evans's route up to the Missouri from Fort Charles, located near modern Sioux City, Iowa, to the Mandan and Hidatsa villages at the mouth of Knife River in present-day North Dakota. The section on display is one section of a seven-sheet copy; it shows the "Grand Detour," a noted feature of the Missouri River. Evans and James Mackay, the expedition leader who remained at Fort Charles, were instructed by Spanish authorities in St. Louis to "discover a passage from the sources of the Missouri to the Pacific Ocean." Although their venture was unsuccessful, this was the first accurate map of the Missouri River above Fort Charles to illustrate in detail major river channels and principal tributaries. Evans's authorship was hidden for nearly forty years until Aubrey Diller noticed that maps credited to William Clark had been misidentified, most notably by a reading of Saint Charles for Fort Charles. This map was carried by Lewis and Clark on their expedition but is a copy that was made for them. Evans's original map has been lost.

Western Americana Collection
Beinecke Rare Book and Manuscript Library
Yale University
New Haven, Connecticut

References: Wood, 39–42; Diller (1946), 516–18.

II.10. *[Map of Spanish Louisiana]* (Illustrated)

Anonymous
Manuscript; 39 × 49 cm
ca. 1797

This is the earliest surviving map of the North American interior to provide a reasonably accurate geographical representation of the middle and upper Missouri River watershed. The middle course of the Missouri from the mouth of the Platte to the Mandan-Hidatsa villages on the Great Bend was based on the journals and maps of James Mackay and John Evans, who were employed by the Spanish-chartered Missouri Company in St. Louis to explore and chart the Missouri River from 1795 to 1797. The upper course beyond the Mandan-Hidatsa villages is depicted with its headwaters located near the latitude of Santa Fe in conformance with contemporary Spanish speculation that New Mexico was close to the Missouri River. The Yellowstone River, also shown and named for the first time, may have been derived from an Indian map given to Mackay by Jacques d'Eglise, the first Spanish-sponsored explorer to reach the Mandan-Hidatsa villages from St. Louis. The delineation of the northern plains and the Manitoba lakes was taken from Soulard's map of 1795 (Map II.8).

This map was found with a collection of other maps once in the custody of William Clark and later transferred to the Bureau of Indian Affairs.

Lewis & Clark Collection "i"
Geography and Map Division
Library of Congress

References: Allen, 43–44; Nasatir, 2: 415–16.

III. British Exploratory Maps

British exploratory mapping of the North American plains dates from 1760 to 1865. Although English seafarers reached Hudson Bay in 1612 and had mapped its general outline by 1632, they did not begin mapping the adjacent Great Plains region until after the collapse of French power in Canada.

Beginning in the 1760s, the Hudson's Bay Company and later the North West Company prepared manuscript sketch maps on the geographical character of portions of the Canadian prairies. This information was obtained from natives and by direct observation. To ensure the accuracy of these maps, the Hudson's Bay Company established the position of Surveyor of the Company's Inland Settlements. Manuscript maps compiled by the Hudson's Bay Company were sent to London, where they were retained for the company's exclusive use until 1791, when the geographical information they contained was incorporated by Aaron Arrowsmith in his large printed maps of North America.

III.1. *A New Map of North America, From the Latest Discoveries, 1778.*

Jonathan Carver
London, 1778
Engraving; 33 × 36 cm

Following the Treaty of Paris in February 1763, French possessions in North America were transferred to Great Britain and Spain, an action that opened the interior plains and prairies to British exploration and mapmaking. Carver's map was published with his popular *Travels Through the Interior Parts of North America in the Years 1766, 1767, and 1768,* which was read widely throughout Europe. The delineation of the lower St. Peter (Minnesota) River was based on his visit to that region in 1766–67 as the first English-speaking explorer to push beyond the upper Mississippi River valley. The western interior, however, was copied from Bellin's map of 1755, which in turn was derived from the maps of La Vérendrye. One innovation introduced by Carver is a dotted line connecting the Mandan River (the Great Bend of the Missouri River) with the imaginary "River of the West," which empties into the Pacific Ocean. This misconception led later

American explorers to believe that the Pacific Ocean could be reached by a direct water route.

G3300. 1778. C32 Vault
Geography and Map Division
Library of Congress
Not in the Library of Congress Exhibition

III.2. *A Map of the North West Parts of America, With The Utmost Respect, Inscrib'd To His Excellency, Sir Guy Carleton, Knight of the Bath: Captain General and Governor of the Province of Quebec: General and Commander in Chief of His Majesty's forces In the Said Province, and Frontiers thereof.*

Alexander Henry
Manuscript, with letterpress title and notes: 102 × 215 cm
1776?

This is the first known map to delimit a portion of the North American plains. It was compiled by Alexander Henry following his visit to the northern plains in 1775–76, and apparently presented to Lord Dorchester, then governor of Canada, just before Henry's departure for England and France in 1776. Henry was one of the first English-sponsored traders from Montreal to reestablish the western fur trade following the British victory that ended French control of the North American interior.

The map encompasses an extensive region stretching from Lake Superior and Hudson Bay west to the headwaters of the Saskatchewan River, and from the Churchill River south to the headwaters of Red River of the North. Henry's configuration of Lake Winnipeg, the placement of *Lake Bourbon* (Cedar Lake), and the delineation of the Saskatchewan (here named the *Posquayaw,* which meant "prairie" or "desert") represent the most accurate rendition of these geographical features to date.

In addition to delineating the natural eastern "Course of the Great Plaines," a line which extends southward from the headwaters of the Churchill River to the Red River of the North, Henry distinguished between northern and southern plains. The former covers the upper Saskatchewan River region; the latter, the region of the Assiniboine and the Red River of the North. The tribal ranges of ten major Indian groups are shown by different color washes.

The map also contains several letterpress notes briefly describing regions, Indian tribes, and customs. With respect to the "Great North Plaines" and its inhabitants, the Assiniboine, Henry wrote: "Of Unknown Extent, Inhabited by the Ousineboins, who, having no fixed Villages live in Tribes like Wandering Scythians, hunt the Buffaloe, which when they destroy all around, they repair to other parts, where these Animals are in greater abundance, few of these Savages use Fire Arms: Which, those Bordering on the River Pasquyai; purchase at Fort des Prairies, Fort Dauphin, and Ousnaboins River. Their Arms chiefly consist of the Bow and Arrow, they cloath themselves with Buffaloe skins, and Expose without shame those Parts, which the generality of Mankind hide with so much care, each Man takes as many Wives as he can maintain." The "Great South Plain," Henry noted is occupied by "THE MANDANES, a Nation different from the Ousinaboins, both in Manners and Language, they inhabit the Southern Part of the Plains, cultivate land, raise Indian-Corn, Squashes, Beans and Tobacco, they have little Intercourse, and no trade with Europeans."

G 3470 1776. H4 Vault
Geography and Map Division
Library of Congress
Not in the University of Nebraska Exhibition

III.3. *[Moses Norton's Drt. of the Northern Parts of Hudson's Bay laid down on Indian Information brot Home by Him]*

Drawn for Moses Norton by Northern Indians
Manuscript on Parchment; 63 × 86 cm
Fort Churchill, 1760

Toward the end of the French regime in 1763, the Hudson's Bay Company began exploratory mapping of the watercourses traversing the Canadian prairies. This is the earliest map which shows a highly generalized western interior extending to the Rocky Mountains, including the North Saskatchewan and the Saskatchewan-Nelson rivers. It was based on Indian information given to Moses Norton, the Fort Churchill factor. Although distances and directions are greatly distorted, this map provided English explorers with their first image of the northern plains.

(HBCA G2/8)
Hudson's Bay Company Archives
Provincial Archives of Manitoba

References: Warkentin and Ruggles, 88.

III.4. *Hudson's Bay's Country* (Illustrated)

Peter Pond; originally copied by J. Hector St. John de Crèvecoeur, and recopied by Johann Georg Kohl
Manuscript
Milford, Connecticut, 1785; copy, 1855–59

This is the most elaborate and detailed map of the western interior drawn by a fur trader since La Vérendrye's maps of the 1740s. It was compiled by Peter Pond, an American trader employed by the North West Company. Three variations of this map were prepared by Pond: the first was presented to Congress in March 1785; the second was given to Lord Hamilton in April 1785; and the third was prepared for presentation to the Empress of Russia in 1787. The map on display was copied by Johann Georg Kohl from a manuscript facsimile in the Archives

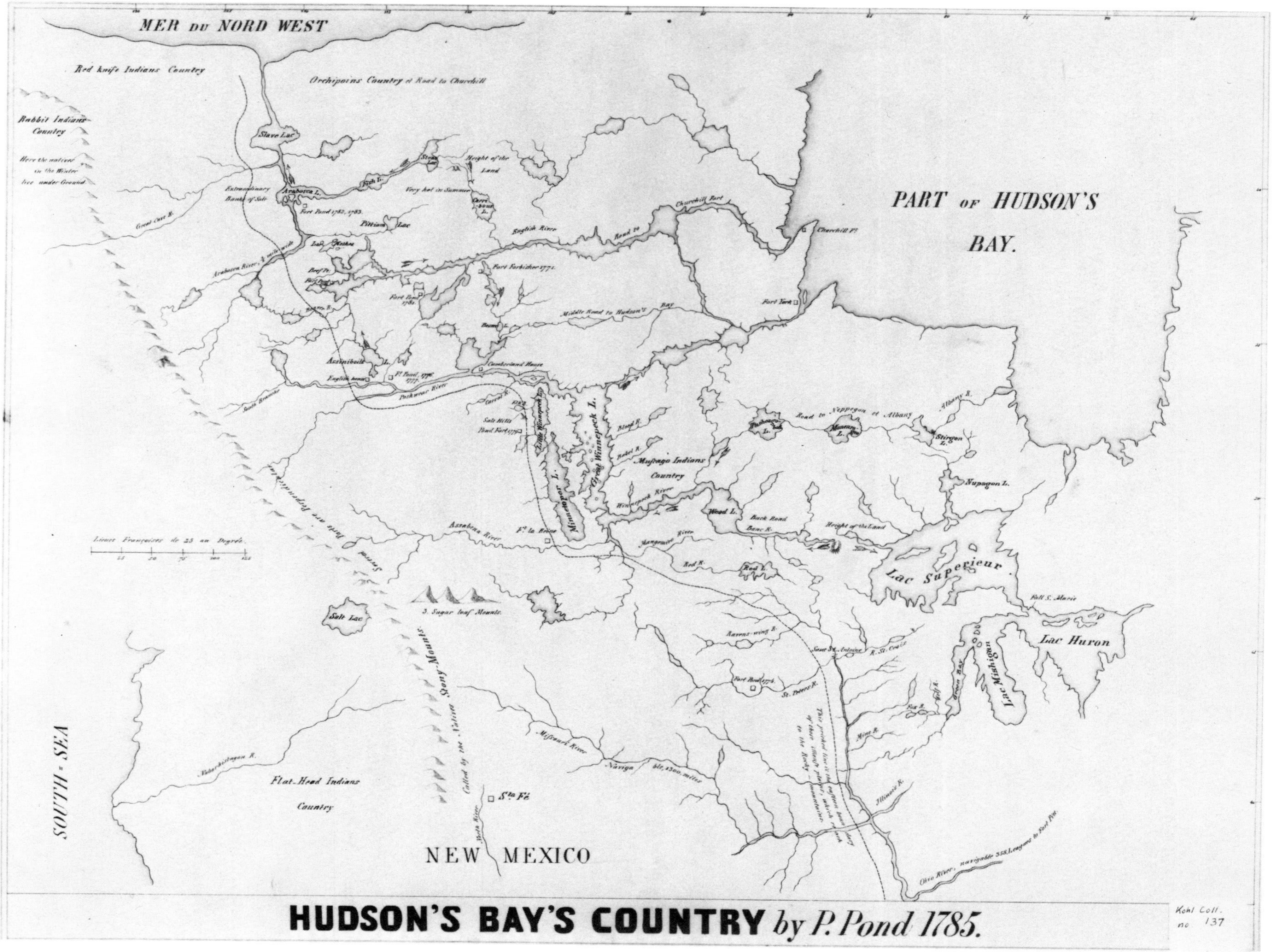

HUDSON'S BAY'S COUNTRY *by P. Pond 1785.*

III.4. "Hudson's Bay Country by P. Pond." (Library of Congress)

of the Hudson's Bay Company (London) made by J. Hector St. John de Crèvecoeur, the French consul in New York. The original manuscript, which has not been found, was presented to Congress in New York City by Pond on 1 March 1785. The title of Crèvecoeur's original copy is *Copy of a Map, presented to the Congress by Peter Pond, a native of Milford in the State of Connecticut. This extraordinary man has resided 17 years in these countries, and from his own discoveries as well as from the reports of Indians, he assures himself of having at last discovered a passage to the North Sea. He is gone again to ascertain some important observations. New York. 1st March, 1785.*

This map's major contribution is the introduction of new hydrographic features north of the Saskatchewan, where Pond explored the Athabasca Lake and River region and was the first non-Indian to discover a river—the Clearwater—flowing to the Pacific Ocean. With respect to the northern Plains, Pond's delineation of the upper Saskatchewan is similar to that of Alexander Henry, although Pond shows that northern and southern branches trending in a more westerly direction and almost touching the Stoney Mountains (Rocky Mountains). The configurations and placements of the Manitoba lakes and the delineation of the Red River of the North, however, represent a notable advance over earlier maps. For the first time Lakes Winnipegosis, Dauphin, and Manitoba are portrayed as separate lakes in their proper positions, and the Sheyenne River (a 325-mile-long western tributary of the Red River of the North) and Devil's Lake (which lies along the upper course of the Sheyenne) are shown as major geographical features of the Dakota prairies. To the west of the Sheyenne River, Pond depicts a three-peaked mountain symbol with the name "3 Sugar loaf Mounts," which may represent the Turtle Mountains, a small range crossing the Canadian-American boundary north of the headwaters of the Sheyenne, which had long served as a familiar prairie landmark.

Another important feature of this map is the delineation of the "eastern boundary of those imense plains, which reach to the Rocky-mountains," which is shown by a dotted line stretching from Slave Lake south to below the Missouri River.

Kohl Collection, No. 137
Geography and Map Division
Library of Congress

Reference: Wagner.

III.5. *[Great Bend of the Missouri River]*

Attributed to David Thompson; copied by Meriwether Lewis, annotated by Thomas Jefferson
Manuscript; 43 × 52 cm
1798

A detailed sketch of the Great Bend of the Missouri, Knife, and Little Missouri rivers drawn by Thompson or based on his surveys of the Upper Missouri in early 1798 and copied by Lewis for the first leg of the Lewis and Clark expedition. Thompson was a British astronomer and surveyor employed by the North West Company. Trained to undertake scientific instrumental surveys, he mapped a large part of the American northwest with detailed accuracy. A notation in Jefferson's handwriting provides information on the latitude and longitude of the great bend of the Missouri: "Bend of the Missouri, Long. 101 25'—Lat. 47 32' by Mr. Thompson, astronomer to the N.W. Company in 1798." This section of the Missouri River was embodied in Aaron Arrowsmith's 1802 map of North America. In addition, Thompson's map shows the number of warriors, houses, and tents of the Mandan and the Pawnee Indian villages in the vicinity of the junction of the Missouri and Knife rivers, and the overland trail to the Souris River.

Lewis & Clark Collection "j"
Geography and Map Division
Library of Congress

III.6. *[An Indian Map of the Different Tribes, that Inhabit the East and West Side of the Rocky Mountains with all the rivers and other remarkable places, also the number of tents]* (Illustrated)

Ac ko mok ki (The Feathers); reduced by Peter Fidler
Manuscript; 46 × 33 cm
Chesterfield House, junction of Red Deer and South Saskatchewan rivers, 1801

One of the many early maps drawn by American Indians to guide European fur trappers and explorers. Ac ko mok ki, a Blackfoot Indian chief, drew this highly stylized map for Peter Fidler to show for the first time in some detail the headwaters of the Missouri and Saskatchewan river systems and adjacent plains. A surveyor, cartographer, and trader, Fidler was responsible for mapping much of the western interior of Canada for the Hudson's Bay Company from 1788 to 1822.

This map represents a composite effort and was derived from the direct observations of Ac ko mok ki and Fidler, from information obtained by Ac ko mok ki from other Indians, and from reports from other company traders. In his accompanying letter to the governor and committee of the company, dated 10 July 1802, Fidler wrote, "This Indian map conveys much information where European documents fail; and on some occasions are of much use, especially as they shew that such & such rivers & other remarkable places are, tho' they are utterly unacquainted with any proportion in drawing them." He further noted that "The source of the Red Deers & Saskatchewan rivers are put down from the accounts of some of our men who has been there. I have regulated their courses & Distances & imagine they will be found to be placed pretty near their true positions."

III.6. "An Indian Map of the Different Tribes, that Inhabit the East and West Side of the Rocky Mountains. . . ." (Provincial Archives of Manitoba)

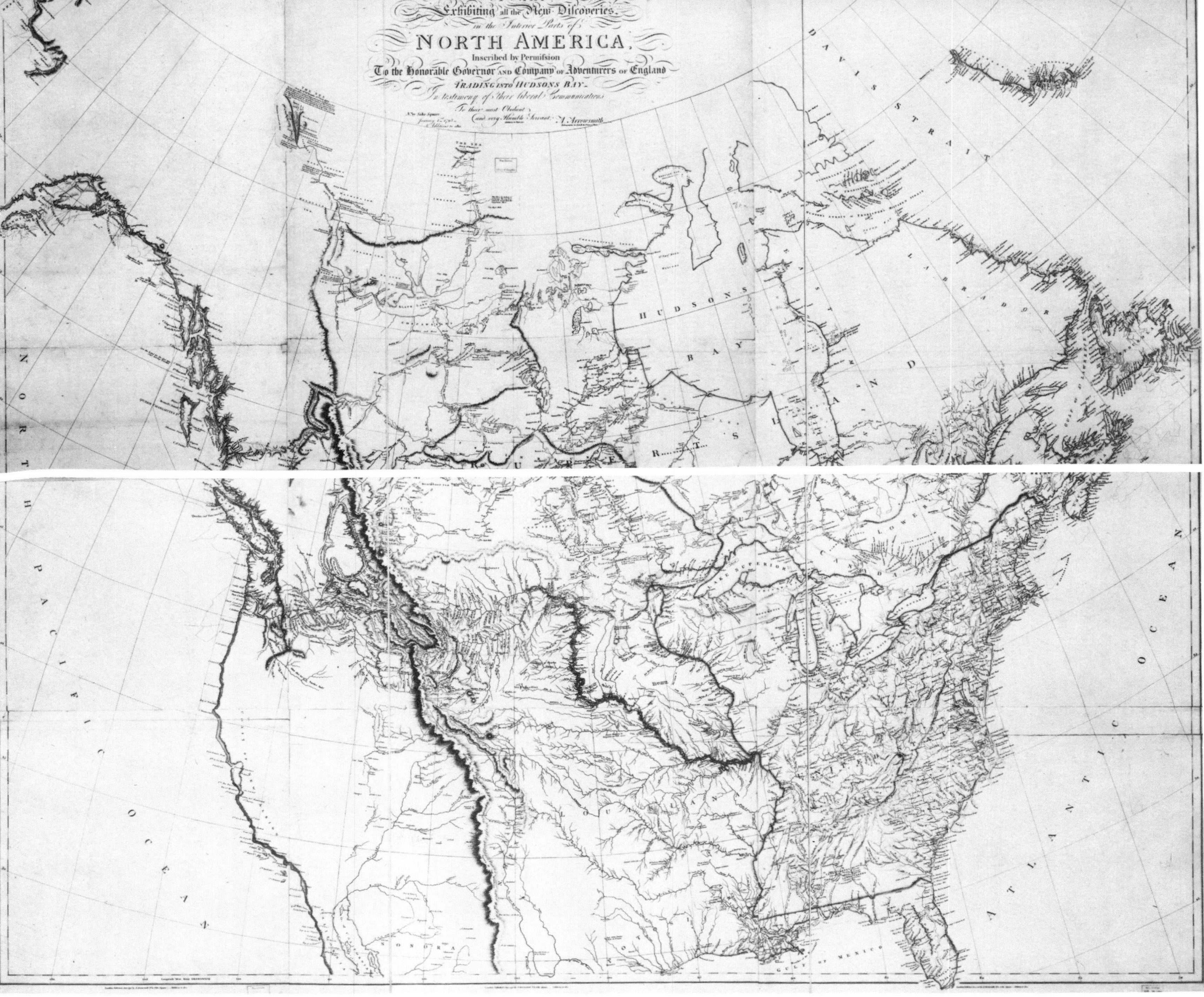

III.7. "A Map Exhibiting all the New Discoveries in the Interior Parts of North America. . . ." (Library of Congress)

In addition to drainage patterns, the map shows the location and census of thirty-two Indian groups (keyed to the map by numbers) and twelve significant mountains and highlands (keyed by letters). The heights of several of these mountains were measured from their base by Fidler with the aid of a sextant. "I am unacquainted with the method of finding the Altitude of mountains above the level of the Sea," he confessed; "probably other Instruments are requisite to ascertain that point."

This remarkable work was immediately incorporated into Aaron Arrowsmith's 1802 map of North America, in which form it provided Lewis and Clark with one of their few cartographic guides west of the Great Bend of the Missouri River.

HBCA E 3/2, 106d–107
Hudson's Bay Company Archives
Provincial Archives of Manitoba

Reference: Moodie and Kaye, 12–15.

III.7. *A Map Exhibiting all the New Discoveries in the Interior Parts of North America, Inscribed by Permission to the Honorable Governor and Company of Adventurers of England Trading into Hudsons Bay* (Illustrated)

Aaron Arrowsmith
Engraved; 2 sheets, each 62 × 146 cm
London, 1814

Arrowsmith's map was the most comprehensive map of the western interior for its time. Geographic information about the American Northwest obtained by Fidler, Thompson, and other Hudson's Bay Company explorers remained restricted until the mid-1790s, when the company made its reports and maps available to Aaron Arrowsmith, a distinguished London mapmaker. From 1795 to 1857 at least nineteen editions or revisions of Arrowsmith's map of North America were published by Aaron or his successors. The 1802 edition was closely consulted by Thomas Jefferson and Meriwether Lewis in preparation for the Lewis and Clark expedition of 1804 to 1806. On the 1814 edition, the Missouri River system is based on Lewis and Clark's published map.

North America 1814
Geography and Map Division
Library of Congress

References: Allen, 78–79; Warkentin and Ruggles, 140.

III.8. *A General Map of the Routes in British North America Explored by the Expedition under Captain Palliser, During the Years 1857, 1858, 1859, 1860.*

John Palliser
Lithograph; 32 × 128 cm
London: Stanford's Geographical Establishment, 1865

A classic work that was to serve as a basic cartographic source for the study of the Canadian prairies, this route map covers the region from Lake Superior to the Rocky Mountains and from the international boundary to the Athabasca and North Saskatchewan rivers. Palliser, who conceived of the idea to explore the British prairies north of the border and arranged support from the British Colonial Office, patterned the expedition after the U.S. Pacific Railroad Surveys, which surveyed the American West from 1853 to 1856. Palliser's detailed map was based on John Arrowsmith's 1854 map of British North America, to which significant corrections and additions were made. The revision was begun by Arrowsmith, a nephew of Aaron Arrowsmith, who had planned to collate the expedition's materials with his own, which he had been collecting over many years. After he had devoted some twenty-eight hundred hours to the task, however, the project was transferred to Edward Stanford in 1863 by the Colonial Office, which became discouraged over the delay. The map was issued as part of Palliser's final *Report* to both houses of Parliament in 1865.

In addition to correcting or improving the positioning of the North and South Saskatchewan; many tributaries of major rivers; and the headwaters of the Qu'Appelle, Pembina, and Souris rivers, Palliser depicts for the first time the three prairie steppes. This map also includes many notes describing the geology, fauna, and flora of the northern plains.

M 71. 185. P17
Nebraska State Historical Society
Not in the Library of Congress Exhibition

References: Spry, 1963; Spry, 1968: xcvi, c–ci, 495–97.

IV. American Exploratory Maps

While the Canadian prairies were mapped primarily by traders and missionaries, the southern and central regions of the North American plains were mapped by military expeditions. The first phase of mapping, from 1803 to 1807, was directed by Thomas Jefferson, who, as president, initiated the exploration of both the trans-Missouri region by Captains Meriwether Lewis and William Clark and the southern plains by Lt. Zebulon Montgomery Pike.

The second phase of American mapping, from 1819 to 1863, was carried out by the War Department's elite Corps of Topographical Engineers as part of their coordinated program to explore and map the trans-Mississippi West. This phase began with Major Stephen H. Long's overland expedition to the Platte River and ended with Lt. Gouverneur K. Warren's reconnaissance of the Dakota country.

IV.1. *[Map of Western North America]*

Nicholas King; annotations in brown ink by Meriwether Lewis
Manuscript; 51 × 78 cm
Washington, D.C., 1803

This composite map reflects the geographical concepts of government leaders on the eve of the Lewis and Clark expedition. It was compiled by King at the request of Thomas Jefferson and Albert Gallatin, secretary of the treasury, from published and manuscript sources. King's source for the Mandan-Hidatsa villages and the Great Bend of the Missouri River was David Thompson's manuscript map of 1798 (Map III.5). The general outline of the headwaters of the Missouri River and the northern plains was taken from Aaron Arrowsmith's map of British North America, which was, in turn, based on the Ac ko mok ki–Fidler map of 1801 (Map III.6). The lower Missouri is generalized and shown by a broken line. It is believed that King's map was carried by Lewis and Clark at least as far as the Mandan-Hidatsa villages, where Meriwether Lewis added additional information obtained from traders. King, a War Department copyist and surveyor of the City of Washington, prepared a number of maps of western exploration for the War Department from 1803 to 1812.

Lewis & Clark Collection "a"
Geography and Map Division
Library of Congress

References: Allen, 103–104; Ehrenberg (1971), "Nicholas King."

IV.2. *A Map of part of the Continent of North America, Between the 35th and 51st degrees of North Latitude, and extending from 89° Degrees of West Longitude to the Pacific Ocean*

Nicholas King
Manuscript; 78 × 111 cm
Washington, D.C., 1805

Compiled in part from direct observations and surveys, this is the first official map of the trans-Mississippi West credited to Lewis and Clark. Although framed by the Mississippi on the east, the headwaters of the Rio Grande and Arkansas on the south, the Pacific Ocean on the west, and the North Saskatchewan and Lake Winnipeg on the north, the map focuses primarily on the Great Plains. The lower Missouri and its immediate vicinity are portrayed quite accurately. The Missouri River from Camp Dubois near St. Louis north to Fort Mandan at the Knife River above present Bismarck, North Dakota, was based on river charts prepared by William Clark during his ascent of the river; the lower Platte was derived from Mackay's maps and reports prepared in 1795–96. To the west of Fort Mandan and the Missouri River, the depiction of the major rivers is quite distorted and fanciful. Reminiscent of eighteenth-century Spanish and French maps, the upper courses of the major rivers of the Great Plains are greatly elongated toward the west, with the headwaters of the Missouri and the Yellowstone curving southward until they almost touch the headwaters of the Rio Grande, Arkansas, and South Platte rivers. The portrayal of the northern plains is from Aaron Arrowsmith's map of 1802.

In addition to hydrographic features, this map contains copious information concerning the location and size of Plains Indians situated between the Platte and the upper course of the Missouri. A noted feature of great interest to Lewis and Clark, "The war path of the Big Bellies Nation," is shown by a dotted line extending from the Mandan-Hidatsa villages to the Columbia River.

The original map was compiled by William Clark at their first winter quarters at Fort Mandan early in 1805. It was forwarded to the secretary of war in Washington just before the pathfinders' departure for the West Coast in order to give "the idea we entertain of the connection of these rivers, which has been formed from the corresponding testimony of a number of Indians who have visited the country." Under the direction of General Dearborn, Clark's map was recopied by Nicholas King "for the inspection of Congress."

University of Nebraska Exhibition
RG77, AMA 21 (1806 Manuscript edition)
Records of the Office of the Chief of Engineers
Cartographic and Architectural Branch
National Archives

Library of Congress Exhibition
G4050 1805.L Vault (1805 Manuscript edition)
Geography and Map Division
Library of Congress

References: Friis (1954); Moulton (1983), "Another Look."

IV.3. *[Sketch Map of the Missouri and Yellowstone Rivers]* (Illustrated as fig. 5.1)

Sheheke (Big White); copied by William Clark
Manuscript; 21 × 30 cm
7 January 1805

During the Lewis and Clark expedition the captains gathered information from the native peoples they met along the way. Occasionally the Indians would draw maps on animal hides or scratch out terrain in the earth. This is one of the better sketches taken from Indian information.

Sheheke, principal chief of a Mandan village, visited Fort Mandan on 7 January 1805 after returning home from a hunting trip. Clark noted that "he gave me a Scetch of the Countrey as far as the high Mountains, & on the South Side of the River Rejone [Yellowstone]." Clark apparently placed the names of the Yellowstone's affluents on the map as he traveled down that stream on his homeward trip in 1806. There is a nearly identical map on the reverse of this sheet.

Western Americana Collection
Beinecke Rare Book and Manuscript Library
Yale University

Reference: Moulton (1983), *Atlas*, 8 and maps 31a and 31b.

IV.4,5. *[Field Draft and Finished Copy for Route of the Lewis and Clark Expedition West of Fort Mandan, April 7–14, 1805]*

William Clark
Manuscripts; field draft sheet, 21 × 30 cm; finished copy, 20 × 15 cm
1805–1806

These two maps trace the course of Lewis and Clark along the Missouri River out of Fort Mandan, where they had wintered during 1804–1805 near the mouth of the Knife River in present-day North Dakota. The original field draft sheet containing course and distance tables represents the earliest extant example of Clark's field mapping. Not only were drafts like these made in the field, but Lewis and Clark also added the daily log of course and distance to the sheets. From those tables they were able to plot their course of travel on the map and use the background grid as an aid in plotting. It is likely that there were such field maps for the initial phase of the trip from near St. Louis to Fort Mandan, but they are lost to us today. The second map is a finished copy of the field draft covering nearly the same area along the Missouri River out of Fort Mandan at a scale of six miles to the inch. Because of the condition of the paper and the clarity of penmanship, it seems likely that this copy was made after the expedition was completed. Clark dropped the course and distance tables from this copy and made minor changes and corrections from the field draft.

Western Americana Collection
Beinecke Rare Book and Manuscript Library
Yale University

Reference: Moulton (1983), *Atlas*, 9 and map 33.

IV.6. *[Route Map of Lewis and Clark Expedition West of Fort Mandan, April 7–14, 1805]*

William Clark; copied for Maximilian, Prince of Wied-Neuwied, by Benjamin O'Fallon and an unidentified copyist
Manuscript; 26 × 40 cm
St. Louis, 1833

This book of maps was made for Prince Maximilian in St. Louis before his trip up the Missouri River in 1833. It contains thirty-four copies of original Lewis and Clark expedition maps, more than half of which are lost today. The maps were copied from Clark's large-scale maps of the course of the Missouri which were made available to Prince Maximilian by Maj. Benjamin O'Fallon, an Indian agent and Clark's nephew. The atlas is particularly important because it illustrates some nine hundred miles of the Missouri River from near St. Louis to Fort Mandan and also some sections of the Yellowstone River known only from the explorers' journals. Upon the completion of his Missouri River expedition, Maximilian, who was accompanied by the Swiss artist Karl Bodmer, returned to his estate on the Rhine River and had the separate maps bound into a small atlas. An English edition of his *Travels in the Interior of North America* was published in London in 1843, without the maps.

Joslyn Art Museum
Omaha, Nebraska

Reference: Wood and Moulton.

IV.7. *[Map of Yellowstone River showing Route of Lewis and Clark, July 25–27, 1806]*

William Clark
Manuscript; 41 × 30 cm
1806

This map by William Clark may be a finished copy from another set of field drafts of the Yellowstone River. It covers a portion of the return route of Lewis and Clark from "Pompy's Tower" to the mouth of the Big Horn River (*Ar sar tas*). Pompey's Pillar, a landmark named by Lewis and Clark for the child of their Indian guide Sacajawea, still bears the signature of William Clark, dated 25 July 1806. The torn portion contained another map of the Yellowstone upriver from this area. That piece, separated from those at Yale University, is at the Missouri Historical Society, St. Louis.

Western Americana Collection
Beinecke Rare Book and Manuscript Library
Yale University

Reference: Moulton (1983), 11 and map 10.

IV.8. *A Map of Lewis and Clark's Track, Across the Western Portion of North America From the Mississippi to the Pacific Ocean; By Order of the Executive of the United States. in 1804.5 & 6.*

William Clark; copied and reduced by Samuel Lewis

Engraved by Samuel Harrison; 30 × 70 cm
Philadelphia, 1814

This was the first published map to incorporate the discoveries of Lewis and Clark. It was compiled from William Clark's large manuscript map of the West by Samuel Lewis, a noted Philadelphia mapmaker, for publication with Nicholas Biddle's *History of the Expedition under the Command of Captains Lewis and Clark.* Appointed superintendent of Indian affairs at St. Louis in 1807 and governor of Missouri Territory in 1813, William Clark continued to update his large master map of the West by interviewing fur trappers and explorers upon their return to St. Louis. The published version of Clark's map is the first to show the existence of the Black Hills of South Dakota and other outlying mountains of the northern plains. The upper Missouri and its major tributaries are depicted in great detail.

G4126.S 12 1806.L4 Vault
Geography and Vault Division
Library of Congress

Reference: Allen, 382–83.

IV.9. *[Map of the 'Santa Fe Trail']*

Zebulon Montgomery Pike
Manuscript: 31 × 20 cm
St. Louis, 1806?

Believed to be the earliest American representation of the Santa Fe trail, this small sketch map was compiled or copied by Zebulon Pike. The information for the map was obtained from earlier French traders before Pike left St. Louis on his 1806–1807 expedition to the Southwest. The map is oriented with Santa Fe on the top. A dotted line depicts the trail of a 1797 trading expedition that began at the Grand Pawnee village on the Platte River and continued overland to Santa Fe, crossing the headwaters of the Kansas, Arkansas, and Red rivers. Distances along the trail are noted in days' journeys. An inscription on the reverse side in Pike's handwriting provides a contemporary image of the southern plains and its inhabitants, the Comanche: "the country level and without wood, except on or near the water courses. The surface being covered with snow; in some places neither wood nor water for 70 or 80 m. . . . The Prairie is high and Dry with short Grass in summer.—met upwards of 3000 Ietans [Comanche]. They are short wellset men all with long Hair. The women are close cropt and are remarkably ugly and filthy. The whole were on Horse back, and are armed with Bows, arrows, and Lances. Their Saddles are made of skin and wood with wooden stirrups, but they procure some Bridles from the Spaniards. They are Erratic raise no corn and have no fixt residence. They hunt only for Buffaloe robes, which is their only dress except a Breech cloute which they procure from the Spaniards, tho many of the men are quite Naked. The women are covered with a Buffalo robe tyed around their necks."

This map was found in Pike's possession when he was arrested near Santa Fe by Spanish authorities, who feared that it showed a military route to their city. Along with eighteen other documents retained by Spaniards, this map was returned to the U.S. State Department by the Mexican government in 1910.

Records of the Adjutant General's Office, 1780s–1917, RG 94
Navy and Old Army Branch
National Archives

References: Jackson (1966), 1: 458–59; Martin and Egli, 18.

IV.10. *A Chart of the Internal Part of Louisiana, Including all the hitherto unexplored Countries, lying between the River La Platte of the Missouri on the N: and the Red River on the S.: the Mississippi East and the Mountains of Mexico West; with a Part of New Mexico and the Province of Texas* (Illustrated)

Zebulon Montgomery Pike; reduced by Antoine Nau
Engraved; Plate I, 44 × 45 cm; Plate II, 43 × 39 cm
Philadelphia: C&A Conrad & Co.; Baltimore: Fielding Lucas, Jr., 1810

Consisting of two separate engraved sheets, this composite map provided nineteenth-century Americans with their first reasonably accurate image of the southern plains. It was compiled by Lieutenant Pike, who had been sent by James Wilkinson, commanding general of the U.S. Army, to explore the Arkansas and Red rivers in 1806–1807; drafted by Sgt. Antoine Nau, a French draftsman on General Wilkinson's staff in St. Louis; and published in part by Fielding Lucas, Jr., who was to become a distinguished publisher of maps and atlases during the next two decades.

For the Osage, upper Kansas, and lower Arkansas river systems, Nau relied upon Lt. James B. Wilkinson's "very elegant protracted sketch" of the explorer's route, which noted "the streams, Hills &c that we crossed—their courses, bearings &c." The son of General Wilkinson, James had accompanied Pike overland as far as the Great Bend of the Arkansas, where the two separated. Pike continued upriver while the younger Wilkinson descended the Arkansas. The lower Platte, Kansas, and Red rivers were derived from existing War Department maps: William Clark's map of 1805–1806 (Map IV.2) and Nicholas King's map of William Dunbar's expedition up the Red River in 1804. The portrayal of the upper Arkansas region was reconstructed from traverse tables that Pike surreptitiously carried out of Santa Fe in the rifle barrels of his men, since his sketch maps were confiscated by the Spanish; the upper Rio Grande and the Red Rivers were taken from Alexander von Humboldt's famous map of New

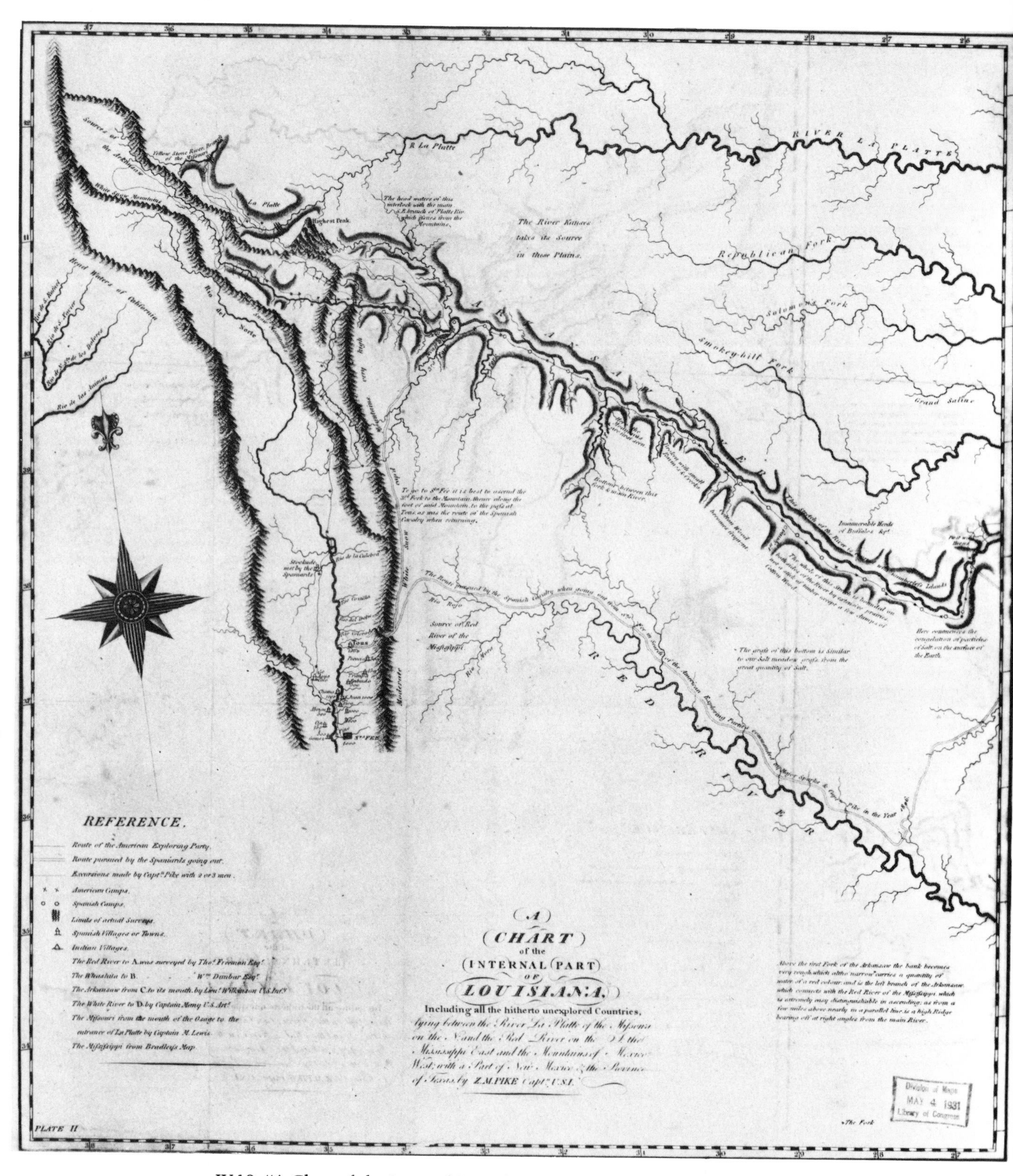

IV.10. "A Chart of the Internal Part of Louisiana. . . ." (Library of Congress)

Spain. The latter was not published until 1811, but a manuscript copy, loaned to Albert Gallatin by Humboldt during his visit to Washington in 1804, was apparently made available to Nau.

Although Pike provided the first detailed treatment of the Arkansas and its tributaries, he badly misplaced the Yellowstone, whose source is shown near the headwaters of the Arkansas and the Rio Grande (thereby perpetuating the erroneous Spanish notion that the Missouri-Yellowstone river system encircled the Great Plains), and the head of the Red River, which is shown too close to Santa Fe. The routes of Pike and his Spanish pursuers, American and Spanish camps, the limits of Pike's actual surveys, Spanish villages and towns, and Indian camps are shown by different symbols. Brief notes describe the "extensive prairies" and vegetation. Hachures or small pictorial drawings of mountains depict watersheds, bluffs, and mountains. Shown near the head of the Arkansas is a rather large perspective view of a mountain, entitled "Highest Peak," which was discovered by Pike on 27 November 1806, and now bears his name.

G1380.P 1810 Copy 2 Vault
Geography and Map Division
Library of Congress

References: Jackson (1966), 1: 459–60; 2: 156; Loomis and Nasatir, 245.

IV.11. *[Sketch Map of the Missouri and Yellowstone River Region]*

Derived from John Dougherty
Manuscript; 34 × 40 cm
1819?

This sketch map of the upper Missouri was derived from John Dougherty, one of a number of American fur trappers and mountain men who entered the region in search of furs shortly after Lewis and Clark's momentous journey. Dougherty's experiences as a fur trapper, which are recounted on the verso of this map, "passed 4 years on the Head waters of the Missouri the Yellowstone & the Columbia Rivers . . . in quest of Beaver." The map appears to have been prepared in 1819 for either Benjamin O'Fallon, Indian agent on the upper Missouri, or Maj. Stephen H. Long, who led an exploring expedition from Council Bluffs to the mouth of the Yellowstone that same year. Dougherty, who spoke several Indian languages, became O'Fallon's subagent and interpreter in 1819.

This map is an absorbing example of an early map prepared by a frontiersman to acquaint someone with the central plains as the area was perceived in about 1811. In addition to delineating the major tributaries of the upper Missouri and Yellowstone, numerous notes assess the quality of the land and soil. The region lying between the Missouri and the Little Missouri, for example, is described as "poor stoney soil, broken & high intersperced with immense plains perfectly level & destitute of timber. Buffaloes."

Records of the Office of the Quartermaster General, RG 92, Post & Reservation File, Map 281.
Cartographic and Architectural Branch
National Archives

Reference: Ehrenberg (1971), "Sketch," 73–78.

IV.12. *Map of Arkansas and Other Territories of the United States Respectfully inscribed to the Hon. J. C. Calhoun, Secretary of War*

Maj. Stephen H. Long, U.S. Army, Topographic Engineers
Engraved by Young & Delleker; 42 × 53 cm
Philadelphia: H. C. Carey & I. Lea, 1822

Derived from Major Stephen H. Long's expedition to the Rocky Mountains in 1820, the first systematic, organized military and scientific survey of the southern plains, this map provided the earliest portrayal of the Platte and Canadian river systems and the "front wall of the Rockies . . . from Long's Peak to the Spanish Peaks." Long, an experienced surveyor and skilled mathematician, was the first of a long line of military topographers sent out by the elite U.S. Army Corps of Topographical Engineers to survey and map the frontier. In addition to the southern plains, the map shows the upper Missouri and Yellowstone rivers, which were based on William Clark's map of 1814 (Map IV.7).

While this map provided the most complete and accurate portrayal of the southern plains for its time, its major legacy is associated with Long's assessment of the southern plains as a "Great Desert." This concept, first introduced by Zebulon Pike, persisted in later maps and presumably hindered settlement of the region for decades.

Carey and Lea, the Philadelphia publishers of the official account of Long's expedition, issued this commercial edition of his map in their *A Complete Historical, Chronological & Geographical American Atlas* one year before they published the official map in two sections.

University of Nebraska Exhibition
Map 17
Don Forke Collection

Library of Congress Exhibition
Arkansas 1822
Geography and Map Division
Library of Congress

References: Friis (1967), 75–87; Nichols and Halley, 176–79.

IV.13. *Hydrographical Basin of the Upper Mississippi River From Astronomical and Barometrical Observations Surveys and Information by J. N. Nicollet. in the Years 1836, 37, 38, 39, and 40;*

assisted in 1838, 39 & 40, by Lieut. J. C. Frémont, of the Corps of Topographical Engineers.

Joseph N. Nicollet; reduced and compiled by Lt. William H. Emory
Engraved by William J. Stone; mountains engraved by E. F. Woodward, Philadelphia; 93 × 78 cm
Washington, D.C.: United State Senate, 1843.

The earliest accurate map of the eastern border of the central plains was based on systematic instrument surveys undertaken by Nicollet, a French mathematician and astronomer, between 1836 and 1840. Nicollet, who was employed as a civilian by the newly reorganized U.S. Army Corps of Topographical Engineers and assisted by Lt. John C. Frémont of the corps, initiated scientific mapping of the trans-Mississippi West by the War Department. Surface relief is conveyed by hachures (short parallel lines that depict degree of slope) and spot heights (elevation figures) based on hundreds of barometric readings taken by Nicollet and Frémont. Nicollet was the first explorer to make much use of the barometer in the North American interior. He was also one of the first to incorporate place names on maps based on systematic analysis of Indian and French place names.

G4042.M5 1843.N5
Geography and Map Division
Library of Congress

Reference: Bray, 246–47.

IV.14. *Map of an Exploring Expedition to the Rocky Mountains in the Year 1842 and to Oregon & North California in the Years 1843–44 by Brevet Capt. J. C. Frémont of the Corps of Topographical Engineers Under the orders of Col. J. J. Abert, Chief of the Topographical Bureau.*

Charles Preuss; annotated by George Gibbs
Lithograph; two sheets, 77 × 65 cm and 77 × 64 cm
Baltimore: E. Weber & Co., 1845

A landmark map re-creating the American West of Jedediah Smith, the legendary explorer. The base map, compiled by Frémont and drawn by his assistant, the German-born surveyor and draftsman Charles Preuss, provided the first accurate continental representation of the West south of the region explored by Lewis and Clark. It was based on personal observation and constructed according to the most advanced surveying and reconnaissance principles of the period. Large areas of the map were left blank in the absence of direct knowledge. Upon this copy of Frémont's map, George Gibbs recorded the travels and discoveries of Smith, including Smith's activities on the central and southern Plains when he was associated with Gen. William Ashley, the St. Louis fur trader of the early 1820s. For nine years, until he was killed by Comanches in 1831, Smith explored more of the West than any other American. His manuscript map, which has been lost, was prepared in about 1830 in St. Louis and consulted by Gibbs when he made these transcriptions in 1851.

American Geographical Society Collection
University of Wisconsin—Milwaukee Library

Reference: Jackson (1970), Map Portfolio, 11–14; Wheat, 2: 119–39.

V. Military Maps

A major impetus for detailed mapping of the North American plains was the frequent warfare in the area during the three decades after the Civil War. Clashes among nomadic Indian tribes, hostilities between Indians and the armed forces of the United States and Canada, and two insurrections by French settlers in Canada contributed to mapmaking. Four major classes of military maps were on display: composite regional or departmental maps prepared for administrative and planning purposes; maps depicting military operations; order-of-battle maps showing field positions of units; and general war maps designed for public use.

V. 1. *[Indian War Map Showing Route of War Party]* (Illustrated)

Gero schunu wy ka (The Man That Is Very Sorry)
Manuscript, 53 × 42 cm
Missouri River, 12 August 1825

Although highly stylized, this map is interesting for providing an image of the central plains as seen through the eyes of a Plains Indian. The central plains are nearly encircled by the Arkansas and Missouri rivers, which frame the map; the interior region is left blank except for some western tributaries of the upper Missouri. The map was compiled by Gero schunu wy ka, an Otoe Indian, to show the route of his "war party of five Ottoes against the Arapahoes accompanied with a few pawnees. . . . Three Arapahoes were killed & five horses taken." The trail of the Otoe war party is depicted as following the Platte upriver and then crossing over to the upper Arkansas, leading to the raid on the Arapaho camp. Otoe and Pawnee villages are shown along the Platte River. In addition, the map depicts the route of Capt. William Armstrong, who led forty mounted soldiers from Council Bluffs up the Missouri to act as cavalry scouts for the Yellowstone Expedition of 1825. Under the command of Gen. Henry Atkinson, the main body of the expedition

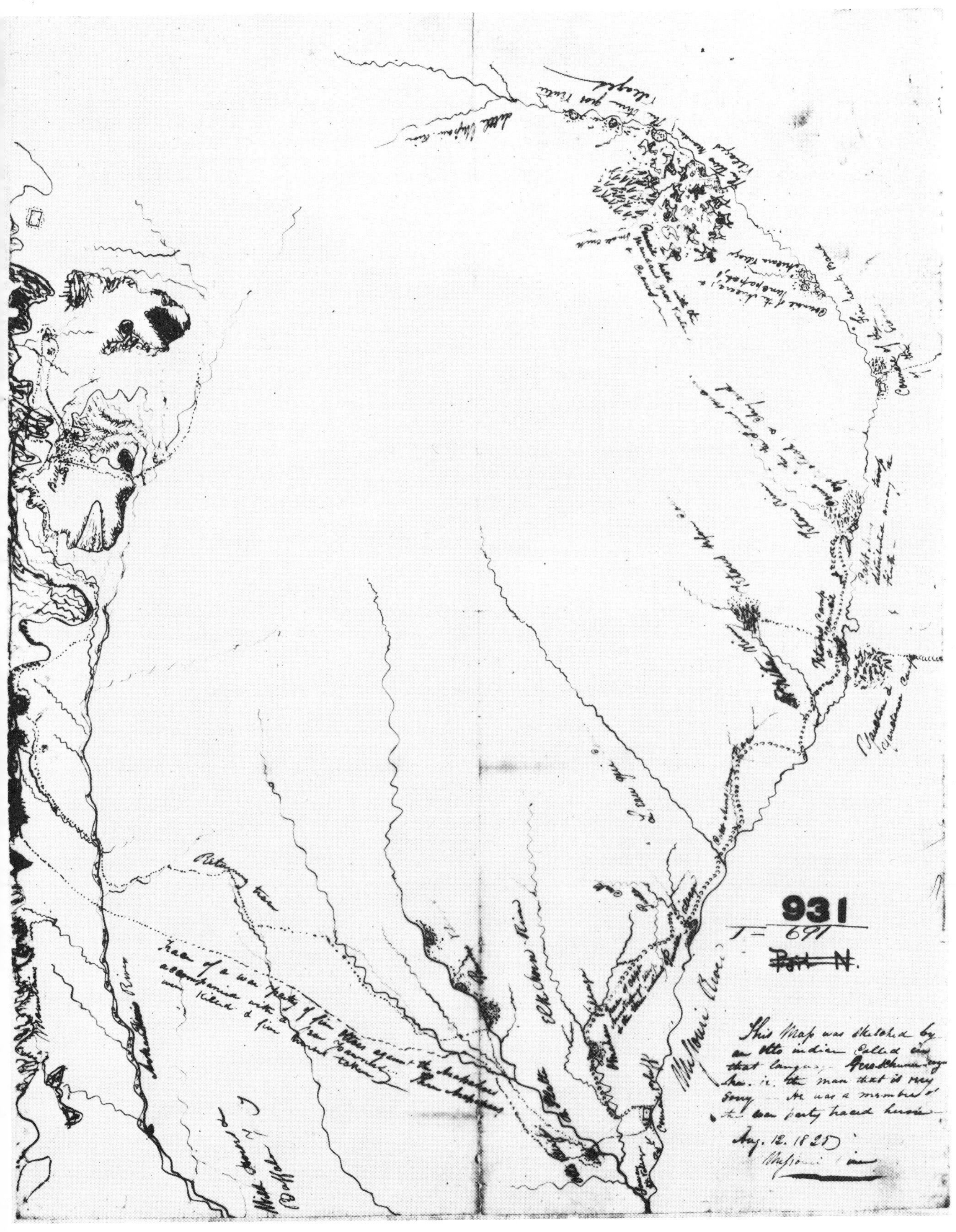

V.1. Indian war map showing route of war party. (National Archives)

ascended the river as far as the Yellowstone in eight keelboats. Their objective was to make treaties with the Indian tribes situated along the Missouri. Twelve treaties were concluded with sixteen bands of Plains Indians during the eighteen-week journey. Many of the treaty sites are located on this map. At the Mandan-Hidatsa villages, a pictograph shows a circle of Indians mounted on horses with the note "Council held with the Mandan, gros Ventres & Crows."

Records of the Bureau of Indian Affairs, RG 75, Map 931
Cartographic and Architectural Branch
National Archives

Reference: Nichols, 90–108.

V.2. *Sketch of the Blue Water Creek embracing the field of action of the force under the command of Bvt. Brig. Genl. W. S. Harney in the attack of the 3rd Sept. 1855, on the "Brule" Band of the Indian Chief Little Thunder.*

Gouverneur K. Warren
a) Manuscript; 25 × 20 cm.
b) Lithograph; P. S. Duval, Philadelphia: 27 × 22 cm.
3 September 1855

Original pen-and-ink sketch by Lieutenant Warren and lithograph copy of the cavalry attack at Blue Water Creek, Nebraska. Warren, who later played a prominent role at the Battle of Gettysburg during the Civil War, observed this conflict from a nearby hill. His report and map detail Gen. William H. Harney's dishonorable tactics which resulted in the killing of many Indian women and children. Under the subterfuge of holding peace talks, Harney met with Little Thunder only long enough to allow time for his cavalry to encircle the Brulé Indian camp. Harney then attacked with his main force, driving the Indians up the Blue Water Creek valley, where the remaining troops were concealed. Warren, who disagreed with this strategy, marked the meeting site on his map with the note "Place where the 'talk' was held with Little Thunder."

The lithograph copy shows that the engraver closely followed Warren's original manuscript.

Records of the Office of the Chief of Engineers, RG 77, Q55 and Q57A
Cartographic and Architectural Branch
National Archives

Reference: Goetzman (1959), 409–10.

V.3. *Reconnoissances [sic] in the Dacota Country* (Illustrated)

Gouverneur K. Warren; Edward Freyhold, draftsman
Manuscript; two sheets, each 106 × 75 cm
Washington, D.C., 1855–57

The northern plains were the focus of several maps prepared by or for Lt. Gouverneur K. Warren, U.S. Army Corps of Topographical Engineers, who undertook extensive military and topographic reconnaissance surveys of the region from 1855 to 1857. The impetus for these military maps was the fear of a general uprising by the Sioux, who felt threatened by continued encroachment on Indian lands by settlers. Following the massacre of thirty soldiers of the Sixth Infantry on 19 August 1854, Warren was assigned to General William H. Harney's punitive expedition into the Sioux country the following year. Warren's assignment was to map the strategic but poorly known badlands country between the North Platte and the upper Missouri. The result of Warren's survey was this large-scale manuscript map, drafted by Edward Freyhold, a civilian employed by the corps, and corrected by Warren. Originally compiled in 1855 in corps headquarters in Washington, D.C., this map illustrates the several steps involved in the preparation of a nineteenth-century map. First, the projection and geographic coordinate system were determined and inked. Next, drainage patterns and topographic features were transferred from existing reliable maps to the new map base by the method of squares (a grid of faint pencil lines which was used to guide the draftsman in the accurate placement of these features). Existing maps included those prepared by Maj. Stephen H. Long (Map IV.12), Joseph N. Nicollet (Map IV.13), Capt. John C. Frémont (Map IV.14), and Capt. Howard Stansbury, who explored and surveyed the Great Salt Lake of Utah in 1850. Third, new information from explorers' field notes and sketch maps was added, using the same method of squares to reduce and transfer the data. Next, names of important features and places were added in pencil and then inked. Finally, the manuscript was reviewed and corrections were suggested.

In 1857, additional information was added to this manuscript map based on Warren's survey of the Niobrara River and Black Hills in that year. The portrayal of the Black Hills is particularly well done, and represents one of the finest examples of the use of hachures to illustrate topographic relief on official maps during the nineteenth century.

Warren's map was lithographed for the War Department by Peter S. Duval & Company in Philadelphia in 1859 but does not include the Black Hills and Niobrara River. A French immigrant, Duval carried out numerous experiments with lithography and zincography and was the first printer in the United States to use a rotary steam press.

Records of the Office of the Chief of Engineers, RG 77, Q579-47
Cartographic and Architectural Branch
National Archives

References: Ristow, 21; Goetzman (1959), 406.

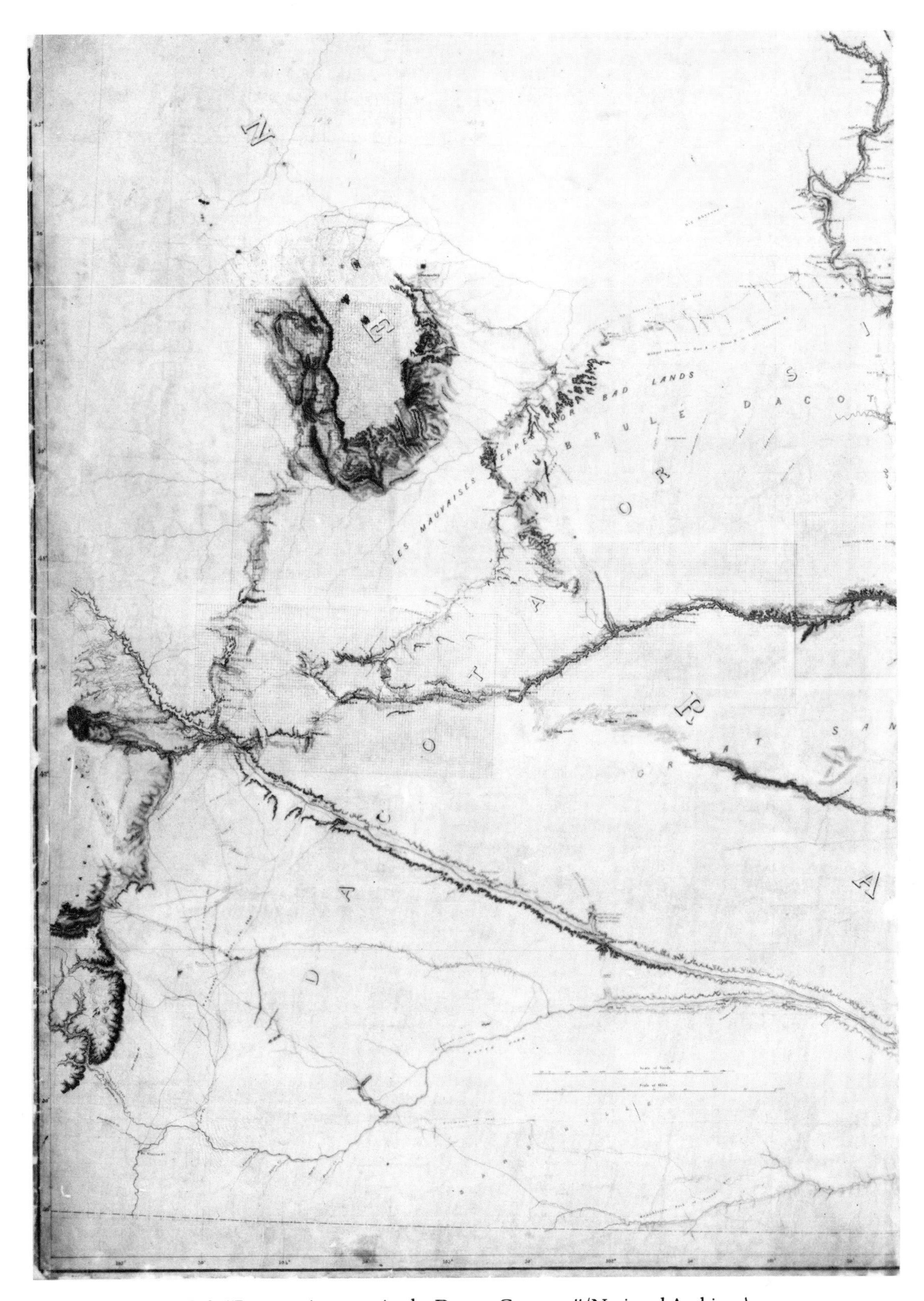

V.3. "Reconnoissances in the Dacota Country." (National Archives)

V.4. *Military Map of Nebraska and Dakota*

Gouverneur K. Warren
Lithograph, 110 × 81 cm
Washington, D.C.: United States Senate, 1857

This remarkable regional map incorporated data from Lt. G. K. Warren's three expeditions to the central plains from 1855 to 1857 and twenty other surveys ranging from Lewis and Clark's to the work of the U.S. General Land Office. Designed for strategic planning, it shows the locations of previous military expeditions, Indian camps, battle sites, springs and water holes, and significant topographic features. The most noteworthy feature is the depiction of the Black Hills, which was based on Warren's 1857 reconnaissance map of Dakota Country (Map V.3).

North Dakota 1857 Warren
Geography and Map Division
Library of Congress

V.5. *Map of the Line of March of the Troops Serving in the Department of Dakota in the Campaign Against Hostile Sioux*

Sgt. James E. Wilson, Battalion of Engineers, U.S. Army
Manuscript on tracing cloth; 48 × 125 cm
Headquarters, Department of Dakota, 1876

Custer's Battle-Field (June 25th 1876)

Sgt. Charles Becker, Battalion of Engineers, U.S. Army
Manuscript; 38 × 44 cm
Headquarters, Department of Dakota, 1876

The Battle of Little Big Horn was the most notable of the central plains military campaigns. Sioux hunting bands under the leadership of Sitting Bull set the stage for the Sioux War of 1876. The Indians had previously contested the advance of settlements across the Dakota prairie, the building of the Northern Pacific Railroad to the Missouri River in 1872, and the exploration of the Black Hills. These actions were viewed by the Sioux as encroachment of their lands. Lt. Col. George Armstrong Custer was the military leader of the Black Hills Expedition of 1874 that discovered gold in a treasured part of the Plains Indians' hunting ground. This earned Custer the particular dislike of the Sioux.

Custer's defeat on 25 June 1876 was a great victory for the northern plains tribes, but it prompted the government to expand its defense system, which led ultimately to the defeat of the Sioux in 1881.

The first of two maps in this display is a general map depicting the route of the Seventh Cavalry during the Sioux campaign of 1876. It was prepared under the direction of Lt. Edward Maguire, chief topographical engineer for the Department of Dakota. This campaign map shows the topography of the region along the line of march of Brig. Gen. A. H. Terry's expedition, the trail of Colonel Custer, the location of the Custer battlefield on the Little Big Horn, and the movements of other cavalry units.

The second map depicts the Custer battlefield. It was also surveyed and prepared under the direction of Lieutenant Maguire on 28 June 1876, two days after the battle. The battle consisted of two separate engagements. One encounter resulted in the complete destruction of Colonel Custer and his immediate command of five troops of the Seventh Cavalry on 25 June. The other engagement was fought by the remaining seven troops and pack train of the Seventh Cavalry under the command of Maj. Marcus A. Reno and Capt. Frederick W. Benteen. It ended on 26 June. The map shows the course of the Little Big Horn, the vast Indian encampment, the line of attack, Major Reno's skirmish line, and the locations of the graves of Custer's officers. A copy of this map was used for the court of inquiry that exonerated Major Reno in 1879.

Records of the Office of the Chief of Engineers, RG 77, Q304-1 and Q288-4
Cartographic and Architectural Branch
National Archives

Reference: Wheeler, 648.

V.6. *[Map of the Battle of the Little Big Horn]*

[Pictograph of George Custer's Column during the Battle of the Little Big Horn]

Shin-ka-kan-sha (Red Horse)
Manuscript on manila paper; each 61 × 66 cm
Cheyenne River Agency, Dakota Territory, 1881

A unique collection of forty-one pictographs and a map of the Battle of the Little Big Horn were drawn by Sioux Chief Red Horse in 1881 for Charles E. McChesney, acting assistant surgeon, U.S. Army. The map and drawings accompanied Red Horses's eye-witness account of the 25 June 1876 battle which he gave to McChesney in sign language. The map shows the location of the Sioux camp, depicted by small pictorial drawings of teepees, before Lt. Col. George Custer's attack; the site of the battlefield, indicated by red dots symbolizing dead soldiers; and prominent features of the surrounding region. The Little Big Horn River is identified by its Sioux name, Greasy Grass Creek.

The pictograph on display shows Sioux warriors and soldiers in direct combat. "Once the Sioux charged right in the midst of the different soldiers [Custer's battalion] and scattered them all," Red Horse reported, "fighting among the soldiers hand to hand."

BAE File 2367-A
National Anthropological Archives
Smithsonian Institution

Reference: Powell, 563–66.

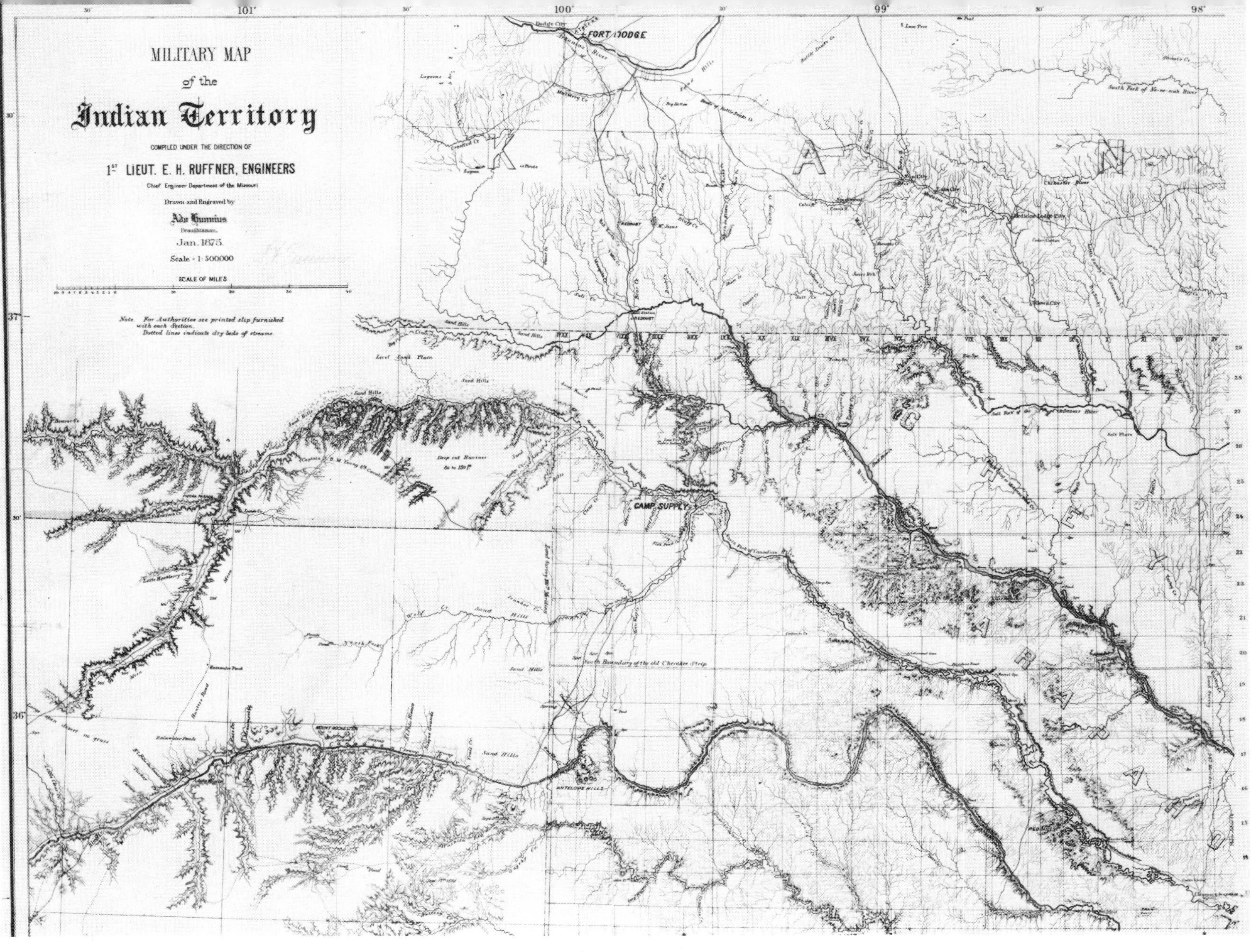

V.7. "Military Map of the Indian Territory." (Library of Congress)

V.7. *Military Map of the Indian Territory* (Illustrated)

E. H. Ruffner; Ado Hunnius, draftsman
Lithograph; four sheets, each 61 × 76 cm
St. Louis, 1875

Following the Civil War, general military maps of the plains were prepared for strategic planning. In the United States, these maps were generally compiled for each of the geographical military commands such as the Department of Dakota, Department of the Platte, and Department of the Missouri. A separate map was issued for Indian Territory. Often large-scale insets of significant military posts were included. Unlike most of the maps that were issued before 1860, these departmental maps were compiled and printed in the West under the direction of departmental engineering officers.

The map on display was compiled and engraved under the direction of Lieutenant Ruffner, chief engineer of the Department of the Missouri. Extremely detailed, it recreates the late-nineteenth-century geographical landscape of the various Indian nations that originally made up present-day Oklahoma. It shows a land of military routes and campsites, Indian agencies and council grounds, cattle trails and ranches, detailed drainage patterns and blank spaces ("Large open desert no grass"), springs and "sand Hills," and stage routes and railroad lines.

Oklahoma 1875
Geography and Map Division
Library of Congress
Not in the Library of Congress Exhibition

V.8. *[Military District No. 10]*

Department of the Militia & Defense
Manuscript; two sheets, each 91 × 53 cm
1878

Military units in both the United States and Canada compiled intelligence maps of portions of the Great Plains for tactical and planning purposes. The map on display provides a military assessment of the eastern Canadian Prairies in southern Manitoba. It shows general topographic features that would aid or hinder military movements (marshes, "water camping grounds," bridges and trails, fordable points along streams, ferries, elevated sites such as ridges, river banks, and hills), settlement patterns ("Settlement Springing up Rapidly," "Mennonite Settlement," "Projected Town") and land appraisal ("Well Fitted for Agriculture," "Extensive And Unbroken Prairie Land [where] Small groves or Islands of woods occasionally occur").

National Map Collection
Public Archives of Canada

Reference: Warkentin and Ruggles, 482.

V.9. *Map of the Country North of Fort Assinniboine, M.T.* (Illustrated)

S. C. Robertson
Manuscript; 50 × 51 cm
1887

This intriguing intelligence map embraces the area from Fort Assiniboine (spelled Assinniboine on the map) on the Milk River in northern Montana north to Medicine Hat and Maple Creek in southeastern Alberta and southwestern Saskatchewan, respectively. It was compiled by 1st Lt. S. C. Robertson of the First Cavalry, U.S. Army, in 1887. This region played an important role in the northern Plains Indians wars. Fort Assiniboine was established in May 1879 as part of the U.S. northern plains defense system in an effort to defeat the Sioux. The Cypress Hills, which are shown in great detail, provided temporary haven in Canada for Sitting Bull and some four thousand refugee Indians from 1879 to 1881.

The map contains a column of extensive notes describing the approaches to two Canadian North West Mounted Police posts at Medicine Hat and Maple Creek "for the purpose of attack."

> Both Maple Cr. and Med. Hat are permanent stations of mounted Police—the garrisons of the two consisting of portions of the same troop (120 men). From these, detachments of about a dozen men are scattered in temporary camps south of Cypress Hills, & from these patrol the vicinity of the [International Boundary] line from about March or April, till about November. The approximate location of those camps for 1887 is given on map. They include all camps belonging to above two posts. It is not difficult to avoid their small patrols on Cypress Road.

To further aid the reader, a larger scale map of the approaches to Medicine Hat is provided in an inset. As the terminus of the Canadian Pacific Railroad in 1887, Medicine Hat served as an important trans-shipment point. Robertson noted that the "town is in deep cut and not tenable if attacked by respectable force."

G3501.R1 1887.R Vault
Geography and Map Division
Library of Congress
Not in the Library of Congress Exhibition

V.10. *Bishop's North-West War Map* (Illustrated)

Lithograph, George Bishop Engraving and Printing Company; 81 × 58 cm
Montreal, 1885

This decorative propaganda map portraying the Saskatchewan Rebellion of 1885 was designed to arouse support of eastern Canadians. In 1869–70 and 1885, Louis Riel led rebellions which briefly took control of parts of the Canadian prairies and established provisional governments to secure the rights of the Métis (descendants of

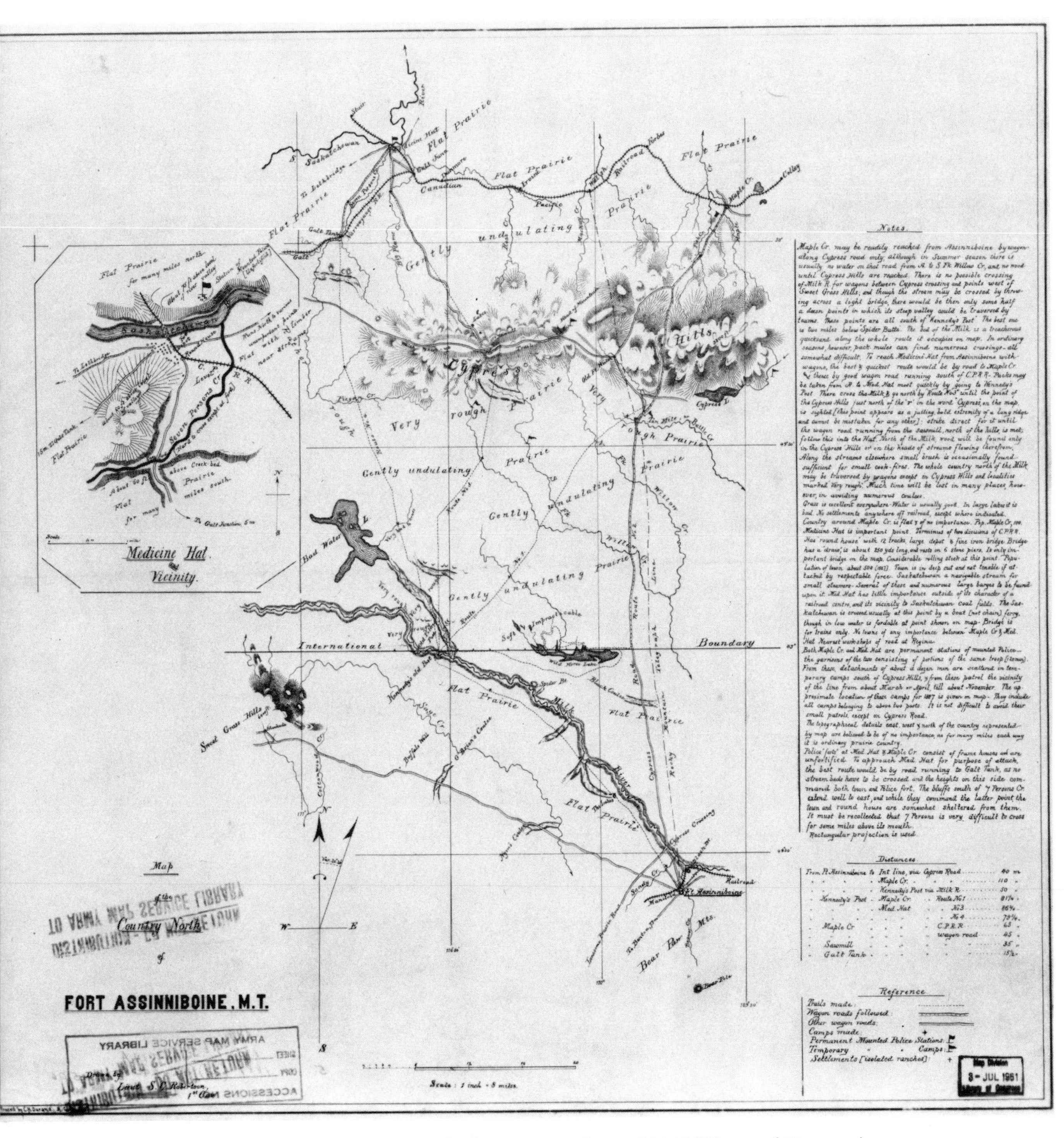

V.9. "Map of the Country North of Fort Assinniboine, M.T." (Library of Congress)

V.10. "Bishop's North-West War Map." (Library of Congress)

French fur traders and Indian women) and Plains Indians. The map, a copy of a general Canadian Pacific Railway map made before 1885, appeared many times in newspapers during this period to aid readers who followed the reports of the Riel Rebellion. An important feature of this map is the inclusion of views and portraits associated with the rebellion. Tranquil panoramas of a Qu'Appelle valley farm scene and the railroad center of Medicine Hat contrast sharply with sketches of heavily armed Cree and Sioux Indians and a depiction of a scalping by a "Savage Indian Warrior." Portraits of Maj. Gen. Frederick Middleton, commander of the Canadian militia, and Riel are also shown.

Georgraphy and Map Division
Library of Congress
Not in the University of Nebraska Exhibition

Reference: Oppen, 94.

VI. Jurisdictional Mapping

Federal agencies in both Canada and the United States were responsible for distributing lands to settlers who moved onto the North American plains, for defining and marking political boundaries, and for surveying Indian lands. In the course of their work, these agencies produced large-scale cadastral plats that recorded land division and ownership, detailed boundary maps, and maps of Indian reserves.

VI.1. *Map Showing the Lands assigned to Emigrant Indians West of Arkansas & Missouri*

Topographical Bureau, U.S. Army
Engraving; with hand coloring; 48 × 46 cm
Washington, D.C., 1836

Early in the nineteenth century Indian tribes were moved from their ancestral lands in the rapidly developing eastern states to the newly acquired and largely empty western lands. Pressure from several state governments, particularly Georgia, led to the enactment on 28 May 1830 of the Indian Removal Act. This map shows the new lands on the southern and central plains assigned to each tribe. Tables indicate the estimated acreage assigned to each tribe, the number of Indians by tribe who have emigrated, and the total number of Indians by tribe. The map is based on the work of Lt. Washington Hood (see fig. 7.3).

U.S. Indian Reservations, 1836
Geography and Map Division
Library of Congress
Not in the University of Nebraska Exhibition

VI.2. *Plat of the survey of the Northern Boundary of Kansas Lands*

John Calvin McCoy
Manuscript; 120 × 43 cm
July 1836

Indian lands located on the Great Plains were systematically mapped by the War Department and later the Bureau of Indian Affairs during the treaty period as tribes, bands, and nations were restricted to certain areas. The lands set aside for Indians became recognized reservations, and their boundaries were surveyed and delineated by contract surveyors. This boundary map was prepared by John Calvin McCoy and his father, Isaac. Together they surveyed and mapped twenty-two major Indian and territorial boundaries in Indian, Kansas, and Nebraska territories from 1830 to 1855. Isaac McCoy, a Baptist missionary, was instrumental in establishing the 1830 act providing for the removal of eastern Indians to the West. The map is typical, showing only the basic drainage pattern and the boundary lines of the Kansas Reserve and the Delaware Outlet.

Records of the Bureau of Indian Affairs, RG 75, Map 333
Cartographic and Architectural Branch
National Archives

Reference: McCoy.

VI.3. *Kickapoo Reservation in Brown County, Kansas Territory, Atchison [County], Calhoun [County]* (Illustrated)

Bureau of Indian Affairs
Manuscript with watercolor wash; 104 × 68 cm
ca. 1856–71

The Kickapoo Nation, an eastern Algonquian tribe from the Great Lakes region, was removed in 1833 to northeastern Kansas, where it was assigned to a small reservation. In this map, the open lands of the reservation contrast sharply with the surrounding townships laid out by the U.S. General Land Office in preparation for settlement. Several features of this map make it somewhat unique among official maps of Indian reservations. First, the native family names, scattered along the two forks of the Grasshopper (Craig) River, a minor tributary of the Kansas, have not been translated into English. Second, the inclusion of trees and buildings, drawn in perspective and enhanced with watercolor wash, reflects a level of artistry not found on most reservation maps. Finally, one or more symbols (dots, squares, rectangles) whose meaning is now lost are found next to the Indian names scattered throughout the map.

The perspective view of the large two-story building probably represents a Presbyterian boarding school which occupied the site from 1856 to 1871.

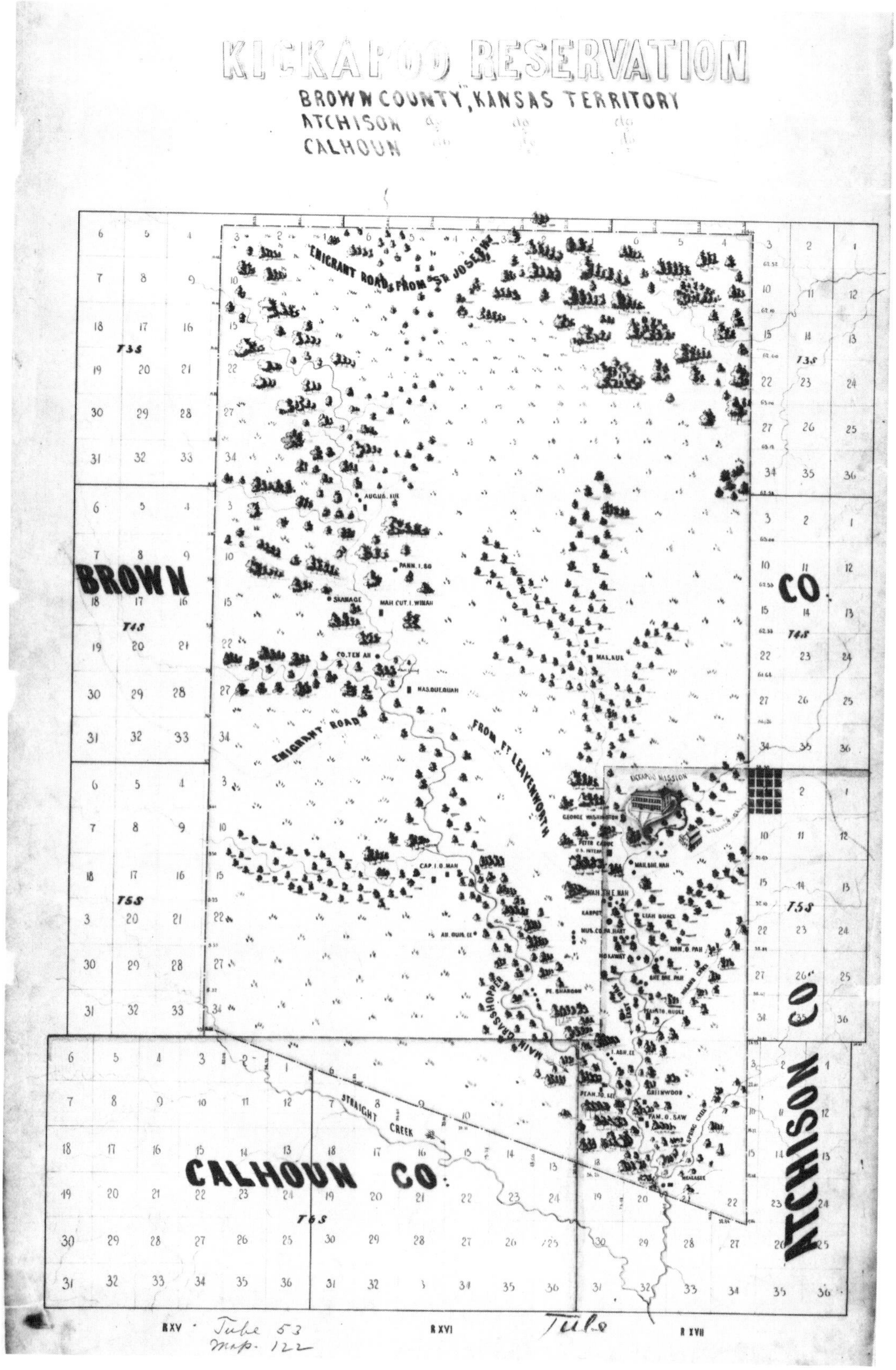

KICKAPOO RESERVATION

BROWN COUNTY, KANSAS TERRITORY
ATCHISON do do do
CALHOUN do do do

EMIGRANT ROAD FROM ST JOSEPH

EMIGRANT ROAD FROM FT LEAVENWORTH

BROWN

CO.

ATCHISON CO.

CALHOUN CO.

KICKAPOO MISSION

STRAIGHT CREEK

GRASSHOPPER

T3S

T4S

T5S

T6S

RXV

RXVI

RXVII

Tube 53
Map. 122

VI.3. "Kickapoo Reservation in Brown County, Kansas Territory. . . ." (National Archives)

Records of the Bureau of Indian Affairs, RG 75, Map 122
Cartographic and Architectural Branch
National Archives
Washington, D.C.

Reference Lutze, 207–11.

VI.4. *Map Showing the progress of the Public Surveys in Kansas and Nebraska, To accompany Annual Report of the Surveyor General 1862*

Surveyor General's Office
Lithograph; 64 × 51 cm
Leavenworth, Kansas, 1862

The first systematic large-scale mapping of the Great Plains was undertaken by the General Land Office. Established in the Treasury Department in 1812 as an outgrowth of the Land Ordinance Act of 1785, the General Land Office controlled the transfer of government land to individual settlers. In order to ensure the orderly distribution of public lands, federally appointed surveyors general supervised the division of these lands into townships measuring six miles square. The townships were arranged in accordance with meridian lines aligned in a north-south direction and parallel lines aligned in an east-west direction. Beginning in 1855, in a series of status maps of Kansas and Nebraska, the General Land Office showed the westward progress of the rectangular land survey.

University of Nebraska Exhibition
Don Forke Collection Map 30
Library of Congress Exhibition
Kansas-Nebraska 1862
Geography and Map Division
Library of Congress

VI.5. *Township No. 32 North, Range No. 8 West of the 6th Principal Meridian* (Illustrated)

General Land Office
Manuscript; 40 × 53 cm
Nebraska City, 20 July 1850

In the rectangular land survey system devised by the General Land Office, townships were divided into thirty-six sections, each measuring one mile square. Following the survey of the township, a plat was drawn to a scale of two inches to a mile on durable paper of the "best quality and of uniform size." Courses and distances of township, section, and subdivision lines are depicted in black ink, swamps by shaded black lines and dots, and prairies by shaded green lines and dots. In addition, land surveyors were required to indicate cultural features such as roads and Indian trails, Indian villages, boundaries of Indian cessions and reserves, and existing settlers. Each township plat was prepared in triplicate; one copy was filed with the Office of the Surveyor General, one with the local land office having jurisdiction over the township, and one with the General Land Office in Washington, D.C.

M297-504
Nebraska State Historical Society
Lincoln, Nebraska

VI.6. *Map of Cuming City, Washington County, N[ebraska] T[erritory]* (Illustrated)

Manucript; 58 × 47 cm
Cuming City, Nebraska Territory, 1855

As a result of the Kansas and Nebraska Act of 30 May 1854, which opened to settlement a large portion of the central plains north of the Missouri River, many towns were hastily platted along the Missouri River in late 1854 and 1855 by speculators who hoped to capitalize on the potential prosperity of these sites. Detailed maps were prepared for each town to advertise the sale of lots. Today, these plats reflect the overly optimistic expectations of the town promoters, who depicted elaborate nonexistent street patterns and neatly numbered city blocks and lots. Most of these towns never came into existence; others existed only briefly as they lost out to towns better situated for trade and transportation.

This map portrays a typical speculative town of the plains frontier. Cuming City, named in honor of Thomas B. Cuming, the twenty-five-year-old acting territorial governor, was located about twenty-four miles north of Omaha near DeSoto, another early town. Reminiscing in 1899, one early settler recalled these Missouri River towns: "We were greatly disappointed at DeSoto—a cluster of cabins in a hollow by the river. . . . There was a place, called Cuming City after Secretary of State Cuming, a few miles above on a fine site, but far from the river. . . . Both the places have departed the earth, and their very sites have been forgotten."

M297-503
Nebraska State Historical Society
Lincoln, Nebraska

Reference: Irvine, 151.

VI.7. *Rough Diagram, Based on Hind's Map, intended to illustrate Report of this date on Township Surveys, Red River Territory* (Illustrated)

J. Stoughton Dennis
Manuscript; 57 × 43 cm
1870

In preparation for the transfer of the vast Canadian prairies to Canada by the Hudson's Bay Company in 1869,

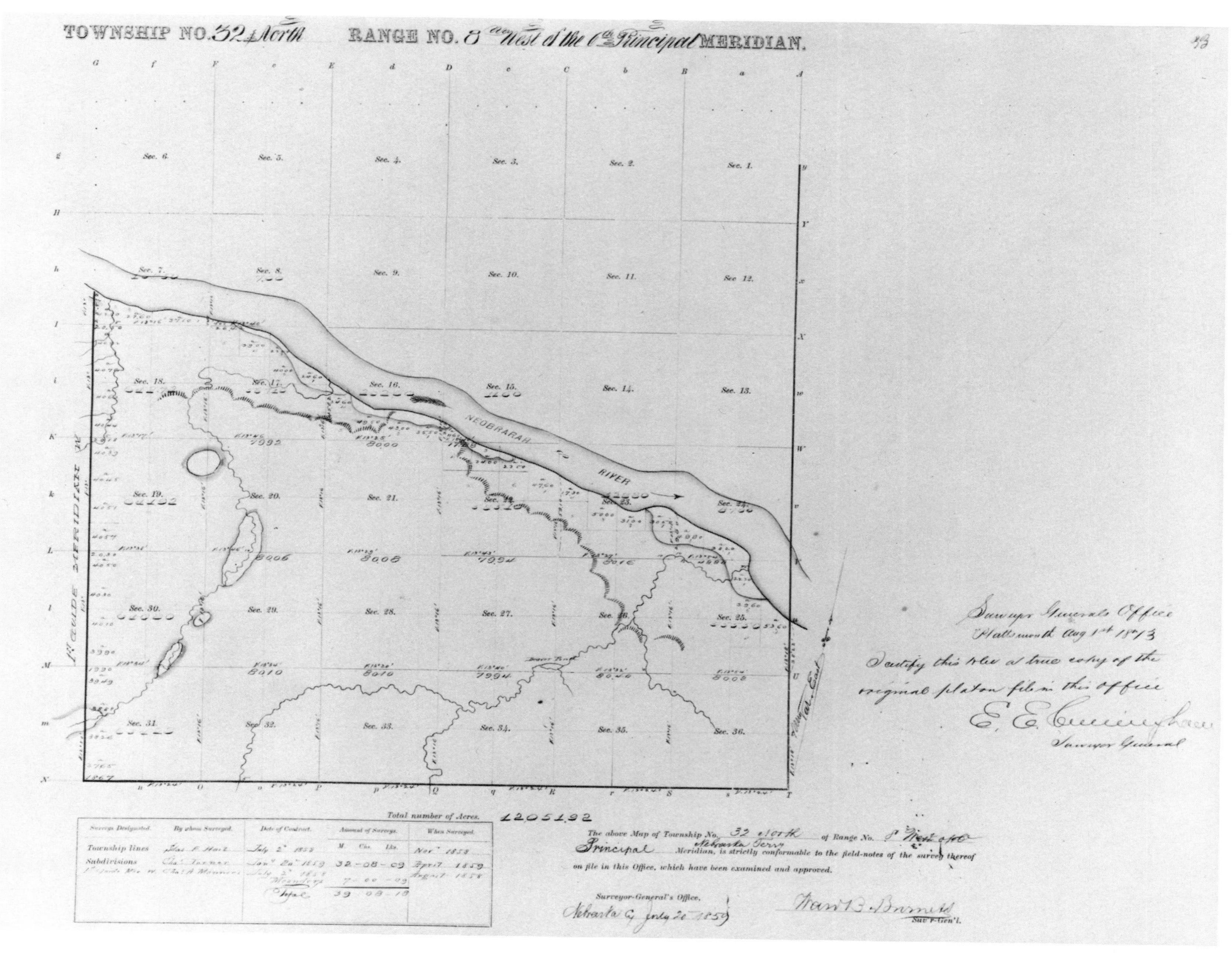

VI.5. "Township No. 32 North, Range No. 8 West of the 6th Principal Meridian." (Nebraska State Historical Society)

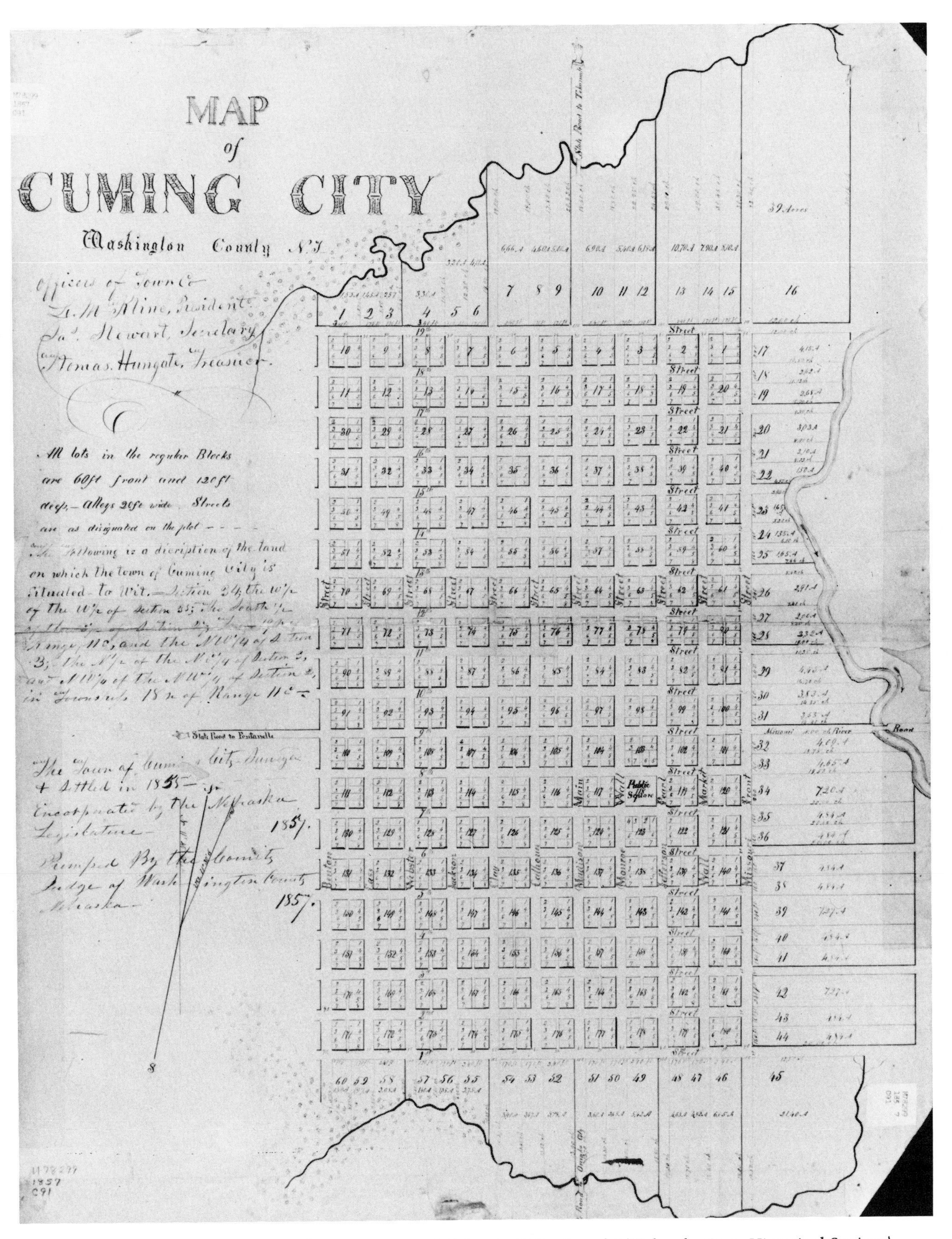

VI.6. "Map of Cuming City, Washington County, N[ebraska] T[erritory]." (Nebraska State Historical Society)

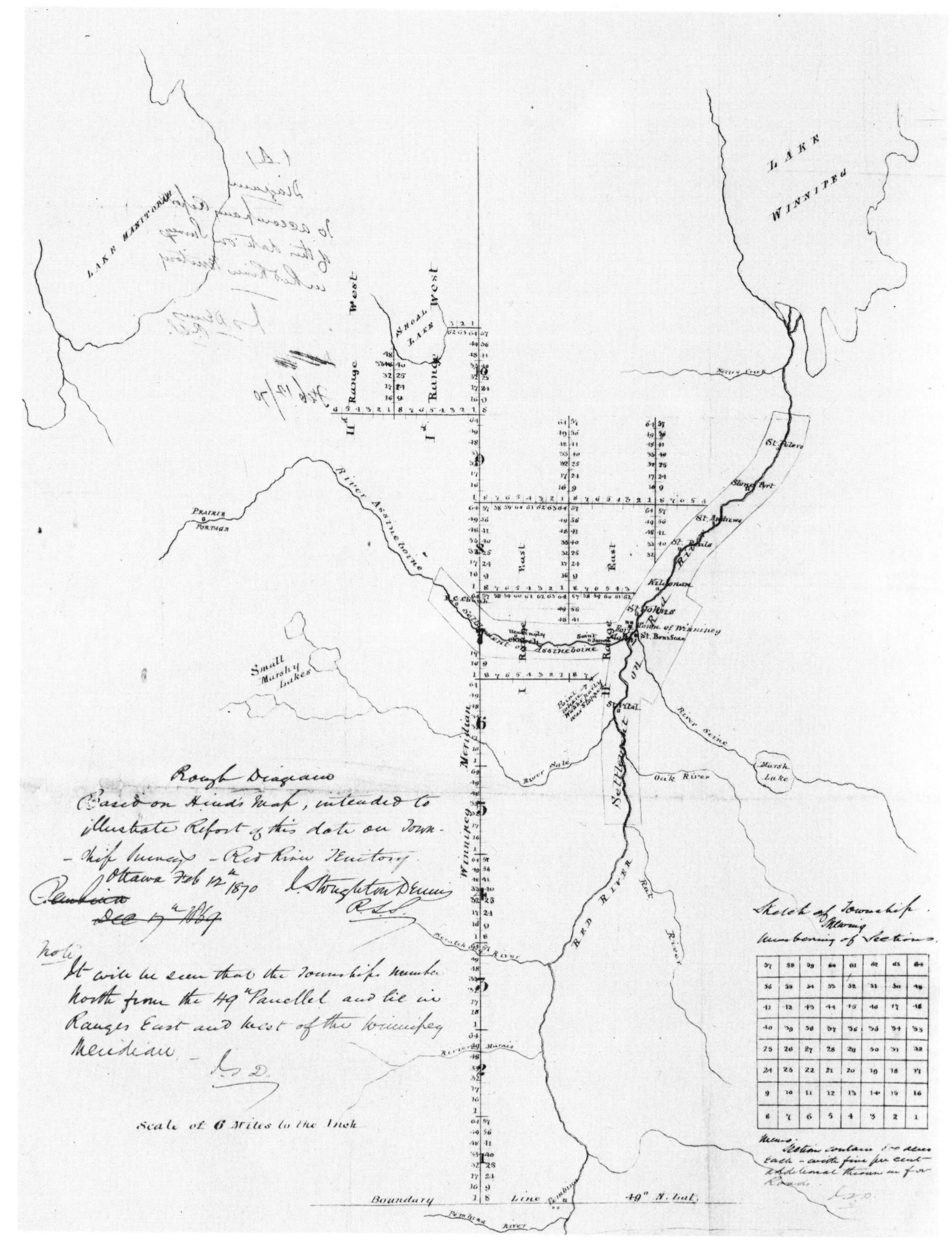

VI.7. "Rough Diagram, Based on Hind's Map, intended to illustrate Report of this date on Township Surveys, Red River Territory." (Public Archives of Canada)

Lieutenant Colonel Dennis, an Ontario land surveyor, devised a rectangular grid system modeled on that of the United States. Dennis prepared this map of his proposed system based on test surveys along the Red River of the North. For a base, he used the map prepared by the explorer Henry Youle Hind, who carefully mapped the Assiniboine and Saskatchewan river county in 1859. An inset shows Dennis's proposed township divided into sixty-four sections of eight hundred acres each. A note near the confluence of the Assiniboine and the Red River mark the point where one of Dennis's survey parties was stopped by Louis Riel on 11 October 1869, which may be considered the beginning of the Riel Rebellion (Map V.12). The Métis population, which was located along the Red River, feared that the rectangular survey system would replace their long river lot landholding system and therefore their way of life.

NMC-7064
National Map Collection
Public Archives of Canada
Not in the Library of Congress Exhibition

References: Oppen, 2–3, 10.

VI.8. *Joint Maps of the Northern Boundary of the United States, From the Lake of the Woods to the Summit of the Rocky Mountains*

U.S. Northern Boundary Comission and Her Majesty's North American Boundary Commission
a) Title Sheet: Lithograph, 34 × 45 cm
b) Index Sheet: Manuscript, 32 × 55 cm
c) Sheet IX: Manuscript, 35 × 55 cm
1876

Boundary map were prepared by official mapping agencies to subdivide the Great Plains into national, territorial, provincial, state, and Indian reserve units. Compiled in the form of strip maps, they show cultural and topographical information adjacent to or crossing the boundary line. Boundary markers, usually set one mile apart, and points from which astronomical and geodetic data were obtained are also shown. The sheets on display are an example of a joint mapping program by the United States and Great Britain to map the international boundary between the United States and Canada from Lake of the Woods to the Rocky Mountains. This map incorporated information on both sides of the boundary lines, compiled from British surveys north of the line and U.S. surveys south of the line.

Records of the Boundary and Claims Commission, RG 76, Series 35
Cartographic and Architectural Branch
National Archives

VI.9. *Plan Muskoday Indian Reserve Chief John Smith South Saskatchewan River Treaty No. 6*

E. Steward; Resurveyed by A. W. Ponton
Manuscript; 43 × 50 cm
1876; resurveyed 1884

Following the transfer of the Canadian plains to Canada by the Hudson's Bay Company in 1869, the Plains Indians agreed to a series of treaties from 1871 to 1921 in an effort to retain their way of life in the face of white settlement. Beginning in 1871, seven treaties were signed establishing reserve lands in the present Prairie Provinces. Reserve lands were platted by Dominion land surveyors to show individual landholdings, which were allocated on the basis of either 160 or 640 acres per family of five. This plan of the Muskoday Indian reserve was originally surveyed according to treaty number 6 by E. Stewart in September 1876. The reserve has been subdivided into individual long lots adjoining the South Saskatchewan River. Each lot is identified by the name of the individual landowner. The land within the reserve is described as "Rolling Prairie Throughout Soil 1st Class."

Indian Affairs Survey Record No. 268 (NMC 8387)
National Map Collection
Public Archives of Canada

Reference: Camponi et al, vii–ix

VI.10. *Map Shewing the Townships Surveyed in the Province of Manitoba and North-West Territory in the Dominion of Canada*

John Johnson
Photolithograph; 64 × 95 cm
Julius Bien, New York, 1877

This composite map shows the progress of Dominion land surveys by the Canadian Department of the Interior based on topographic data obtained from land surveyors' field books. Each township is subdivided into thirty-six one-mile-square sections, except for the lands along the Red River and lower course of the Assiniboine. These were excluded from the survey as part of the treaty with the Métis. Other townships were reserved for the Canadian Pacific Railroad, Menonites, English, and Icelandics. A feature that distinguishes Canadian and U.S. land survey maps is the depiction of vegetation. While both Canadian and American surveyors recorded the type of vegetation found along section boundary lines, American surveyors did not normally include this information on their maps. In this map, three types of vegetation are distinguished: woodlands, swamp or "muskeg," and prairies or meadowland.

National Map Collection
Public Archives of Canada
Ottawa, Canada

Reference: Warkentin and Ruggles, 268.

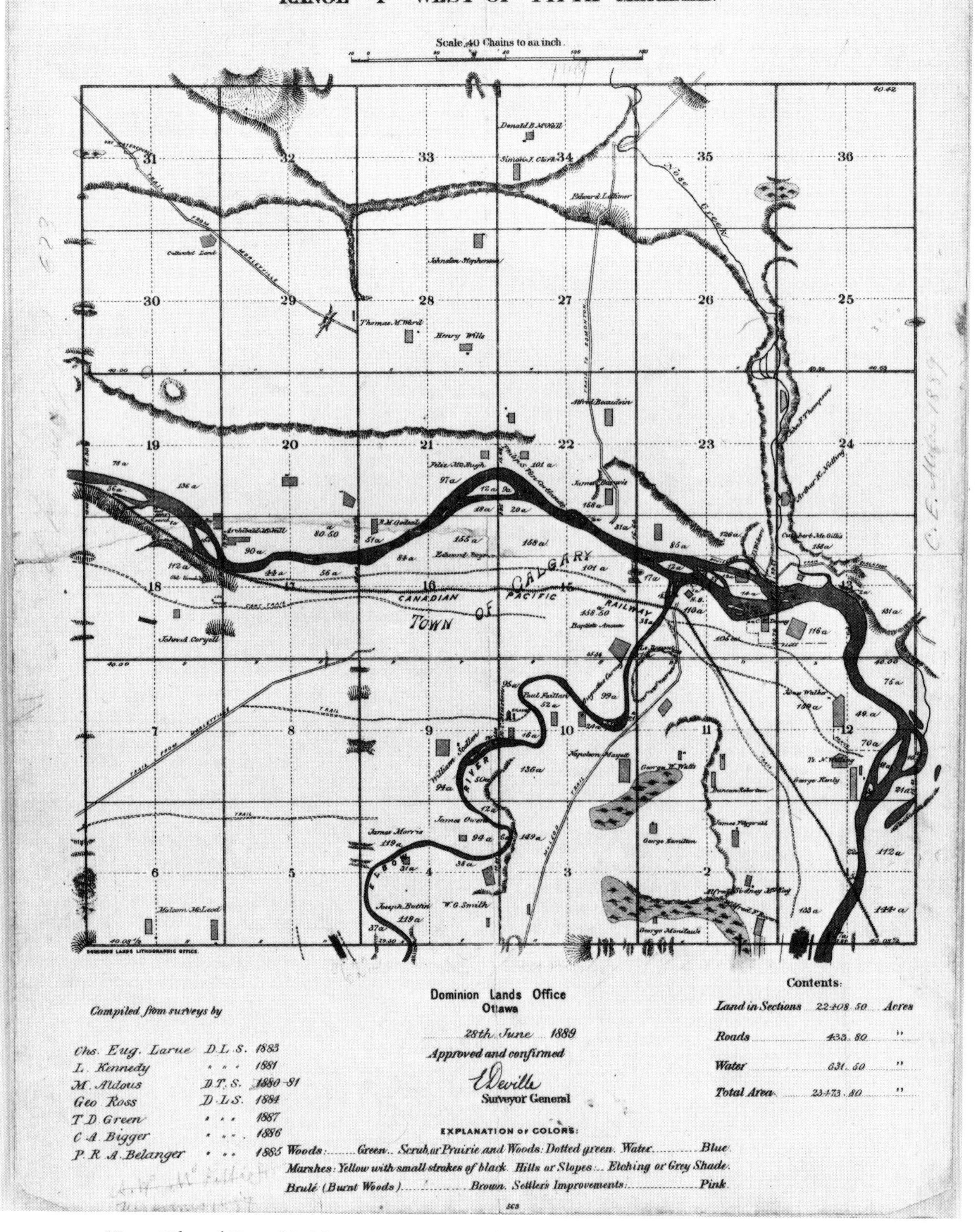

VI.11. "Plan of Township No. 24 Range 1 West of Fifth Meridian. (Public Archives of Canada)

VI.11. *Plan of Township No. 24 Range 1 West of Fifth Meridian* (Illustrated)

Dominion Lands Office
Lithograph; 48 × 35 cm
Ottawa, 1889

The Canadian cadastral survey system was established in 1869 to distribute farmlands to settlers who moved onto the prairies in the years following confederation. Canadian township plats were modeled closely on the rectangular grid system adopted in the United States in which the land was divided into six-mile-square townships and one-mile-square sections. In a Canadian township, however, sections were divided by six ninety-nine-foot-wide roads. Woodlands, marshes, creeks, lakes, and improvements such as roads, trails, and field patterns were included by the surveyors from both nations as townships were platted. A Canadian township plat also showed topographic features such as hills or slopes.

NMC-26423
National Map Collection
Public Archives of Canada
Ottawa, Canada

Reference: Warkentin and Ruggles, 266.

VI.12. *Qu'Appelle Sheet West of Second Meridian*

Photolithograph; 60 × 86 cm
Revised 23 March 1894

The first extensive map series to record the western landscape was the Three-Mile Sectional Map of the Canadian Prairies. This series superseded the survey status maps on a scale of six miles to one inch in 1892. These maps show the Dominion land survey pattern, drainage, roads, railroads, topographic features, and settlements. Progress of settlement was shown by the lands patented, those entered, and those reserved for various purposes. From 1892 to 1967, a total of 134 sheets of this series were issued.

National Map Collection
Public Archives of Canada
Not in the Library of Congress Exhibition

Reference: Nicholson and Sebert, 19.

VI.13. *Map of the Indian Territory. Showing the Progress and Status of Townsite Surveys, under the Direction of The United States Indian Inspector for the Indian Territory.*

Bureau of Indian Affairs
Manuscript on tracing paper; 53 × 49 cm
30 June 1901

Town of "Alabama" Creek Nation, Ind. Territory

W. E. Winn
Lithograph; 50 × 41 cm
Sapulpa, Indian Territory, 1901

Several hundred manuscript and published plats of townsites located on Indian reservations in Kansas, North and South Dakota, and Oklahoma were prepared by the Bureau of Indian Affairs between the 1880s and 1916. These reservations were subdivided in accordance with the rectangular land survey system devised by the U.S. General Land Office. The accompanying annual status map shows the progress of townsite surveys in the Chickasaw, Choctaw, Creek, Cherokee, and Seminole nations of Indian Territory. Like the towns in the rest of the Great Plains, most of the 116 towns shown on this map are located along railroad lines.

Records of the Bureau of Indian Affairs, RG 275, Map CA 497 and Map 2019
Cartographic and Architectural Branch
National Archives
Washington, D.C.

VII. Geological and Topographic Maps

In response to growing economic and scientific interests following the gradual settlement of the North American plains, special-purpose maps began to appear after the Civil War that recorded and assessed the natural resources and topography of the region. Prepared by trained scientists and topographic engineers from systematic field surveys, these include maps of geological structures, mineral deposits, land classification, and topography.

VII.1. *Carte Géologique des États-Unis et des Provinces Anglaises de L'Amérique du Nord* (Illustrated)

Jules Marcou
Chromolithograph; 41 × 59 cm
Paris, 1855

Jules Marcou based this geological map on data he obtained while serving as geologist to one of the U.S. Pacific Railroad expeditions. Marcou was a Frenchman and protégé of the scientist Louis Agassiz. In the spring of 1853 Marcou joined Lt. Amiel Weeks Whipple's exploration of a railroad route that took him from Fort Smith, Arkansas, to the Pacific Coast. Secretary of War Jefferson Davis, who insisted that Marcou's report be completed in the United States, had the scientist's official notebooks seized upon his return to Paris. As a result, Marcou's account of the

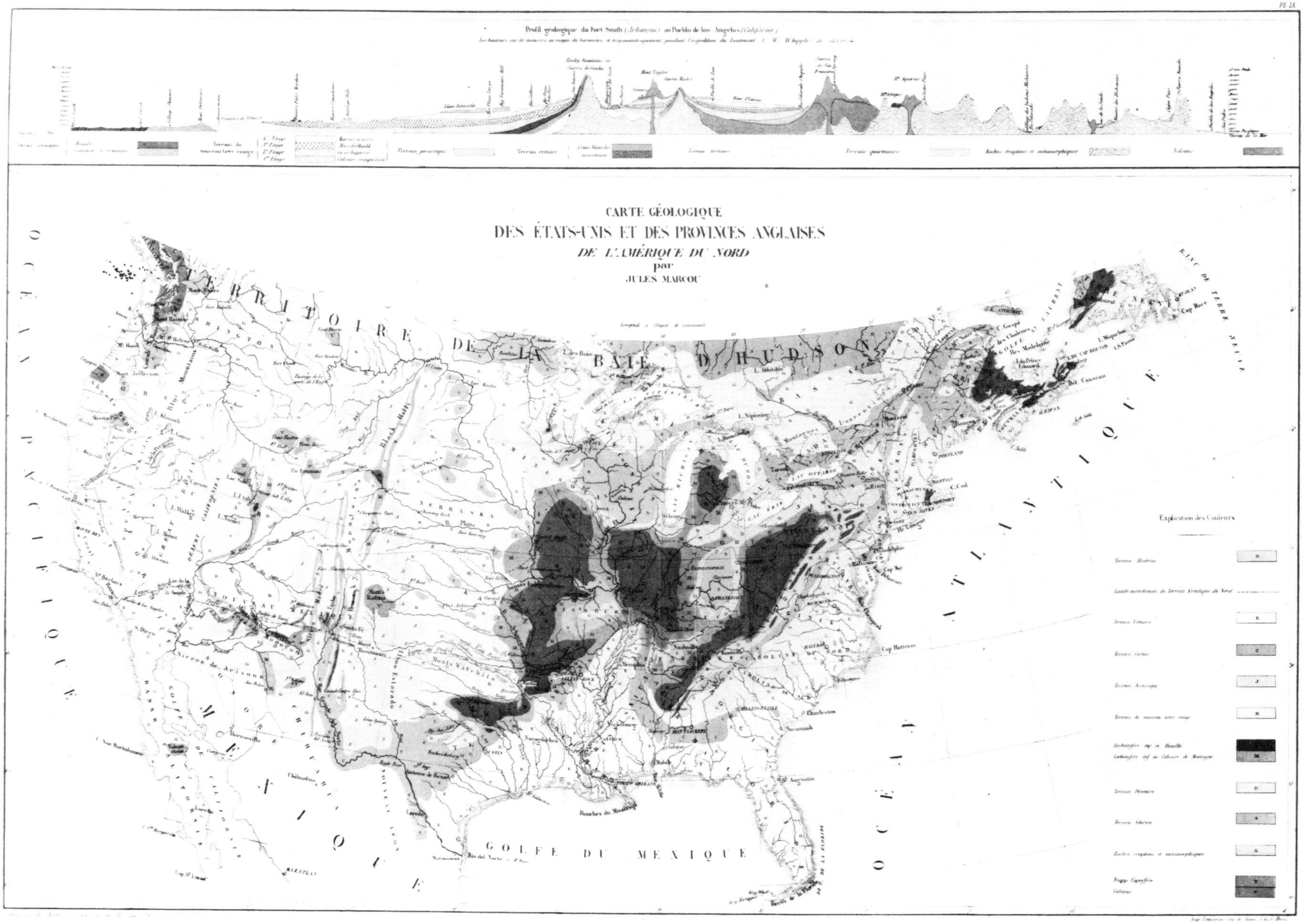

Profil géologique du Fort Smith (*Arkansas*) au Pueblo de los Angeles (*Californie*)

CARTE GÉOLOGIQUE
DES ÉTATS-UNIS ET DES PROVINCES ANGLAISES
DE L'AMÉRIQUE DU NORD
par
JULES MARCOU

OCÉAN PACIFIQUE

OCÉAN ATLANTIQUE

GOLFE DU MEXIQUE

Explication des Couleurs

Terrain Moderne	Q
Limite méridionale du Terrain Erratique du Nord	- - - -
Terrain Tertiaire	E
Terrain Crétacé	C
Terrain Jurassique	J
Terrain du nouveau Grès rouge	R
Carbonifère sup. ou Houille	
Carbonifère inf. ou Calcaire de Montagne	M
Terrain Dévonien	D
Terrain Silurien	S
Roches éruptives et métamorphiques	G
Trapp Cuprifère	T
Volcans	V

trip and his geological map were compiled frm memory and published initially in French instead of English. This map contains the first geological cross-section of the United States and is one of the earliest maps of the United States to be printed in color.

U.S. Geology 1855
Geography and Map Division
Library of Congress

Reference: Goetzman (1959), 323–24.

VII.2. *North America*

Arnold Guyot
Lithograph, with water color wash; 77 × 66 cm
New York: C. H. Scribner & Co., 1865

This is one of the first maps of North America depicting surface relief by generalized hachures and watercolor washes to be prepared by a trained scientist. Guyot, a native of Switzerland, was a student of the celebrated German geographers Alexander von Humboldt and Karl Ritter. He introduced geography as an academic discipline to the United States when he taught at Princeton from 1854 to 1884. The "Great Western Plains" are indicated on both the map and cross-section.

North America, Physical, 1865
Geography and Map Division
Library of Congress

VII.3. *Diagram of Nebraska Exhibiting character of the Soil and Timber and Projected and completed Railroads* (Illustrated)

H. F. Greene
Manuscript; 40 × 50 cm
Plattsmouth, Nebraska, 1869

This composite map of Nebraska was prepared in 1869 for the commissioner of the General Land Office to show the status of subdivided townships. In addition, it showed timber and coal deposits, several other soil types, and mineral deposits. This interest in the vegetation and soil not only reflected the General Land Office's long tradition of requiring its surveyors to record the character of the land in their township survey field notes, but more specifically the General Land Office's sponsorship of Ferdinand Hayden's geological surveys in Nebraska in 1867 and 1868 (Map VII.5).

Records of the Bureau of Land Management, RG 49, Nebraska 7
Cartographic and Architectural Branch
National Archives
Washington, D.C.

VII.4. *Bird's Eye View of the Black Hills to Illustrate the Geological Structure* (Illustrated)

Henry Newton
Lithograph; 56 × 81 cm
New York: Julius Bien, 1879

A beautiful example of scientific art, this panoramic view portrays the Black Hills as seen through the eyes of a geologist. It was published in Newton's *Topographic and Geological Atlas of the Black Hills of Dakota,* which accompanied his official *Report on the Geology and Resources of the Black Hills,* compiled with Walter P. Jenney. The leaders of one of the most thorough scientific expeditions conducted by the federal government during the nineteenth century, Newton and Jenney were sent to the Black Hills in 1875 to dispel growing rumors about gold. Instead, their expedition and report confirmed the existence of gold and spurred the gold rush.

The artist's touch can be seen in this view even in the use of symbols, where drawings of birds in flight are used to identify the different rock formations.

G1447. B5U5 1880 fol.
Geography and Map Division
Library of Congress

Reference: Goetzmann (1966), 421–24.

VII.5. *General Geologic Map of the Area Explored and Mapped by Dr. F. V. Hayden, and the Surveys under his Charge 1869–1880*

Ferdinand V. Hayden
Chromolithograph; 63 × 93 cm
New York: Julius Bien, 1880

One of several geological maps prepared under the direction of Hayden, who had first explored and mapped part of the Great Plains as a geologist with Lt. G. K. Warren's Dakota expeditions of 1856 and 1857 (Map V.3). After the Civil War, Hayden returned to the western plains as head of the U.S. Geological Survey of the Territories. The latter organization was later consolidated with three other post–Civil War western surveys to form the U.S. Geological Survey. Through his surveys and maps, Hayden helped create the earliest fundamental order in western American geology.

In this map, which provides one of the earliest geological assessments of a portion of the plains based on direct observation, Hayden employed a combination of watercolor tints and letters to show nine different rock formations lying to the west of the Missouri River.

U.S. West
Geography and Map Division
Library

Reference: Goetzman (1966), 492.

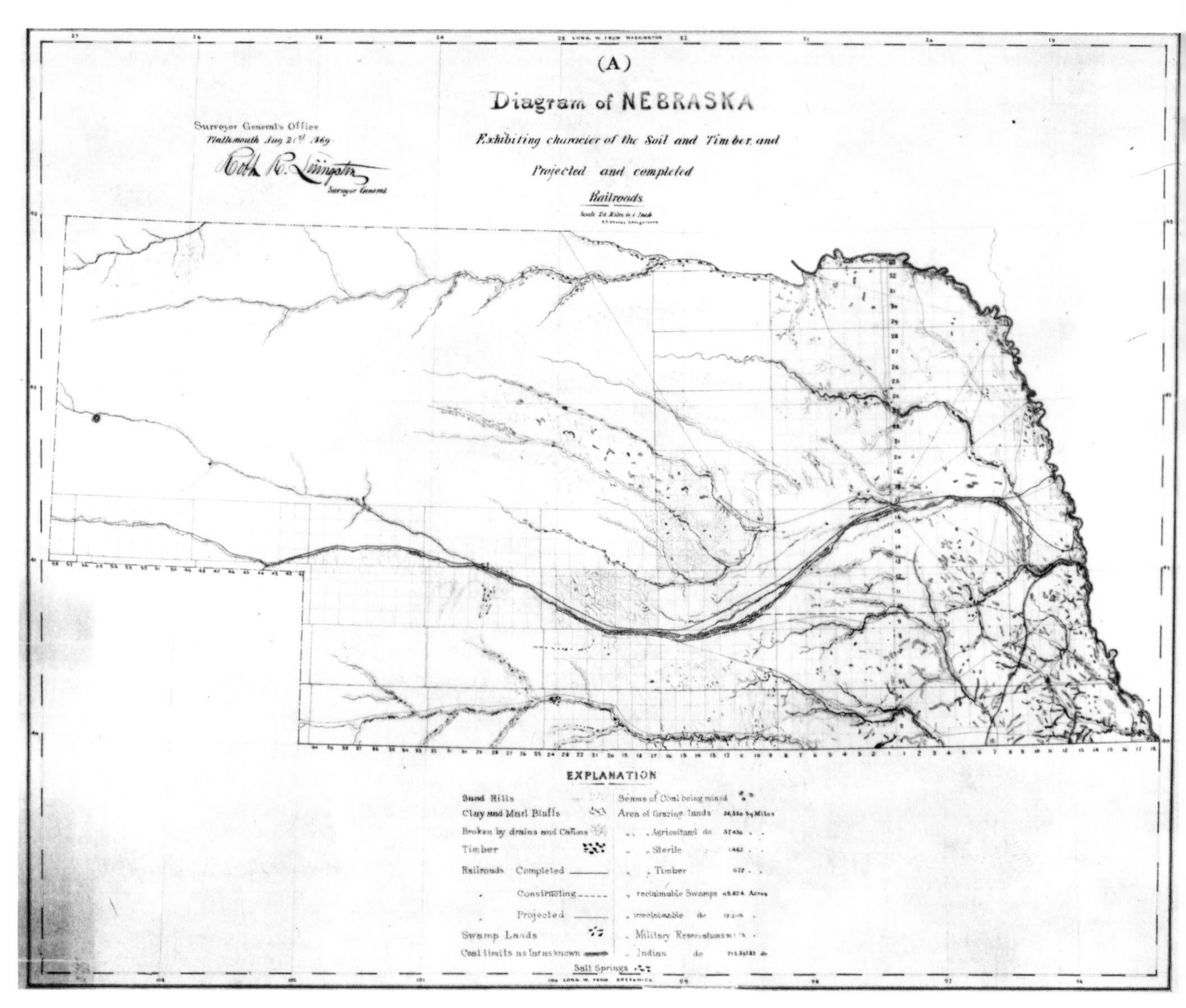

VII.3. "Diagram of Nebraska Exhibiting character of the Soil and Timber and Projected and completed Railroads." (National Archives)

VII.4. "Bird's Eye View of the Black Hills to Illustrate the Geological Structure." (Library of Congress)

VII.6. *Kansas El Dorado Sheet*

A. H. Thompson; topography by E. T. Perkins, Jr.
a) Manuscript; 94 × 75 cm
b) Lithograph; 49 × 42 cm
1885

The U.S. Geological Survey, established in 1879, began the systematic topographic mapping of the Great Plains about 1883. The initial quadrangle map series was compiled in a thirty-minute format at a scale of 1:62,360 and then reduced to 1:125,000 for the published sheet. The thirty-minute format represents about 780 to 1,030 square miles. Standard symbols, colors and place names were adopted to simplify map interpretation and to encourage map reading by the general public. The most innovative aspect of the topographic mapping program was the decision to depict surface features by contour lines (lines of equal elevation), the first time this was done for an entire country on a uniform series of maps.

Records of the U.S. Geological Survey, RG 57
Cartography and Architectural Branch
National Archives

VIII. Commercial Maps

Commercial map publishers synthesized and disseminated geographic information about the North American plains obtained from explorers, military expeditions, and official surveyors. Beginning with the earliest printed maps of North America, private map printing establishments provided the general public with decorative maps of new discoveries and settlements. This process accelerated during the nineteenth century as private printing firms in England, Canada, and the United States gained access to manuscript maps in the custody of the Hudson's Bay Company and official government bureaus. With the opening of the Great Plains to settlers in the late 1850s and the discovery of gold in the Black Hills in the 1870s, railroad companies, land speculators, and mining companies commissioned maps promoting their activities. At the same time, publishing firms issued detailed illustrated county wall maps and atlases of settlements and farms designed to appeal to the settlers' regional and community pride.

VIII.1. *Map of the United States of America With Its Territories & Districts. Including also a part of Upper & Lower Canada and Mexico*

Engraving, with watercolor wash; 72 × 108 cm
Hartford: B. B. Barber and A. Willard, 1833

With the publication of Lewis and Clark's map by Samuel Harrison and Samuel Lewis in 1814, American commercial map publishers began to dessiminate geographic information obtained from official surveys (Map IV.7). The depiction of the central plains on this map is based on Lewis and Clark's map. Designated as the "Mandan District," it is described as "a vast wilderness of immense plains and meadows interspersed with barren hills and almost destitute of wood except in the neighborhood of streams. It is traversed by numerous herds of Buffaloes and wild Horses and by a few roving tribes of Indians. Occasional bands of white hunters and trappers range this country for Furs."

Enhancing the map are statistical tables providing comparative information about the twenty-eight states and territories (acreage; number of "Whites," "Free Colored," and "Slaves"); distances of cities from Washington, D.C., and from each other; and the heights of major mountains in feet above sea level. The map is also decorated with patriotric views of the Capitol building and the "President's House" in Washington.

U.S. 1833
Geography and Map Division
Library of Congress

VIII.2. *Topographical Map of the Road From Missouri to Oregon Commencing at the Mouth of the Kansas in the Missouri River and Ending at the Mouth of the Wallah-Wallah in the Columbia. Section II* (Illustrated)

Charles Preuss
Lithograph; 40 × 65 cm
Baltimore: E. Weber & Co., 1846

The U.S. Senate ordered ten thousand copies of this topographical road map of the Oregon Trail, which followed the Kansas and Platte rivers across the Great Plains. A German cartographer, Preuss had compiled John C. Frémont's earlier maps based on his surveys and the explorer's journal and field notes (Map. IV.14). The map was lithographed by E. Weber and Company, a Baltimore firm that later, under the name of August Hoen and Company, became one of the principal publishers of government reports and maps. Edward Weber and his cousin, August Hoen, were also natives of Germany. Weber had studied lithography in Munich under the watchful eye of Aloys Senefelder, who had developed the process.

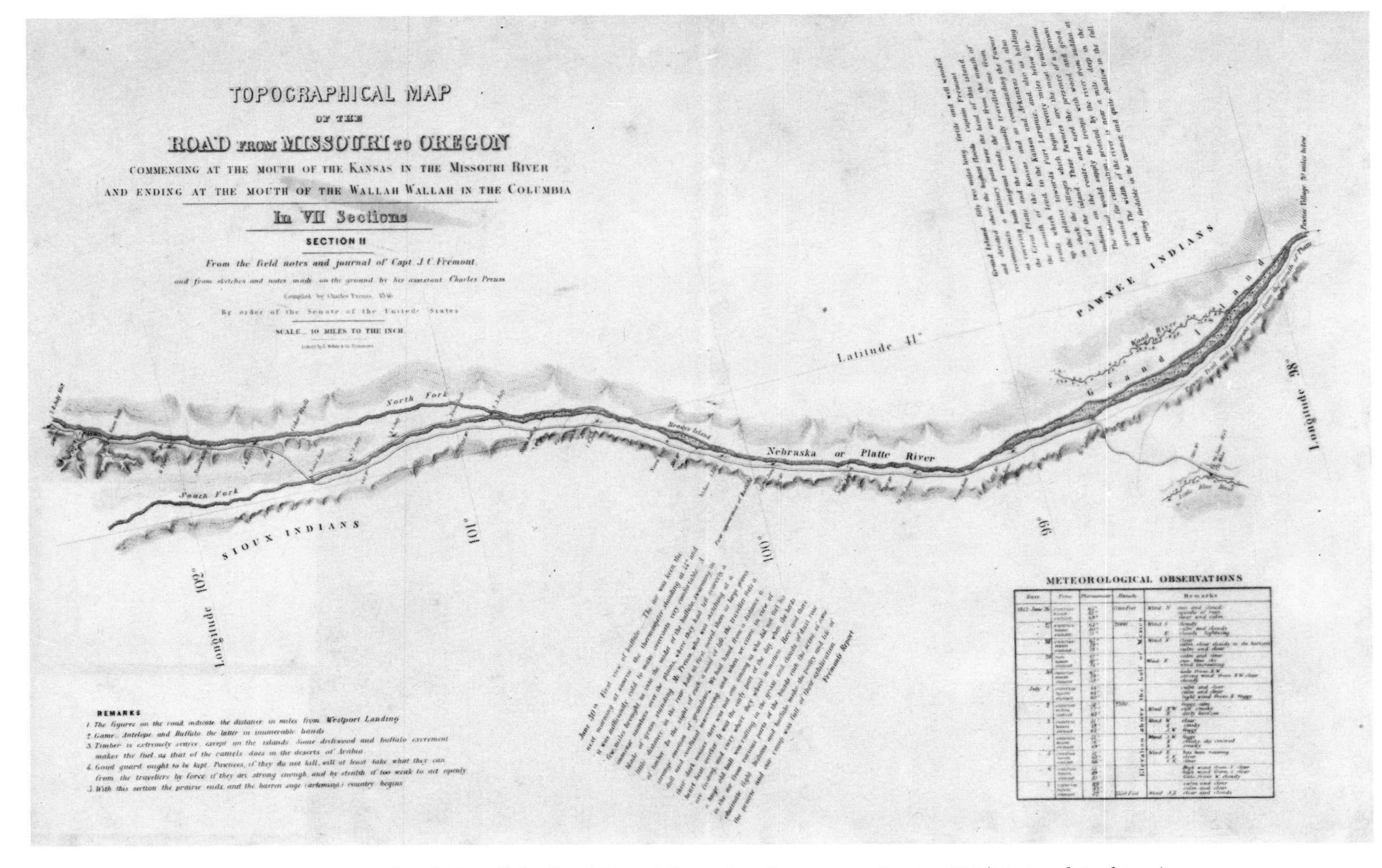

VIII.2. "Topographical Map of the Road From Missouri to Oregon. . . . Section II." (National Archives)

Preuss's map was issued in seven sections so that it could easily be read by wagonmasters under the most adverse conditions. Each section was drawn at a scale of one inch to 10 miles and covered approximately 250 miles. A table of meteorologic observations and notes extracted from Frémont's report provided information about the region. The section on display shows that portion of the Oregon Trail from its junction with the Platte River at Grand Island to beyond the confluence of the north and south forks. A note, marking the point of Frémont's "First view of Buffalo," captures the nineteenth-century romanticism of the plains: "We had heard from a distance a dull and confused murmuring and when we came in view of their dark masses, there was not one among us who did not feel his heart beat quicker. . . . Indians and buffalo make the poetry and life of the prairie and our camp was full of their exhiliration."

Like later road maps, Preuss's map contains helpful travelers' aids. Distances in miles from Westport Landing (Kansas City) are marked along the trail. The map also provides comments on the availability of wild game for food, fuel ("Some drift wood and buffalo excrements makes the fuel, as that of the camels does in the deserts of Arabia"), and safety ("Good guard ought to be kept. Pawnees, if they do not kill, will at least take what they can from the travellers by force if they are strong, and by stealth if too weak to act openly.")

Records of the Office of the Chief of Engineers, RG 77, U.S. 155
Cartographic and Architectural Branch
National Archives
Not in the Library of Congress Exhibition

Reference: Jackson (1970), Map Portfolio Commentary, 14–15.

VIII.3. *A New Map of Texas Oregon and California with the Regions Adjoining.*

S. Augustus Mitchell
Engraving; 57 × 53 cm
Philadelphia, 1846

Exquisitely hand-colored general maps such as this one provided Americans with their images of the changing political geography of the Great Plains during the first half of the nineteenth century. In this map the central and southern plains are divided into four large political units: Iowa Territory, extending from the Mississippi to the Missouri rivers; Missouri Territory lying between the Missouri and the Rocky Mountains and between the Canadian boundary and the north fork of the Platte; Indian Territory, situated between Missouri Territory and the State of Texas; and Texas, with its Panhandle pushing northward as far as the Arkansas River in present Kansas. Traversing the central plains are the Oregon Trail and John C. Frémont's routes of exploration, which captured the imagination of easterners during the 1840s. The lingering belief that the southern plains were a desert is suggested by the cartographer's designation of the Santa Fe Trail as a "Caravan route." As an aid to potential travelers, the map also shows the "Summer range of the Comanches," which is situated to the south of the Santa Fe Trail.

A native of Scotland, Mitchell was a prolific publisher of travelers' guides and general maps during the 1830s and 1840s. Unlike some American map publishers, Mitchell continued to rely upon copperplate or steel engraving rather than the faster and less expensive lithography in order to compete with the high-quality copperplate maps then being produced in Europe.

G4050 1846.M5
United States Western
Geography and Map Division
Library of Congress
Not in the Library of Congress Exhibition

VIII.4. *Nebraska and Kansas* (Illustrated)

Joseph H. Colton
Lithograph, with watercolor wash; 77 × 58 cm
New York, 1854

This decorative wall map is representative of the maps prepared by the Colton firm of New York City, a rival of S. Augustus Mitchell (Map VIII.3). Beginning with large wall maps of the United States, Joseph H. and George W. Colton expanded their operations to include gazetteers and atlases and were a major force in map publishing from 1847 to 1884. Colton's maps are characterized by borders embellished with interweaving vines. This map incorporates the latest military surveys of the central and southern plains, extending from the Canadian border in the north to the Staked Plains and Cross Timbers regions of Texas. For the potential western traveler, it shows the Oregon and Santa Fe trails as well as the proposed transcontinental railroads. A large vignette of a horse-mounted Plains Indian shooting a buffalo with his bow and arrow is placed north of the Platte River to evoke the image of the Great Plains.

United States 1854
Geography and Map Division
Library of Congress
Not in the Library of Congress Exhibition

VIII.5. *Colton's New Sectional Map of Nebraska*

G. W. & C. B. Colton
Lithograph
New York, 1866

This large pocket map was designed "for Travelers and Tourists." Based on information obtained from the Gen-

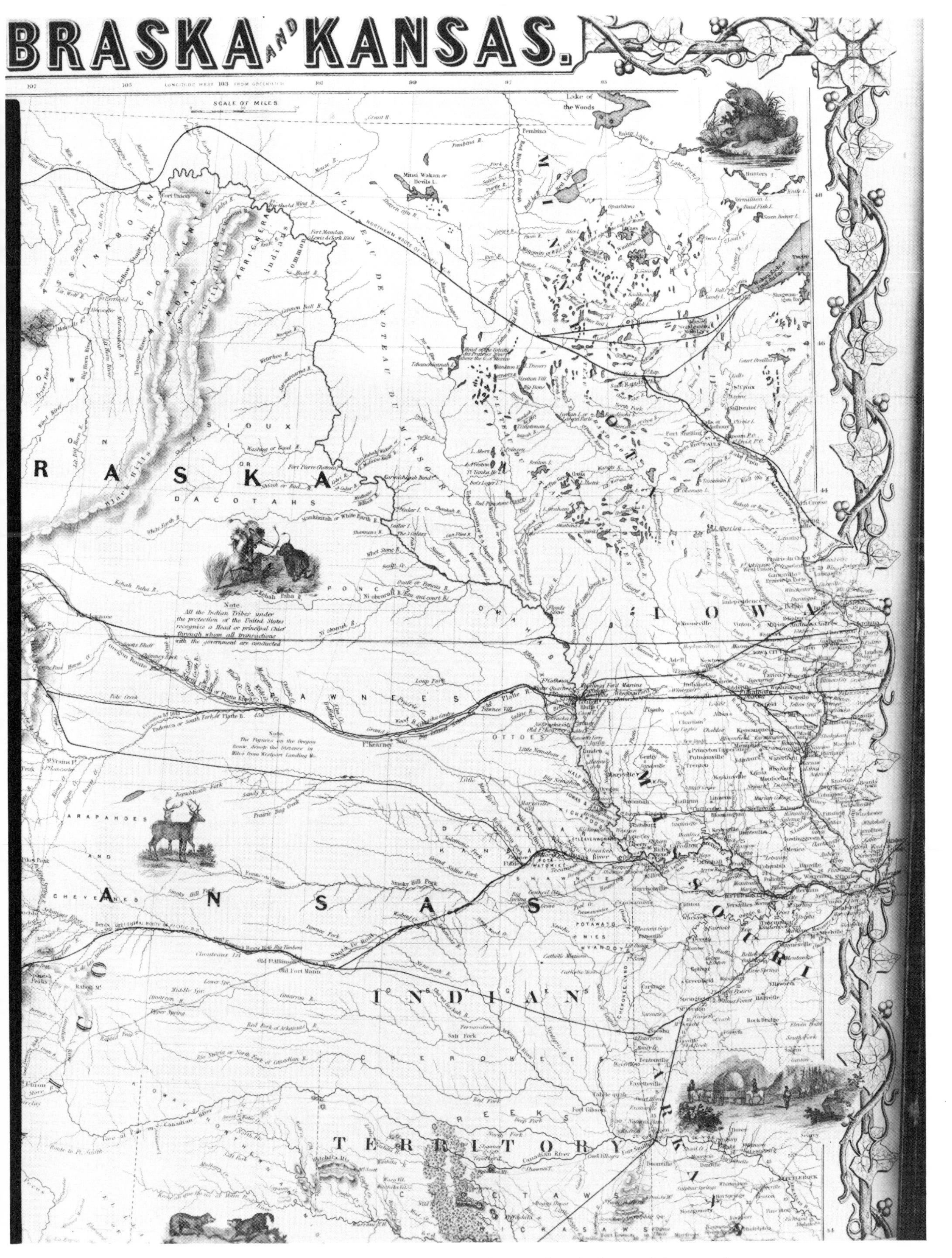

VIII.4. Nebraska and Kansas, by Joseph H. Colton (Library of Congress)

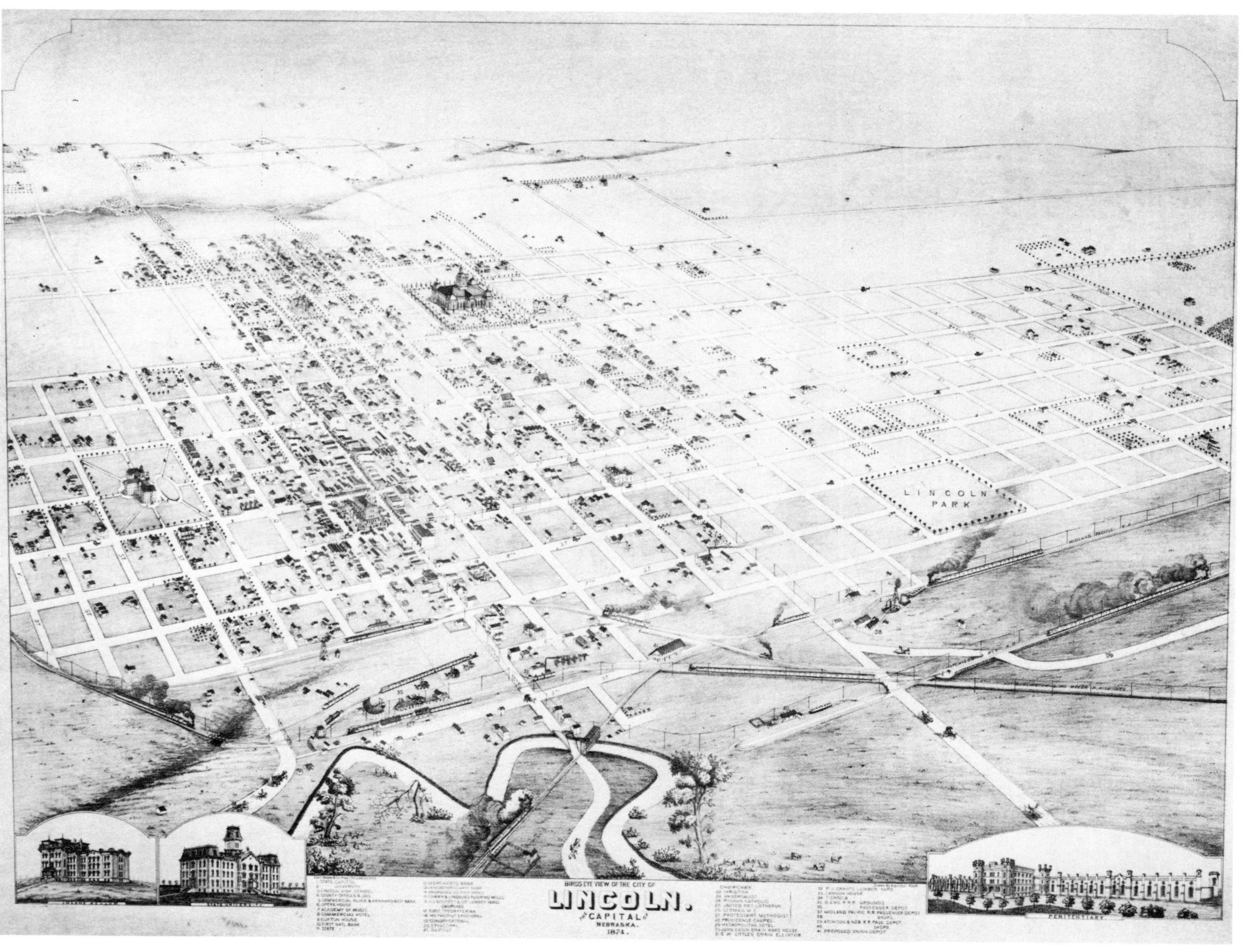

VIII.6. "Birds Eye View of the City of Lincoln. . . ." (Bennett Martin Public Library, Lincoln, Nebraska)

eral Land Office, travelers' maps were often updated in order to show the latest railroads, stage routes, towns, villages, and other internal improvements.

Don Forke Collection, Map 36
Lincoln, Nebraska

Not in the Library of Congress Exhibition

VIII.6. *Birds Eye View of the City of Lincoln. The Capital of Nebraska* (Illustrated)

Augustus Koch
Lithography
Chicago: Chas. Shober & Co., 1874

Panoramic or perspective views recorded the development and expansion of many of the towns and cities that dot the map of the Great Plains. A unique American art form following the Civil War, these views were marketed through door-to-door sales campaigns and newspaper advertisements and were popular with individuals as wall hangings and as promotional aids for real estate speculators and chambers of commerce.

This view of Lincoln, Nebraska, is typical. Lincoln was selected as the site of the state capital in 1867, shortly after Nebraska Territory became a state. In the same year, the undeveloped capital site was designed, surveyed, and platted by August F. Harvey, an accomplished surveyor, lawyer, editor, and politician. Harvey's plan, which is considered one of the most successful city plans in the American West, provided for ample public sites, open spaces, picturesque street vistas, and varying street and lot sizes.

The panoramic view on display captures Lincoln seven years later. It shows a thriving community spreading across the rolling prairie landscape, guided by the rectangular grid pattern that epitomizes western towns.

Bennett Martin Public Library
Lincoln, Nebraska
Not in the Library of Congress Exhibition

Reference: Reps, 417–20.

VIII.7. *New Map of the Black Hills*

Rand, McNally & Company
Wax Engraving; 88 × 112 cm
Chicago, 1877

Discovered by the Vérendryes in the 1730s, the Black Hills, rising abruptly from the northern plains, served as little more than a familiar western landmark until gold was reported in August 1874 by a U.S. Army expedition headed by Lt. Col. George A. Custer. By the end of that year, some fifteen thousand prospectors had entered the region, establishing mining camps at Deadwood and other suitable sites. Commercial maps, including this map by the Rand McNally Company, were printed to take advantage of the gold fever. Designed to promote the region, Rand McNally artists not only identified 125 gold-bearing quartz lodes, but they also decorated their work with picturesque border insets illustrating spectacular natural features, the civilizing impact of the coming of the railroad, recreational activities such as hunting, and the improbable view of an Indian warrior assisting a prospector in his search for gold. A major publisher of railroad maps, Rand McNally adopted the wax engraving process for map printing in 1872 (Map VIII.9).

South Dakota Black Hills 1877
Geography and Map Division
Library of Congress
Not in the University of Nebraska Exhibition

VIII.8. *An Illustrated Historical Atlas of Osage County, Kansas. Compiled, drawn and published from Personal Examinations and Surveys by Edwards Brothers* (Illustrated)

Edwards Brothers
Lithograph; three sheets, each 44 × 35 cm
a) Title Page
b) Township 17 South, Ranges 13 and 14 East of the Sixth Principal Meridian
c) Lithographic Views
Philadelphia, 1879

No nineteenth-century documents evoke a more nostalgic notion of the pastoral virtues of the Great Plains than the lithographic views found in illustrated county atlases like the one on display. The idealized scenes depicted here show children romping with hoops and young adults playing croquet on neatly manicured lawns, fine carriages passing before homesteads, beautifully landscaped gardens and orchards bordering farmhouses, and large herds of cattle roaming tree-lined pastures. Little evidence of the frontier is present.

Accompanying these views are large-scale maps of each township, laid out in the typical grid pattern of the midwest imposed by the U.S. General Land Office. These maps were drawn from information obtained from the U.S. General Land Office, county records, and direct surveys. In addition to showing various cultural and physical features, these maps also gave the names of individual landowners.

County atlases were marketed on a subscription basis by salesmen who canvassed the county. They cost about ten dollars each. An additional fee was charged for the inclusion of the subscriber's biography, portrait, or view of his farmstead or house. Between 1873 and 1900, atlases appeared for most of the eastern counties of Kansas, mirroring the settlement pattern of the state. Altogether some 750 county atlases were published during the last half of the nineteenth century and the early twentieth century, limited primarily to the northeastern and midwestern states.

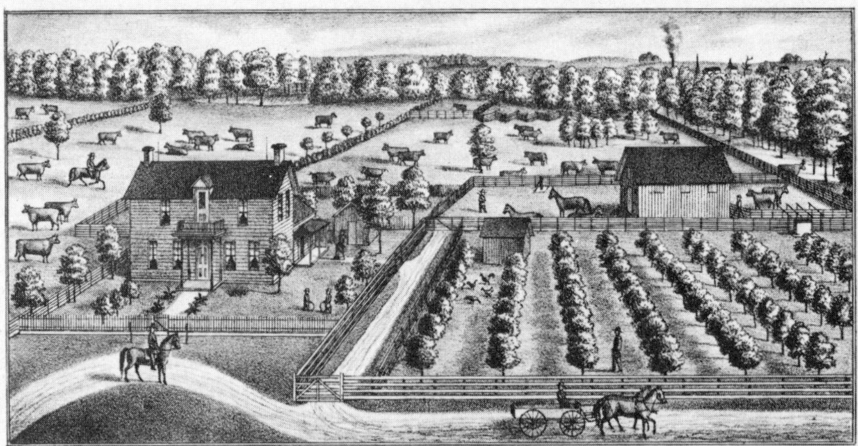

FARM RESIDENCE OF D. NICKEL, 8 MILES. S.W. OF OSAGE CITY & 1½ M. N.E. OF READING, OSAGE CO. KAS.
FARM CONTAINING 615 ACRES. SETTLED IN 1868.

FARM RESIDENCE OF D.G. GRISWOLD, BURLINGAME, OSAGE CO. KAS.

PALACE HOTEL,
CURRY & McELFRESH, PROPRIETORS, OSAGE CITY,
KANSAS.

FARM RESIDENCE OF GEO. MITCHELL, 3 MILES W. OF OSAGE CITY, OSAGE CO. KAS.

FARM RESIDENCE OF M.F. BARR, 3½ MILES S.W. OF ARVONIA, OSAGE CO. KANSAS.
FARM CONTAINING 480 ACRES. SETTLED IN 1874.

VIII.8. "An Illustrated Historical Atlas of Osage County, Kansas. . . ." (Library of Congress)

G1458.06E2 1879 fol.
Geography and Map Division
Library of Congress

Reference: Stephenson, xvii–xix.

VIII.9. *Correct Map of the Burlington and Missouri River R.R.*

Rand McNally and Company
Wax engraving; 43 × 94 cm
Chicago, 1882

An innovative advertising technique was employed by the Rand McNally Company on this map to emphasize the spreading network of the Burlington and Missouri River Railroad in the central plains. The map was designed in such a way that the central states are shown in greater detail and at a larger scale than the coastal states. As a result, the reader focuses on the central plains and the Burlington and Missouri River Railroad.

Located in Chicago, the railroad hub of the United States, Rand McNally began printing railroad maps in 1872 employing wax engraving, a special printing process developed in 1841 by Sydney Edwards Morse and adopted by most American commercial mapmakers during the last decades of the nineteenth century. With the adoption of the wax engraving process and color printing shortly thereafter, Rand McNally became one of the world's leading commercial mapmakers. In wax engraving, the initial image is etched on wax rather than on copper or stone. The engraved wax serves as a mold from which a duplicate printing surface in the form of a thin metal plate in relief is made by a process of electrodepositing. This method allowed the engraver to insert metal letterpress type in the wax instead of laboriously engraving by hand each letter or figure. The letterpress type, however, led to overuse and gave American maps a mechanical appearance.

Railroad 354
Geography and Map Division
Library of Congress
Not in the Library of Congress Exhibition

Reference: Woodward, 1–3.

VIII.10. *Sectional Map Showing the Lands of the Company From Winnipeg to the Second Principal Meridian, Including the Choice Wheat Growing Districts of Pelican & White Water Lakes & Souris River, Also Lands in the Main Line Belt of the Canadian Pacific Railway* (Illustrated)

Canada North West Land Company
Lithograph; 72 × 110 cm
Toronto, ca. 1883

As part of the Canadian Pacific Railway's agreement to construct a railway from Ontario to the West Coast, the company received extensive lands along the planned route. The North West Land Company, a subsidiary of the railway, was responsible for the sale of these lands and undertook a vigorous promotional program, since success in the settlement of the land would bring more business. The two inset views showing prairie forms during planting season and a bountiful autumn harvest are perhaps overly optimistic in the success they promised. The railways are highlighted in red on this map, and available lands are marked in blue. At the right is a diagram informing settlers of how each township was laid out in thirty-six sections.

614.2 gbbd ca. 1883c
Provincial Archives of Manitoba Map Collection
Winnipeg, Canada
Not in the Library of Congress Exhibition

Reference: Warkentin and Ruggles, 326.

VIII.11. *Dodge City, Kansas*

Sanborn Map and Publishing Company
Lithograph with watercolor wash; 64 × 54 cm
New York, October 1884

One measure of the development of the new towns that sprung up along the transcontinental railroads and rivers of the Great Plains during the latter half of the nineteenth century was the fire insurance map. Designed to aid insurance underwriters in computing fire risks on individual buildings, these maps also marked a town's coming-of-age in terms of wealth and importance. Detailed information was provided for each building, including the number of stories and windows. The types of construction material are indicated by watercolor wash (yellow denotes frame construction, red denotes brick, blue denotes stone, and green denotes special materials). Commercial, industrial, and public buildings are identified by function or name.

The Sanborn Map and Publishing Company, which was established in 1867, mapped 238 towns and cities in Kansas between 1884 and 1949, each an average of five different times. One of the first Kansas towns to be mapped was Dodge City. Following the opening of the Western Cattle Trail from Texas and the construction of a stockyard by the Atchison Topeka & Santa Fe Railroad in 1877, Dodge City briefly became the West's leading cattle town. This map captures Dodge City near the height of its prosperity. The numerous hotels, boarding houses, restaurants, saloons and livery stables attest to its title as the "Cowboy Capitol of the West."

Sanborn Collection
Geography and Map Division
Library of Congress
Not in the Library of Congress Exhibition

Reference: Reps, 556.

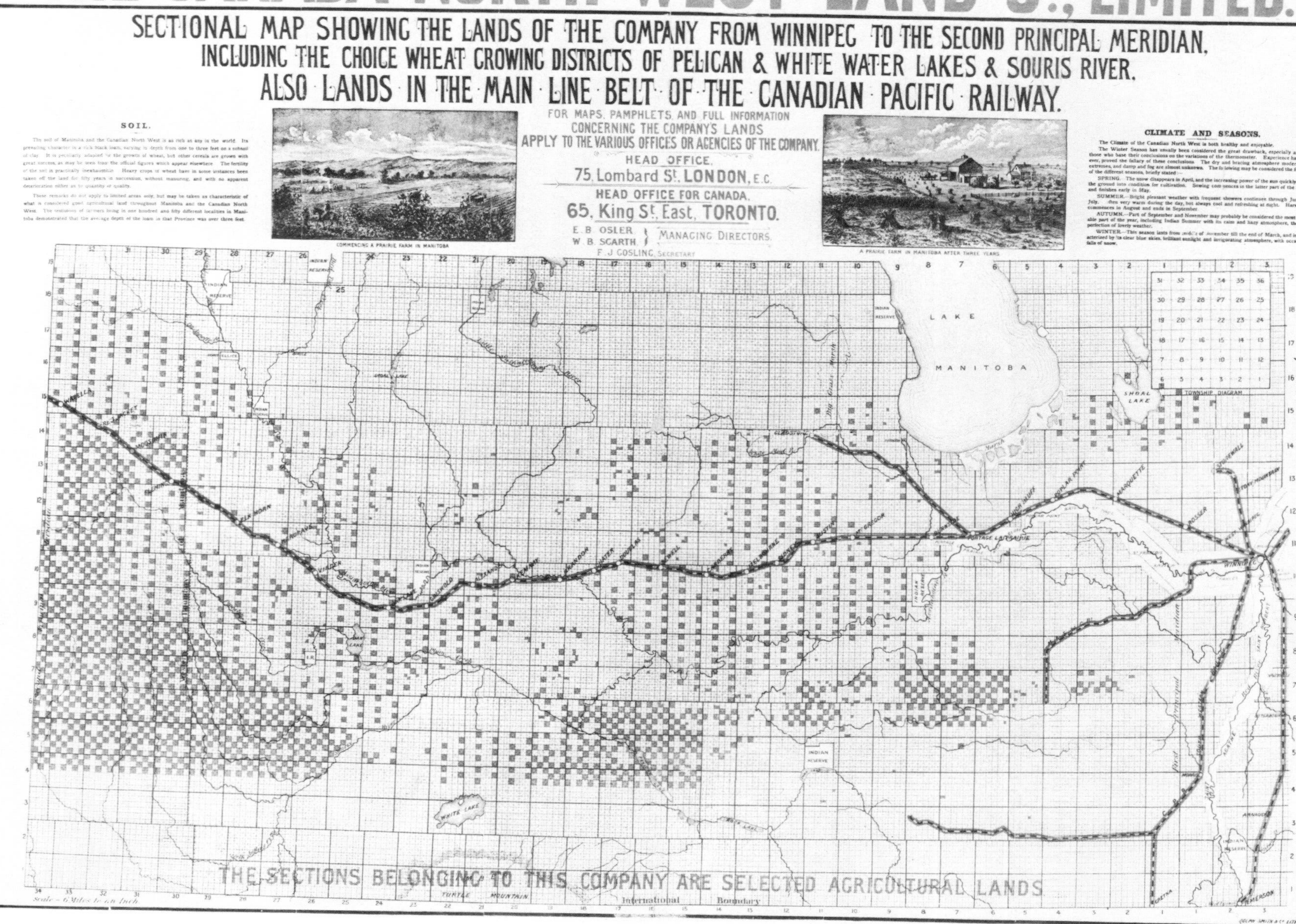

VIII.10. "Sectional Map Showing the Lands of the Company From Winnipeg to the Second Principal Meridian. . . ." (Provincial Archives of Manitoba)

VIII.12. *Map of Wyandotte Co. Kansas*

G. M. Hopkins
Lithograph, with watercolor wash; 69 × 88 cm
Philadelphia, 1887

County wall maps were derived from the same surveys and marketed in the same manner as county atlases (Map IV.12). Although popular as wall hangings in the northeast before the Civil War, wall maps were superseded to some extent by county atlases during the last half of the nineteenth century since the latter were easier to use and store. This was particularly true in the Great Plains, where the settlement process coincided with the emergence of the county atlas.

Hopkins issued this wall map and an atlas of Wyandotte County in this same year.

Kansas, Wyandotte 1887
Geography and Map Division
Library of Congress

Reference: Stephenson, xvii–xix.

Bibliography

Allen, John L. *Passage through the Garden: Lewis and Clark and the Image of the American Northwest.* Urbana: University of Illinois Press, 1975.

Bray, Martha Coleman. *Joseph Nicollet and His Map.* Philadelphia: American Philosophical Society, 1980.

Camponi, Linda; Diane Tardif-Cote; and Guy Poulin. *Maps of Indian Reserves and Settlements in the National Map Collection/Cartes des Reserves et Agglomerations Indiennes de la Collection Nationale de Cartes et Plans.* Vol. 2. Ottawa: Public Archives of Canada, 1981.

Crouse, Nellis M. *Contribution of the Canadian Jesuits to the Geographical Knowledge of New France, 1632–1675.* Ithaca, N.Y., 1924.

Cumming, William P. *The Southeast in Early Maps, with an Annotated Checklist of Printed and Manuscript Regional and Local Maps of Southeastern North America during the Colonial Period.* Chapel Hill: University of North Carolina Press, 1962.

Delanglez, Jean. *Life and Voyages of Louis Jolliet, 1645–1700.* Chicago: Institute for Jesuit History, 1948.

———. "The Sources of the Delisle Map of America 1703." *Mid-America* 25 (1943): 275–98.

———. "Tonti Letters." *Mid-America* 21 (July 1939): 209–38.

———. "The Voyage of Louis Jolliet to Hudson Bay in 1679." *Mid-America* 26 (July 1944): 221–50.

De Voto, Bernard. *The Course of Empire.* Boston: Houghton Mifflin, 1952.

Diller, Aubrey. "Maps of the Missouri River before Lewis and Clark." In *Studies and Essays in the History of Science and Learning Offered in Homage to George Sarton.* Ed. M. F. Ashley Montagu. Pp. 505–19. New York, 1946.

———. "A New Map of the Missouri Drawn in 1795." *Imago Mundi* 12 (1955): 175–80.

Ehrenberg, Ralph. "Nicholas King: First Surveyor of the City of Washington, 1803–1812." *Records of the Columbia Historical Society, 1969–70,* 1971, pp. 31–65.

———. "Sketch of Part of the Missouri & Yellowstone Rivers with a Description of the Country, etc." *Prologue,* 1971, pp. 73–78.

Friis, Herman R. "Cartographic and Geographic Activities of the Lewis and Clark Expedition." *Journal of the Washington Academy of Sciences* 44 (November 1954): 338–51.

———. "Stephen H. Long's Unpublished Manuscript Map of the United States Compiled in 1820–1822 (?)." *The California Geographer* 8 (1967): 75–87.

Goetzmann, William H. *Army Exploration in the American West, 1803–1863.* New Haven: Yale University Press, 1959.

———. *Exploration and Empire: The Explorer and the Scientist in the Winning of the American West.* New York: Alfred A. Knopf, 1966.

Hamilton, Raphael N. "The Early Cartography of the Missouri Valley." *American Historical Review* 39 (1933–34).

Irvine, C. "Recollections of Omaha, 1855–61." *Nebraska State Historical Society* 5 (1902): 150–60.

Jackson, Donald. *The Expeditions of John C. Frémont.* Urbana: University of Illinois Press, 1970.

———, ed. *The Journals of Zebulon Montgomery Pike, with Letters and Related Documents.* Norman: University of Oklahoma Press, 1966.

Kavanagh, Martin, *La Vérendrye: His Life and Times.* Brandon, Manitoba: Martin Kavanagh, 1967.

Klemp, Egon. *America in Maps Dating from 1500 to 1856.* New York and London: Holmes & Meir, 1976.

Loomis, Noel M., and A. P. Nasatir. *Pedro Vial and the Roads to Santa Fe.* Norman: University of Oklahoma Press, 1967.

Lutze, J. J. "The Methodist Missions among the Indian Tribes In Kansas." *Transactions of the Kansas State Historical Society, 1905–1906* 9 (1906): 160–235.

McCoy, John C. "Survey of Kansas Indian Lands." *Transactions of the Kansas State Historical Society* 4 (1890): 298–311.

Martin, Lawrence, and Clara Egli. *Noteworthy Maps No. 3, Accession 1927–28.* Washington, D.C.: Library of Congress, 1928.

The Medical Repository 3 (1806): 315–18.

Moodie, D. W., and Barry Kaye. "The Ac Ko Mok Ki Map." *The Beaver: Magazine of the North,* Spring, 1977, pp. 4–15.

Moulton, Gary E. "Another Look at William Clark's Map of 1805." *We Proceeded On: The Official Publication of the Lewis & Clark Trail Heritage Foundation* 9 (March 1983): 19–22.

———, ed. *Atlas of the Lewis & Clark Expedition.* 126 plates. Lincoln & London: University of Nebraska Press, 1983.

Nasatir, A. P., ed. *Before Lewis and Clark: Documents Illustrating the History of the Missouri, 1784–1804.* 2 vols. St. Louis: St. Louis Historical Documents Foundation, 1952.

Nichols, Roger L. *General Henry Atkinson: A Western Military Career.* Norman: University of Oklahoma Press, 1965.

———, and Patrick L. Halley. *Stephen Long and American Frontier Exploration.* Newark: University of Delaware Press, 1980.

Nicholson, N. L., and L. M. Sebert. *The Maps of Canada.* Hamden, Conn.: Archon Books, The Shoe String Press, 1981.

Oppen, William A. *The Riel Rebellions: A Cartographic History/Lé récit cartographique des affaires Riel.* Cartographica Monograph no. 21–22/1978, Supplement Nos. 1–2 to *Canadian Cartographer* 15 (1978). Toronto: University of Toronto Press in association with the Public Archives of Canada and the Canadian Government Publishing Centre, 1978.

Powell, John W. *Tenth Annual Report of the Bureau of Ethnology to the Secretary of the Smithsonian Institution, 1888–89.* Washington, D.C.: GPO, 1893.

Reps, John W. *Cities of the American West: A History of Frontier Urban Planning.* Princeton: Princeton University Press, 1979.

Ristow, Walter W. *Maps for an Emerging Nation: Commercial Cartography in Nineteenth-Century America.* Washington, D.C.: Library of Congres, 1977.

Skelton, R. A. *Decorative Printed Maps of the 15th to 18th Centuries.* London: Spring Books, 1952.

Spry, Irene M. *The Palliser Expedition: An Account of John Palliser's British North American Expedition, 1857–1860.* Toronto: Macmillan of Canada, 1963.

———, ed. *The Papers of the Palliser Expedition, 1857–1860.* Toronto: Champlain Society, 1968.

Stephenson, Richard W. *Land Ownership Maps: A Checklist of Nineteenth Century United States County Maps in the Library of Congress.* Washington, D.C.: Library of Congress, 1967.

Swanton, John R. *The Indian Tribes of North America.* Smithsonian Institution Bureau of American Ethnology, Bulletin 145. Washington, D.C.: GPO, 1952.

Thwaites, Reuben Gold, ed. *New Voyages to North America by the Baron De Lahontan.* New York: Burt Franklin, 1970.

Utley, Robert M. *Frontier Regulars: The United States Army and the Indian, 1866–1891.* New York: Macmillan Publishing Co., 1973.

Verner, Coolie, and Basil Stuart-Stubbs. *The Northpart of America.* Don Mills, Ont.: Academic Press Canada, ca. 1979.

Wagner, Henry. *Peter Pond: Fur Trader & Explorer.* New Haven: Yale University Library, 1955.

Warkentin, John, and Richard I. Ruggles. *Historical Atlas of Manitoba: A Selection of Facsimile Maps, Plans and Sketches from 1612 to 1969.* Winnipeg: Manitoba Historical Society, 1970.

Wheat, Carl I. *Mapping the Transmississippi West, 1540–1861.* 5 vols. San Francisco: Institute of Historical Geography, 1957–63.

Wheeler, George. *Report upon United States Geographical Surveys West of the One Hundredth Meridian, . . . vol. 1, Geographical Report.* Washington, D.C.: GPO, 1889.

Wolter, John A. "Johann Georg Kohl and America." *Map Collector* 17 (December 1981): 10–14.

Wood, W. Raymond. "The John Evans 1796–97 Map of the Missouri River." *Great Plains Quarterly* 1 (Winter 1981): 39–53.

———, and Gary E. Moulton. "Prince Maximilian and New Maps of the Missouri and Yellowstone Rivers by William Clark." *Western Historical Quarterly* 12 (October 1981): 372–86.

Woodward, David. *The All-American Map: Wax Engraving and Its Influence on Cartography.* Chicago and London: University of Chicago Press, 1977.

CONTRIBUTORS

John L. Allen is a Professor and Chairman of the Department of Geography at the University of Connecticut. His major publications are in historical geography and the environment and include *Passage through the Garden: Lewis and Clark and the Image of the American Northwest* (1975).

Silvio A. Bedini, Keeper of Rare Books of the Smithsonian Institution, is a historian of science specializing in scientific instrumentation. He is author of many books and articles, including *Thinkers and Tinkers: Early American Men of Science* (1975).

Ralph E. Ehrenberg is Assistant Chief of the Geography and Map Division in the Library of Congress. His special field of interest is historical cartography, and he has published numerous articles in that field.

John B. Garver, Jr., is Chief Cartographer for the National Geographic Society and an editor of *National Geographic Magazine.* Formerly an instructor at the U.S. Military Academy at West Point, Dr. Garver blended his interest in geography and cartography with a military career.

Ronald E. Grim is Bibliographer in the Geography and Map Division of the Library of Congress. Dr. Grim's publications include *Historical Geography of the United States: A Guide to Information Sources* (1982) and a contribution to *A Guide to Information Sources in the Geographical Sciences* (1983).

Frances W. Kaye is Associate Professor of English at the University of Nebraska—Lincoln and Editor of *Great Plains Quarterly.* A specialist in midwestern and plains literature, she has published essays in *Southern Review, Agricultural History, Canadian Review of American Studies,* and other publications.

G. Malcolm Lewis is Senior Lecturer in Geography at the University of Sheffield, England. A specialist in historical geography, he has concentrated on the Great Plains. His numerous articles have appeared in *Transactions of the Institute of British Geographers, Annals of the Association of American Geographers, Cartographia, The Map Collector,* and other journals.

Frederick C. Luebke is Professor of History and Director of the Center for Great Plains Studies at the University of Nebraska—Lincoln. He is the author of many books and articles in the field of ethnic history; his edited works include *The Great Plains: Culture and Environment* (1979), *Ethnicity on the Great Plains* (1980), and *Vision and Refuge: Essays on the Literature of the Great Plains* (1982).

Gary E. Moulton is Associate Professor of History at the University of Nebraska—Lincoln and Editor of the *Journals of the Lewis and Clark Expedition,* a long-term multivolume project of the Center for Great Plains Studies. He is the author of *John Ross, Cherokee Chief* (1978) and editor of *The Papers of Chief John Ross* (1984) and *Atlas of the Lewis and Clark Expedition* (1983).

James M. Richtik is Professor and Chairman of the Department of Geography at the University of Winnipeg. He is author of articles in *Agricultural*

History, Regina Geographical Studies, Great Plains Quarterly, and other scholarly publications.

James P. Ronda is Professor of History at Youngstown State University. He is the author of several books and many articles on the history of American Indians; his most recent book is *Lewis and Clark among the Indians* (1984).

Richard I. Ruggles is Professor of Geography at Queen's University, Kingston, Canada. One of his many publications, *Historical Atlas of Manitoba* (1970), coauthored with John Warkentin, won a National Award of Merit.

W. Raymond Wood is Professor of Anthropology at the University of Missouri at Columbia. A former editor of *Plains Anthropologist*, Wood has published his work in various journals, including *Missouri Historical Review* and *Great Plains Quarterly*.

INDEX

Boldface numbers indicate pages where illustrations appear.